INTEGRATED LOGISTICS SUPPORT HANDBOOK

JAMES V. JONES

LOGISTICS MANAGEMENT ASSOCIATES
IRVINE, CALIFORNIA

Dedicated to my wife, Kim, and my daughter, Catherine,
whose love, encouragement, and understanding made this book possible.

FIRST EDITION
FOURTH PRINTING

© 1987 by **McGraw-Hill, Inc.**

Library of Congress Cataloging-in-Publication Data

Jones, James V.
   Integrated logistics support handbook / by James V, Jones.
     p.   cm.
   Includes index.
   ISBN 0-8306-2921-1
    1. United States—Armed Forces—Procurement—Handbooks, manuals,
etc. 2. United States—Armed Forces—Supplies and stores-
-Handbooks, manuals, etc. I. Title.
UC263.J67 1987
355.6′212′0973—dc19   87-19408
                                                 CIP

For information about other McGraw-Hill materials, call 1-800-2-MCGRAW in the U.S.
In other countries call your nearest McGraw-Hill office.

Contents

Acknowledgments

Thanks to the many students who have attended my lectures and seminars over the years and taught me so much.

A special thanks to Global Engineering Documents, Santa Ana, California for allowing the use of drawing illustrations from the *Drawing Requirements Manual, 6th Edition,* that are reproduced in Chapter 2.

I would also like to extend a special thanks to those friends who have assisted in researching and collecting source information for this text. I couldn't have finished this project without them.

Introduction

Integrated Logistics Support is the foundation of the national defense of the United States of America. This is a strong statement, and some might disagree, but I believe it with all my heart. The strength and security of our nation depends on the readiness of our military systems. Readiness is dependent on having the right logistics support available at the right place at the right time. The lack of adequate spare parts and support equipment, trained maintenance personnel, accurate technical documentation, or other support resources will eventually render any military system useless. Therefore, without proper integrated logistics support, readiness will suffer, thereby threatening the ability of our military forces to provide for the national defense.

Integrated Logistics Support is a dynamic and challenging profession. Logistics engineers work hand-in-hand with designers to create military systems that are on the leading edge of technology. The many-faceted problems that face logisticians require an in-depth knowledge of support concepts and the ability to develop alternatives that provide the best support possible. Compounding the problems of logisticians is the fact that most decisions made affecting logistics are made by nonlogisticians. Therefore, it is imperative that all individuals who are associated with the development and production of military systems have a basic understanding of support concepts and philosophies.

As a young logistician, I was eager to learn everything I could about my new profession. I soon found, however, that the world of logistics is not as simple as it first appeared. Within the logistics arena there are many separate disciplines, each having its own unique purpose; in order to understand the overall logistics process, one must first understand each of these internal disciplines. I also realized that these disciplines had developed relatively independently, and they were still separated, in most cases, as semi-independent organizations, each trying to plan and implement support for military systems. This state of affairs was not conducive to learning about logistics. As my career advanced, it became even more apparent that there was no single source available for obtaining detailed information about the overall logistics process. Realizing this need, I began compiling information on all aspects of logistics several years ago. This repository of information, gleaned from practical experience working on military systems, has served as the reference for this text.

The purpose of this text is to provide a single source of information about Integrated Logistics Support. Contained herein are discussions of the concepts and philosophies that guide the activities of logisticians. Each discipline within the Integrated Logistics Support organization is addressed in as much detail as the limited space of the text will allow. Whenever possible, I have also provided two key items: examples of the topic being discussed to aid in understanding, and references to government documents that are the controlling authority for the subject.

Don't let my use of acronyms confuse or frustrate you, because you have to learn to speak the language if you want to make it in logistics. I have tried to use the building-block approach in preparing the text, which starts with a discussion of the field of Integrated Logistics Support in Chapter 1 and basic topics encountered when dealing with the U.S. Government in Chapter 2. Chapters 3 through 17 address specific areas of Integrated Logistics Support. The remaining chapters discuss management issues related to the Integrated Logistics Support program and other subjects of which the logistician should have a basic understanding.

And last, a message to those logisticians and other decision makers who plan and implement support for our military systems. Always remember, the decisions you make today will affect our national defense for years to come, and the future of our nation depends on our national defense.

Chapter 1

Introduction to Logistics

Logistics is the science of planning and implementing the acquisition and use of resources necessary to sustain the operation of military forces. Without logistics, military forces could not operate.

Integrated Logistics Support (ILS) is the disciplined and unified management of the technical logistics disciplines that plan and develop support for military forces. In general, this means that ILS is the management organization that plans and directs the activities of many technical disciplines associated with the identification and development of logistics support requirements for military systems.

PRINCIPAL ELEMENTS OF ILS

The ILS organization contains technical disciplines that specifically address the support aspects of maintenance planning; manpower and personnel; supply support; support and test equipment; training and training devices; technical documentation, computer resources support; packaging, handling, storage, and transportability; facilities; and reliability and maintainability. These areas, as illustrated in Fig. 1-1, are commonly referred to as the principal elements of ILS. Each element is the responsibility of an ILS discipline that is staffed with logistics engineers trained in that particular specialty.

Maintenance Planning

Much of the support of military systems is centered around maintenance of equipment. A primary function of ILS is to develop a concept for the maintenance

| MAINTENANCE PLANNING |
| MANPOWER AND PERSONNEL |
| SUPPLY SUPPORT |
| SUPPORT AND TEST EQUIPMENT |
| TRAINING AND TRAINING DEVICES |
| TECHNICAL DOCUMENTATION |
| COMPUTER RESOURCES |
| PACKAGING, HANDLING, STORAGE, AND TRANSPORTABILITY |
| FACILITIES |
| RELIABILITY AND MAINTAINABILITY |

Fig. 1-1. Principal elements of ILS.

program to support a military system and then to plan the detailed maintenance actions that must occur to support the system. The requirements for maintenance then drive the decisions for resources necessary to support maintenance actions. Logistics engineers are responsible for maintenance planning and analysis through the maintenance engineering process.

Manpower and Personnel

Systems cannot operate and maintain themselves. ILS is charged with the responsibility of identifying the number and skills of military and civilian personnel needed to support operations and maintenance. This is accomplished by maintenance engineers and personnel specialists who participate in the design and analysis process as the system is being developed.

Supply Support

Operation and maintenance actions require material in the form of spare and repair parts. Identification and acquisition of the materials necessary to support operation and maintenance of military systems is another key responsibility of the ILS organization. The disciplines of provisioning and supply support fulfill this requirement.

Support and Test Equipment

Most military systems require additional items of equipment to support operations or maintenance. Support equipment specialists and test engineers conduct analyses to identify and develop these requirements as a portion of the overall maintenance planning process.

Training and Training Devices

Trained and qualified operators and maintenance personnel are required to support military systems. Within the ILS organization are training specialists who participate in the planning process to identify training requirements and develop appropriate training courses for operation and maintenance personnel. Necessary devices and equipment to support training are also developed by this group.

Technical Documentation

The user of equipment needs instructions to operate and maintain it. Technical documentation is prepared by the technical publications discipline that accompanies the system. This documentation describes all the actions required for system operations and maintenance.

Computer Resources Support

Computers are used to operate and maintain many military systems. The facilities, hardware, software, documentation, and personnel needed to operate and maintain these computers are identified through the analysis of operations and maintenance of the system by logistics engineers. The resources to support computers become an integral part of the support package for the system.

Packaging, Handling, Storage, and Transportability

The physical movement of a system must be accomplished in a manner that does not reduce its effectiveness. Logistics engineers plan and implement the procedures and measures necessary for packaging, handling, storing, and transporting military systems.

Facilities

Operating and maintaining most military systems, and training personnel, requires some type of facilities. ILS is responsible for identifying the needs for facilities, planning facility utilization, and developing the justification for acquisition.

Reliability and Maintainability

The areas of reliability and maintainability address how long a system will operate without failing and how long it will take to fix an item when it fails. These disciplines are sometimes within the scope of the ILS organization and sometimes they are not, depending on a company's organization. In either case, reliability and maintainability play an important role in determining the support that will be needed when a system is used. Logistics engineers use information from the analyses performed by reliability and maintainability personnel to develop system support requirements.

GOALS OF ILS

The goals of the ILS organization, shown in Fig. 1-2, are (l) to have logistics support considerations influence the design of a system, (2) to identify and develop support requirements that are related to and supportive of readiness objectives of the system, (3) to acquire the necessary support, and (4) to provide the required support for the minimum cost. The ILS organization is an integral part of the engineering effort that designs military systems. Logistics engineers work hand-in-hand with other engineers to ensure that support is considered in the design process. Logistics analyses are conducted to identify ways that a design can be changed to improve support or supportability. Additional analyses are performed to identify the resources that will be required to support the system when it is used. Logistics support resources are the biggest expense associated with a military system over its useful life, so it is imperative that ILS plan for the most economical use possible of these resources.

ILS PARTNERS

The overall ILS organization that develops the support for a military system is composed of two partners who must work as a team to achieve the goals of the ILS program. These partners are an element of the Department of Defense (DoD), which establishes a need for the system, and the government contractors who design and produce the system. The actual element of DoD that acts for the government depends on the specific system. It could be any branch of the military or an organization within a branch that is charged with the responsibility of acquiring military weapon systems.

For the purposes of this text, the terms "DoD" and "the government" will be used to generically signify any government or DoD organization acquiring military equipment. The success of the ILS organization depends on the ability of both partners to work together to achieve the goals of the ILS program.

THE TWO PHASES OF ILS

There are actually two phases to ILS. Phase I is everything that is done to plan and acquire support before a military system is delivered to the user, and Phase II

- Support influence design
- Develop resource requirements
- Acquire resources
- Provide support for minimum cost

Fig. 1-2. Integrated logistics support goals.

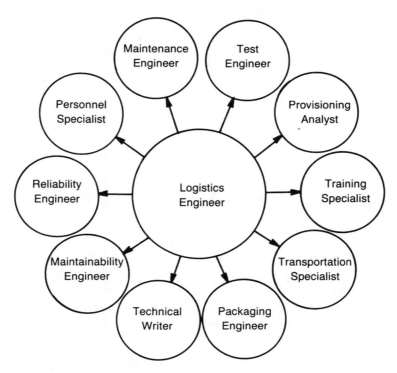

Fig. 1-3. Roles of a logistics engineer.

includes the things that are done to support the equipment while it is being used. Phase I occurs during the design and production of a system, and its duration might be only a few years. Phase II, the useful life of a system, can last up to 20 or 30 years.

The important thing to remember is that the actions that occur during Phase I dictate how well the system will be supported during Phase II. If planning is inadequate during the first phase, then the support will be inadequate in the second phase. During Phase I the majority of ILS planning is accomplished by the government contractor with direction from DoD. Phase II, the support of systems, is normally accomplished by DoD with assistance, in some instances, from government contractors.

THE LOGISTICIAN'S RESPONSIBILITY

The success or failure of military systems rests with those who design, develop, produce, and support them. Logisticians participate in each of these activities and are, therefore, responsible singly and as a group for the success or failure of military systems. The topics in this text are provided to give an insight to better shouldering of this responsibility. Figure 1-3 shows that a logistics engineer must be capable of performing a myriad of tasks, all within the scope of the ILS organization.

Chapter 2

A Foundation for Logistics

As stated in Chapter 1, ILS participates in the design of equipment that is purchased by the government; therefore, the logistician must understand the basic terminologies of government contracts, the acquisition cycle, military equipment, and engineering documentation. Although these topics might seem unrelated, they are the foundation for all ILS activities. Government contracts state the requirements for ILS, and the acquisition cycle dictates when ILS activities are to occur. Delivery of military equipment and the engineering documentation that describes the final design are the result of this contracting process.

GOVERNMENT CONTRACTS

The Department of Defense buys everything used by the military through documents called "government contracts." The government contract is a legally binding document. Its contents describe the equipment that the government is buying, the associated data (information) that is also being procured, and the things that a contractor is required to accomplish to produce both. Every government contract is different. Each one is tailored to fit the specific requirements of what the government is buying. However, for the sake of brevity, they can be grouped into two generic categories: (1) contracts for equipment and (2) contracts for services.

Contracts for equipment deal with the government buying something that is an identifiable quantity, whether it is bolts, beans, or battleships. Contracts for services are issued when the government buys manpower or effort, and can range from providing janitorial services to conducting studies on the feasibility of making a spaceship land on Mars. The government issues contracts for ILS in both of these categories. The difference is that, in the case of equipment contracts, the ILS activity pertains to the equipment being procured, and, in the case of services, the ILS activity will be performed on equipment being procured through a different contract.

The bottom line is that ILS always pertains to equipment. Conversely, many government contracts, such as those procuring bolts, beans, fuel, paper, etc., have no ILS requirements. Therefore, this handbook only deals with contracts for equipment that require ILS. The theories and philosophies discussed herein are applicable to all ILS requirements, whether for contracts procuring equipment or services. ILS participation throughout the acquisition process is required and controlled by government contracts.

The Basic Contract

Typical contracts for equipment have three basic sections that direct ILS activities:

- Product Specification
- Statement of Work
- Data List

Each of these sections has its own purpose, but the result is a single set of requirements that guide the contractor in producing what the government wants.

Product Specification. A complete technical description of the item being procured by the government is contained in the Product Specification. This specification will include performance standards, design limitations, desired characteristics, environmental considerations, and anything else that describes the finished product. The Product Specification for an aircraft would normally tell how fast it must fly, how high it must go, the number of crew members it requires, and the types of things it must do. Even such things as packaging requirements and colors are in the specification. Key ILS information found in the specification are the required equipment mean time between failure, mean time to repair, and proposed maintenance concept. Built-in test, built-in test equipment, support equipment, and other support areas may also be addressed in the specification. Each of these topics is discussed in later chapters. For the time being, however, they should be considered the foundation for all ILS activity, because they are part of the total performance requirements of the final product, and represent the reliability, maintainability, and supportability goals that the item must meet. Figure 2-1 contains example excerpts from a typical Product Specification.

Statement of Work. In order for a contractor to produce the item described in the Product Specification, many actions must occur. These actions are contained in the Statement of Work (SOW). The SOW describes the things that the contractor must do to fulfill all the requirements of the contract. It details all engineering and manufacturing tasks, reporting requirements, reviews, and schedules. In a nutshell, the SOW includes everything that a contractor must do. Specifically, each ILS discipline, as described in Chapter 1, is addressed. The SOW states what ILS tasks will be performed by the contractor and when they will be accomplished. The SOW provides

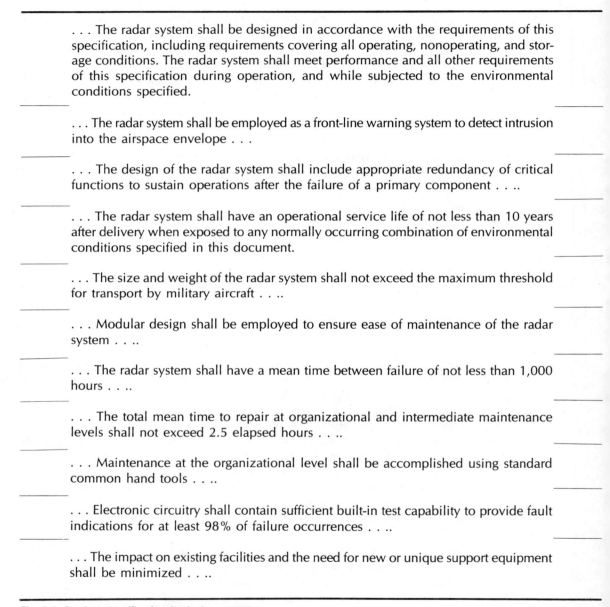

. . . The radar system shall be designed in accordance with the requirements of this specification, including requirements covering all operating, nonoperating, and storage conditions. The radar system shall meet performance and all other requirements of this specification during operation, and while subjected to the environmental conditions specified.

. . . The radar system shall be employed as a front-line warning system to detect intrusion into the airspace envelope . . .

. . . The design of the radar system shall include appropriate redundancy of critical functions to sustain operations after the failure of a primary component . . ..

. . . The radar system shall have an operational service life of not less than 10 years after delivery when exposed to any normally occurring combination of environmental conditions specified in this document.

. . . The size and weight of the radar system shall not exceed the maximum threshold for transport by military aircraft . . ..

. . . Modular design shall be employed to ensure ease of maintenance of the radar system . . ..

. . . The radar system shall have a mean time between failure of not less than 1,000 hours . . ..

. . . The total mean time to repair at organizational and intermediate maintenance levels shall not exceed 2.5 elapsed hours . . ..

. . . Maintenance at the organizational level shall be accomplished using standard common hand tools . . ..

. . . Electronic circuitry shall contain sufficient built-in test capability to provide fault indications for at least 98% of failure occurrences . . ..

. . . The impact on existing facilities and the need for new or unique support equipment shall be minimized . . ..

Fig. 2-1. Product specification (typical excerpts).

detailed instructions on the governing documents that will be followed and the standard practices that will be used; it is an all-encompassing document. Figure 2-2 provides excerpts from a typical Statement of Work.

Data List. The third contract section is the Data List. The formal title for this list is the *Contract Data Requirements List* (CDRL). (For subcontractors, it is called a Subcontract Data Requirements List (SDRL).) The CDRL provides a complete list of the data items that a contractor must submit to the government during the performance of the contract. CDRL items are normally reports, plans, engineering drawings, and other documents related to the contract. In other words, the CDRL identifies all the documentation that a contractor must submit to the government in addition to the

---

. . . This Statement of Work defines the contractor's responsibilities, tasks, deliverable equipment, and data requirements for the procurement of the radar system. This document, in conjunction with the Product Specification and the Contractor Data Requirements List, constitutes the total technical, management, and data requirements for the radar system acquisition . . ..

. . . The contractor is required to design, develop, fabricate, test, and support the radar system . . ..

. . . The contractor shall conduct an Integrated Logistics Support program . . ..

. . . The contractor shall ensure that reliability is a primary design influencing factor in the development of the radar system . . ..

. . . The contractor shall conduct a Reliability Program to include accomplishment of Tasks 101, 102, 104, 201, 202, 203, 204, and 301 of MIL-STD 785B . . ..

. . . The contractor shall conduct a Maintainability Program to include accomplishment of Tasks 101, 102, 201, 202, 203, 204, and 301 of MIL-STD 470A . . ..

. . . The contractor shall conduct a Logistics Support Analysis program to include accomplishment of Tasks 101, 102, 103, 201, 205, 301, 303, 401, and 501 of MIL-STD 1388-1A . . ..

. . . The contractor shall prepare technical documentation for use by operator and maintenance personnel . . ..

. . . The contractor shall identify requirements for spare and repair parts necessary to support operation and maintenance of the radar system . . ..

. . . The contractor shall prepare and submit data in accordance with the Contractor Data Requirements List . . ..

Fig. 2-2. Statement of work (typical excerpts).

Integrated Logistics Support Plan	Personnel Requirements
Common Bulk Item List	Program Progress Report
Computer Operator Manual	Provisioning Parts List
Computer Resources Support Data	Provisioning Screening
Depot Maintenance Work Requirement	Provisioning Technical Documentation
Facilities Plan	RD/GT Procedure
Facilities Recommendations	RD/GT Report
Failure Analysis Reports	Reliability Allocations
Failure Mode and Effects Analysis	Reliability Centered Maintenance
Failure Reports	Reliability Growth Tests
Firmware Manual	Reliability Models
Hazard Analysis Report	Reliability Predictions
Human Engineering Plan	Reliability Program Plan
Instructor Guide	Repair Level Analysis
Life Cycle Cost Data	Repair Parts and Special Tools List
Logistics Support Analysis Plan	Safety Assessment Report
Logistics Support Analysis Records	Software Supportability Plan
Long Lead Time Items List	Student Material
Maintainability Allocations	Supplementary Packaging Data
Maintainability Demonstration	Support Equipment Requirements Data
Maintainability Predictions	Support Materials List
Maintainability Program Plan	System Safety Program Plan
Maintenance Manuals	Technical Manual Plan
Maintenance Plan	Technical Manuals
Maintenance Task Analysis	Test Requirements Document
Maintenance Tools List	Training Course Outlines
Operator Manuals	Training Material
Packaging Data	Training Plan
Part Stress Analysis	Training Requirements
Parts Control Plan	Use Study

Fig. 2-3. ILS data items (examples).

equipment being procured. The CDRL is very significant to ILS because the majority of ILS activities result in some kind of documentation. Typical ILS data items are program plans, technical manuals, spares lists, analysis reports, and the results of other ILS efforts. In many cases, ILS generates the bulk of data items required by a CDRL. Figure 2-3 is a list of typical ILS data items, each of which is discussed in later chapters.

It is important to understand that these three sections are interdependent. The SOW tells what a contractor is required to do in order to produce the item described in the Product Specification. It also tells what activities will be done to produce the CDRL items. The Product Specification defines what the product will do, how it must perform, and how it is to be built. The Data List identifies the documents the contractor must provide to confirm that the effort described in the SOW resulted in an item of equipment that meets the requirements of the Product Specification.

Related Government Documents

The Department of Defense publishes standard guidance documents for most of its activities. Contracts refer to government publications that are applicable to the particular item or service being procured, and the referenced documents become part of the contract. These documents allow standardization of requirements over all contracts and makes contractors' jobs much easier. The most common documents referenced are standards, handbooks, specifications, and directives. Each branch of the military also publishes its own supplementary regulations; such as Army regulations, Air Force regulations, etc. Normally, the military documents are the controlling documents and are supplemented by service regulations.

Standards. Military standards (MIL-STDs) and DoD standards (DoD-STDs) provide guidance. Essentially, they are policy-oriented documents that describe what will be done. Standards are written for specific subjects and normally have DoD-wide application. Extensive references are made to MIL-STDs and DoD-STDs throughout this text.

Handbooks. Handbooks (MIL-HDBKs) and DoD handbooks (DoD-HDBKs) are how-to documents. They provide detailed instructions on methods and techniques and are, many times, extensive reference documents for analyses and mathematical calculations. Many MIL-HDBKs are used by ILS disciplines and are referenced in later chapters.

Specifications. Military specifications (MIL-SPECs) are much like product specifications in that they provide standard specifications for military items and processes. MIL-SPECs form the basic building blocks for most items procured by the government because the parts used to build these items are manufactured to MIL-SPECs. Unlike MIL-STDs and MIL-HDBKs that are identified by the acronym forming the first part of the document number, MIL-SPECs are identified by MIL- (some alpha character), such as MIL-C or MIL-S.

Directives. Directives are issued by DoD to address specific issues in the areas of policies or procedures. These directives can be identified by the DoD-D at the beginning of the document number. These documents are the least referenced of the four types described, because they provide general DoD policy rather than specific guidance.

THE ACQUISITION CYCLE

ILS begins with the acquisition planning for an item of equipment and continues throughout its useful life. This evolutionary process is the same for all items, whether a small piece of equipment or a major weapon system. The acquisition cycle is divided into seven distinctly different phases, and ILS is involved in each phase. These phases are as follows:

- preconcept
- concept

- demonstration and validation
- full-scale development
- production
- deployment and operation
- disposal

Each phase has a definite start and end, although there might be overlap between phases. Figure 2-4 illustrates the phases of the acquisition cycle.

Preconcept Phase

The acquisition cycle begins with the preconcept phase in which the need for new equipment is identified. This need might be based on an evaluation of existing equipment that can no longer perform its mission or on a new type of mission for which equipment does not exist. The purpose of the preconcept phase is to fully define the new need, develop a complete mission profile for the new equipment, identify in gross terms the resources that exist or must be developed to fulfill the need, and establish priorities for continuing the acquisition process.

Concept Phase

When the need has been fully defined, resources identified, and priorities estab-

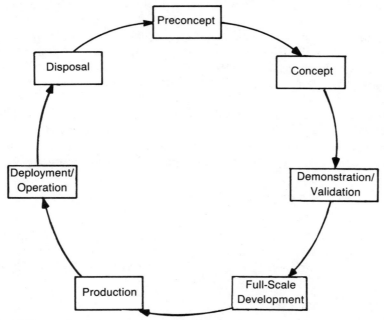

Fig. 2-4. Acquisition cycle.

lished, the next step is to develop alternative approaches for fulfilling the need. The concept phase is where this occurs. During the concept phase, all possible approaches to fulfilling the need are identified. The positive attributes and risks involved with each alternative are evaluated to ensure that the alternatives selected are capable of fulfilling the need. The result of the concept phase is selection of the most feasible alternative, or alternatives, for further study. If an alternative cannot be identified, then the cycle reverts to the preconcept phase for redefinition of the need.

Demonstration and Validation Phase

The alternative, or alternatives, developed during the concept phase must be completely explored to determine if the alternative will actually fulfill the need. To this point, the alternative has been a general approach to solving the need. After all, that's why it is called a "concept." The purpose of demonstration and validation (DEMVAL) is twofold; (1) to transform the concept into a functioning item and demonstrate that it actually works and (2) to validate that the item will fulfill the need defined during the preconcept phase. Failure of all alternatives to pass DEMVAL returns the process to the concept phase for identification of more alternatives.

Full-Scale Development Phase

The alternative, or alternatives, that pass DEMVAL proceed to full-scale development (FSD), also called full-scale engineering development (FSED). During FSD the proposed equipment undergoes a complete engineering process to develop an equipment design that meets all the requirements of the need and that will perform in the field. The purpose of FSD is to produce an equipment design that is reliable, maintainable, producible, and supportable. As is shown in later chapters, a large percentage of ILS activity occurs during FSD.

Production Phase

Actual manufacturing of new equipment occurs during the production phase. This is the first appearance of the complete, operational equipment. The design of the equipment is frozen at the start of production and cannot be changed thereafter without formal approval of the government.

Deployment and Operation Phase

After the equipment is manufactured, the government assumes ownership and the deployment and operation phase begins. The equipment is fielded (deployed) and starts fulfilling the need identified during the preconcept phase. As the equipment functions in its intended environment, its capability to fulfill the need is continually evaluated. The equipment's performance is also evaluated as new needs arise. This ends the circle of one acquisition cycle and begins another.

Disposal Phase

When a new item of equipment is fielded, the item being replaced is phased out and the disposal phase begins for the old item. This phase continues until all of the old equipment is purged from the government inventory or redistributed to fulfill other needs.

As stated above, the acquisition cycle is a never-ending circle of events that continually addresses new needs as they are identified. ILS participates in each phase of acquisition. Subsequent chapters address how ILS provides input to the planning for and use of support resources required to sustain equipment operational capabilities. Figure 2-5 shows typical ILS activities in each acquisition phase.

MILITARY EQUIPMENT

Subsequent chapters refer to weapon systems and items of military equipment as generic "systems," "equipment," or "items" for simplicity and to avoid confusion. However, it is important to recognize that there are different levels of equipment and that these levels are used by ILS disciplines in analyses and for documenting support resource requirements. Also, as the equipment design progresses, different types of models are built to develop and test the design to ensure that the final product will meet the requirements of the product specification. ILS participates in the evaluation of each of these different developmental models.

Equipment Levels

Equipment is developed through the design process using the building-block method of combining pieces in the manner required to produce an item that will perform a required function or mission. MIL-STD 280, Definitions of Item Levels, Item Exchangeability, Models, and Related Terms, defines the levels of equipment as part, subassembly, assembly, unit, group, set, subsystem, and system. These terms are standard definitions used to describe pieces of military equipment and their components. The complexity of the design dictates the levels used to describe an item of equipment. A simple item might have only two or three levels, where a major weapon system might have more than 20.

Part. A part is the lowest level contained in an item of equipment. It refers to one piece, or a group of pieces, that cannot be disassembled without destruction. Examples of a part are a nut, bolt, resistor, or microcircuit. Parts are normally discarded when they fail.

Subassembly. When two or more parts are combined in a manner that will allow disassembly, they form a subassembly. A subassembly is the lowest level where maintenance can be performed by the replacement of parts.

Assembly. A combination of subassemblies and parts is an assembly. The difference between an assembly and a subassembly is the use of the item. An assembly is normally capable of performing a specific function, while a subassembly is dependent on other subassemblies to perform a function. In one application an item might be called an assembly; and in another, a subassembly. Examples of an assembly are a power supply, a circuit-card assembly, or a fuel pump.

Unit. A unit is the lowest level of equipment that is normally able to operate

Pre-concept Phase

- Develop preliminary logistics support requirement estimates.
- Develop preliminary logistics support concepts.
- Prepare preliminary availability predictions.
- Prepare preliminary logistics plans.
- Conduct limited logistic support analysis.

Concept Phase

- Prepare preliminary plans for reliability and maintainability.
- Prepare preliminary predictions for reliability and maintainability.
- Develop preliminary maintenance concept.
- Develop design criteria for ILS disciplines.
- Continue development of logistics support requirements.
- Expand logistic support analysis effort.

Demonstration/Validation

- Update plans developed in previous phases.
- Develop plans for support equipment, training, personnel, facilities, and technical manuals.
- Update predictions developed in previous phases.
- Expand logistic support analysis effort.
- Develop design guidance for logistics concerns.

Full-scale Development

- Update plans and predictions developed in previous phases.
- Conduct full logistic support analysis.
- Generate and maintain logistic support analysis record data base.
- Complete identification of all logistics support requirements.
- Prepare technical manuals.
- Prepare training courses and conduct initial training.
- Conduct initial spares provisioning.
- Participate in design activities to ensure equipment supportability.
- Participate in equipment testing and evaluations.

Production

- Update all plans from previous phases.
- Maintain logistics data base.
- Participate in design changes to upgrade supportability.
- Verify previous predictions using operational data.

Fig. 2-5. ILS activities by acquisition phase (examples).

independently in more than one application. It consists of parts, subassemblies, and assemblies that are combined to perform a distinct function. An electric motor, a computer disk drive, and a truck transmission are examples of a unit.

Group. The next level, a group, is more nebulous, because it relates to the capability of the item, or items, to perform a function. A group is a combination of assemblies and units that is not capable of performing a completely independent operation, but has the capability to perform a role in the accomplishment of an operational function. This level is normally found only in complex items and is used to segregate functions within the equipment.

Set. A set is a combination of units and groups that is capable of performing an independent operational function. Radar sets, radio sets, and test sets are examples of this equipment level.

Subsystem. A major division of a complex item is called a subsystem. It is made up of a combination of assemblies, units, groups, or sets. It performs a function within the overall system. An aircraft might contain, among other things, a radar subsystem, an avionics subsystem, and a propulsion subsystem.

System. A system is the top level of equipment. It contains the total package of parts through subsystems and is capable of performing a complete operational mission. The concept of a system can also include the personnel required to operate and maintain the equipment and the supporting equipment required to sustain operations. An aircraft, a tank, a missile system, and a truck are examples of systems. The term "major weapon system" applies to complex equipment that fulfills a significant tactical or strategic role.

Equipment. Although not defined in MIL-STD 280, the term "equipment" is commonly used to differentiate between big things and smaller things. Equipment is an item that is of lesser scale than a system, normally less expensive, and requires fewer resources to sustain operations. Examples of equipment could be a compressor, a typewriter, or a cargo trailer.

Figure 2-6 shows the relationship between the levels described above. Remember

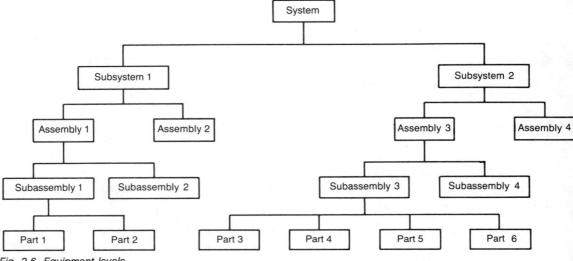

Fig. 2-6. Equipment levels.

Level	Indenture
System	A
Subsystem	B
Assembly	C
Subassembly	D
Part	E

Fig. 2-7. Indenture levels.

that this detail of description does not apply to all items of equipment. The complexity of the design dictates the number of levels used.

Indenture Levels

The method for identifying the relationship between the levels is called the *indenture*. This is accomplished by identification of what makes up an item, starting from the top level and going down. For example, a system is made up of subsystems, a subsystem made up of sets, etc. Many times the same levels of set and group might not be used on the same item of equipment. The term "next higher assembly" (NHA) is commonly used to show indenture relationships. Figure 2-7 illustrates the indenture levels that

INDENTURE	ITEM	NEXT HIGHER ITEM
A	System	None
B	Subsystem 1	System
C	Assembly 1	Subsystem 1
D	Subassembly 1	Assembly 1
E	Part 1	Subassembly 1
E	Part 2	Subassembly 1
D	Subassembly 2	Assembly 1
C	Assembly 2	Subsystem 1
B	Subsystem 2	System
C	Assembly 3	Subsystem 2
D	Subassembly 3	Assembly 3
E	Part 3	Subassembly 3
E	Part 4	Subassembly 3
E	Part 5	Subassembly 3
E	Part 6	Subassembly 3
D	Subassembly 4	Assembly 3
C	Assembly 4	Subsystem 2

Fig. 2-8. Indentured parts list (reference Fig. 2-6).

would apply to Fig. 2-6. Indentures can be signified by either alphabetic (alpha) or numeric characters, but not both. Figure 2-8 is an indentured list of part of Fig.2-6 that shows the NHA relationship of items. Indenture levels are used extensively in ILS planning for the determination of repair levels, maintenance tasks, and spare parts, discussed in later chapters.

Equipment Models

As the design of an item evolves, the transformation of the design from paper to actual working hardware is accomplished by building different types of equipment models. The models can be related to the acquisition phases of equipment development. One of the key phrases to describe the status of these models is the "form, fit, and function" of the item as compared to the required final design. Early models are concerned with developing the required functional capabilities necessary to meet the desired performance criteria. Later models take on the dimensions and mechanical qualities necessary to integrate the final item design.

Preliminary Development Model. The first type of model built is the preliminary development model (PDM). This model is used to evaluate design alternatives in rough experimental form. The term "breadboard" refers to a hand-made electrical/electronic PDM. A PDM isused to evaluate the function being performed. It does not conform to the required form or dimensions (fit) of the final item. The PDM is typically developed during the Demonstration/Validation (DEMVAL) acquisition phase.

Advanced Development Model. As an item nears the end of DEMVAL, an advanced development model (ADM) may be built to demonstrate the feasibility of a design and to show that it meets the desired performance requirements. It is used by ILS to develop initial predictions of the reliability, maintainability, and related support resources requirements. The ADM is much closer to the physical requirements of the final item than the PDM, and provides engineering information to be used during full-scale development (FSD).

Engineering Development Model. The first model built during FSD is the engineering development model (EDM). The EDM typically approximates what the final design will look like and how it will function, and it is used as a test vehicle to see how the final design will operate in its intended environment. ILS uses the EDM to validate and update the logistics predictions developed on the ADM.

Preproduction Model. The preproduction model meets all the requirements of form, fit, and function of the final design. It is the last model built prior to starting full production of the equipment. In many cases, the only difference between preproduction and production models is that the preproduction model is not built using standard manufacturing procedures and processes. The preproduction model is fully capable of performing all required functions and operations.

Production Model. The final design of the equipment is known as the production model. Design activity for the production model is limited to changes that are approved by the government. It is produced using standard manufacturing processes and procedures.

ENGINEERING DOCUMENTATION

Everything that the government buys must be completely described by engineering documentation. This documentation is comprised of narrative information, engineering drawings, or both. The purpose of the documentation is to identify what the final designed equipment looks like, parts and materials used to make the item, how it is assembled, and any special processes required in the manufacturing process. DOD-D-1000B, Engineering Drawings and Associated Lists, provides a basic description of types of engineering documentation and when they are generated. A detailed description of the types of engineering documentation and how they must be prepared is found in DOD-STD 100C, Engineering Drawing Practices. The set of documentation for an item of equipment may be referred to as a *drawing package* or *technical data package* (TDP).

Drawing Classification

DOD-D-1000B divides engineering drawings into three classifications: Level 1—Conceptual and Development Design; Level 2—Production Prototype and Limited Production; and Level 3—Production. The level of drawings is commensurate to the acquisition phases and stages of equipment development. The complexity of drawings increases as the equipment design matures. Figure 2-9 shows the typical relationship between the acquisition phases and drawing levels.

Level 1. Conceptual and development design drawings must be detailed enough to allow evaluation of an engineering concept. The basic requirement for the preparation of level 1 drawings is that they must be legible and provide information in an understandable manner. In most cases, this level of drawings cannot be used to build a deliverable piece of equipment. Typical level 1 drawings can include sketches, undimensioned layouts, and line drawings.

Level 2. As the design matures, level 2 engineering drawings are prepared for use in building production prototypes and limited production equipment. The requirements for format and content of level 2 drawings are much more stringent than for level 1. A level 2 drawing package normally contains parts lists, diagrams, assembly drawings, detail drawings, interface drawings, and special manufacturing and test information.

Level 3. A production drawing package completely describes the equipment to the point that any manufacturer, having the required production capabilities, could take the drawing package and produce equipment that has the same form, fit, and

ACQUISITION PHASE	DRAWING LEVEL
Concept	Level 1
Demonstration/validation	Level 1
Full-scale development	Level 2
Production	Level 3

Fig. 2-9. Drawing Level by acquisition phase.

function as that manufactured by the original designer. A level 3 drawing package is also called a *reprocurement drawing package* because it is used by the government to buy equipment from several sources. Level 3 drawings are also developed by the government for *piece parts*. These parts are referred to as MIL-SPEC parts and have part numbers assigned by the government.

Drawing Types

Standard types of engineering drawings are described in DOD-STD 100C. The most common types are detail, assembly, control, diagram, layout, and special-purpose drawings. Each type has a specific purpose. The specific drawings used in a TDP depend on the complexity of equipment being designed, the phase of the design, and the requirements of the government. All of these drawings, except the layout drawing, can be prepared to level 1, 2, or 3. Figure 2-10 identifies the basic information found on all drawings. Understanding this information makes research of drawings easier.

Detail Drawings. A detail drawing shows all the requirements for the complete end-item. This type of drawing is commonly used to document nonreparable parts such as screws, bolts, fabricated parts, etc. *Monodetail* drawings describe a single part. *Multidetail* drawings are used for uniquely identified parts. Figure 2-11 is an example of a monodetail drawing.

Assembly Drawings. An assembly drawing shows the relationship between two or more parts, subassemblies, or assemblies. Assembly drawings are the most common drawings used to depict an item of equipment. The use of an assembly drawing allows visual comprehension of what the equipment looks like. ILS disciplines use assembly drawings extensively to research information for determining support requirements. An essential part of assembly drawings is the parts list that identifies the parts shown on the drawing. Parts lists can be integral (a part of the drawing) or separate. Separate parts lists have the same part number as the corresponding assembly drawing except that the parts list has a "PL" prefix assigned to the part number, i.e., the number of the parts list for assembly 12345 would be PL12345. An assembly drawing with a tabulated parts lists is shown in Fig. 2-12.

Control Drawings. Drawings that provide information to install specific items with related items, or that provide a complete item description to allow procurement or development of an item commercially, are *control* drawings. (Note: This is where the acronyms get confusing. These next acronyms are commonly mixed up, which can cause a misunderstanding of requirements.) A *specification control* drawing (SCD) describes the specifications to which an item must be manufactured. This type of SCD can be used to buy items from any qualified source of supply. A *source control* drawing (SCD) is used to describe an item that, because of unique characteristics or qualification requirements, can be bought from only specific sources. These are two very different types of drawings that are, unfortunately, both called an SCD. Figure 2-13 shows a specification control drawing that contains suggested sources of supply. Figure 2-14 is an example of a source control drawing showing approved sources of supply. *An interface control* drawing (ICD) provides physical and functional requirements for the interface between two or more related items. This type of ICD is normally used on

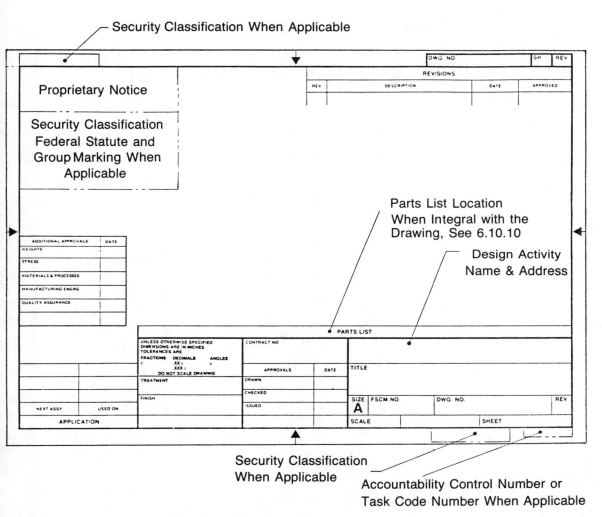

Security Classification When Applicable

Proprietary Notice

Security Classification
Federal Statute and
Group Marking When
Applicable

Parts List Location
When Integral with the
Drawing, See 6.10.10

Design Activity
Name & Address

Security Classification
When Applicable

Accountability Control Number or
Task Code Number When Applicable

Fig. 2-10. Basic drawing information.

equipment levels higher than assembly level. An *installation control* drawing (ICD) describes the physical requirements that must be met for an item to be installed in its intended location. (The same problem exists with ICD that was described for SCD.)

Diagram Drawings. Diagram drawings illustrate the functional relationship among items that are combined to make an assembly or higher level item. This type of drawing uses lines and symbols to describe how the functions are performed. Schematic diagrams show electrical functions of a specific circuit arrangement of electric or electronic parts. Connection and wiring diagrams show the electrical connections of an item or an installation. Interconnection diagrams explain the exterior electrical connections between two or more items. Figure 2-15 illustrates a schematic diagram.

Integrated Logistics Support Handbook

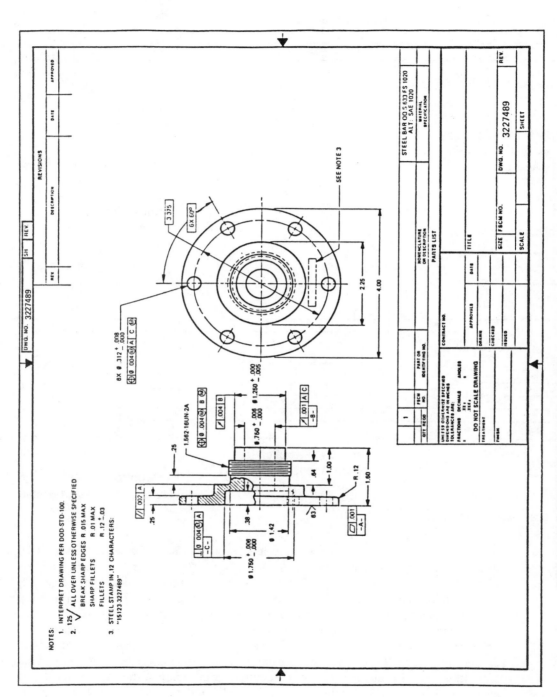

Fig. 2-11. *Mono-detail drawing.*

22

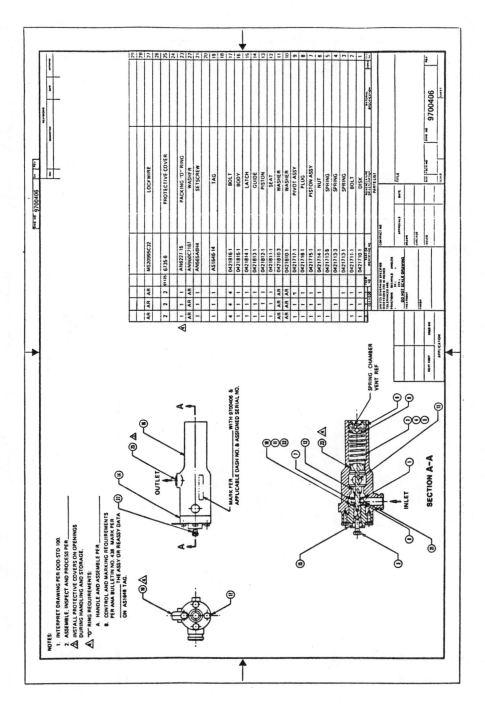

THIS SAMPLE DRAWING IS INFORMATIONAL ONLY AND COMPLETE TO THE
DEGREE NECESSARY TO ILLUSTRATE THE TYPE OF DRAWING BEING DESCRIBED.

Fig. 2-12. Tabulated assembly drawing.

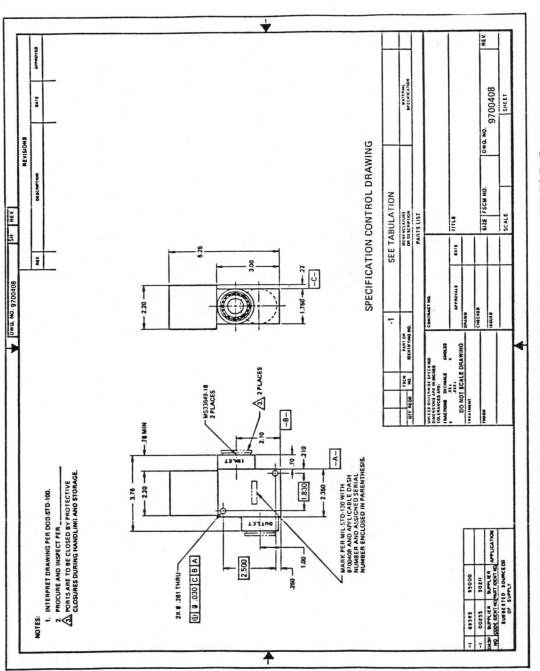

Fig. 2-13. Specification control drawing.

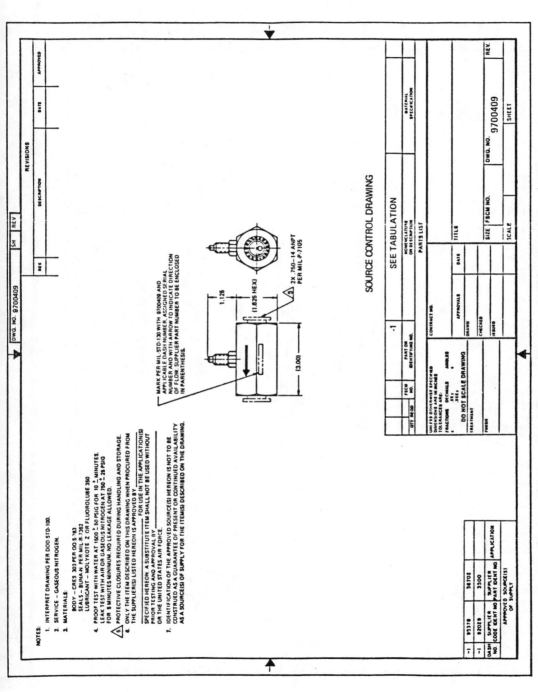

Fig. 2-14. Source control drawing.

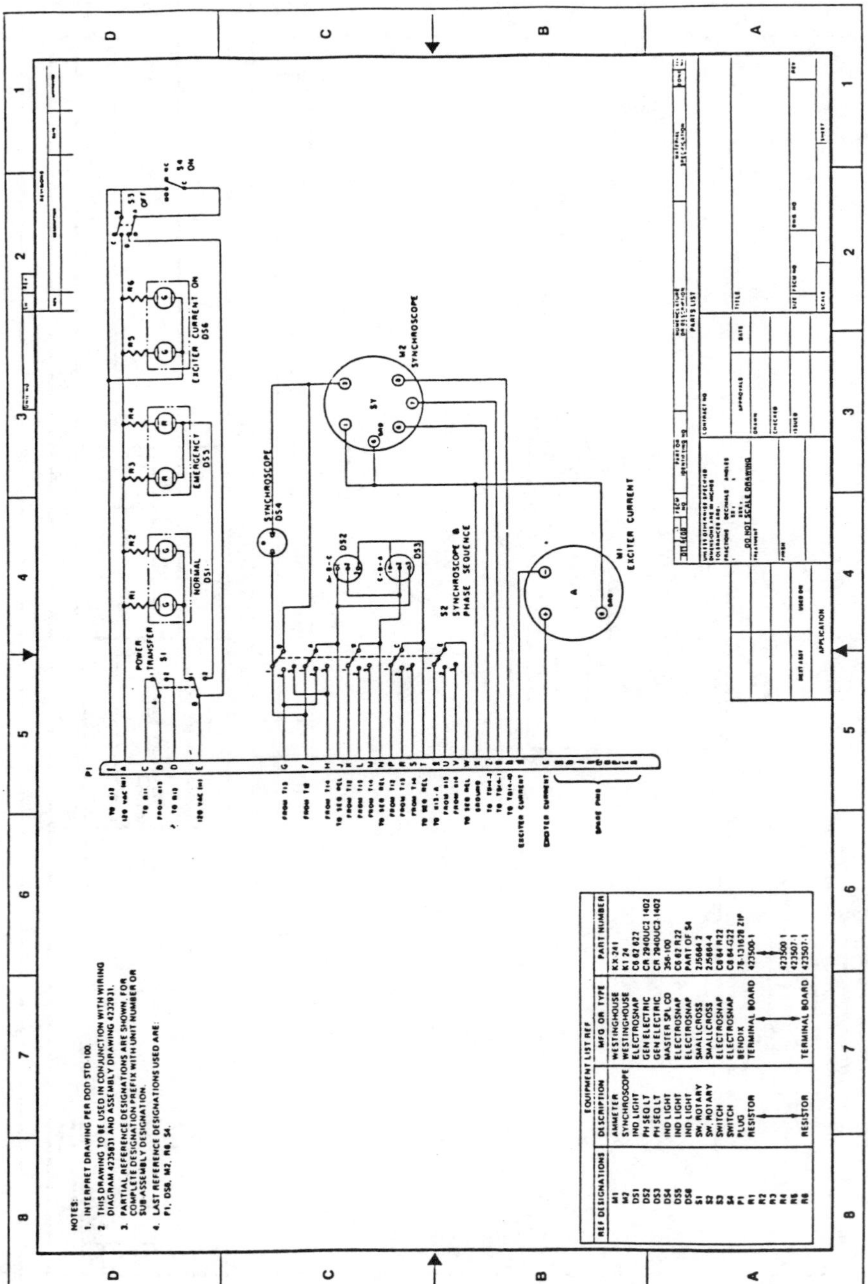

THIS SAMPLE DRAWING IS INFORMATIONAL ONLY AND COMPLETE TO THE
DEGREE NECESSARY TO ILLUSTRATE THE TYPE OF DRAWING BEING DESCRIBED.

Fig. 2-15. Schematic drawing.

Special-Purpose Drawings. Drawings used to supplement those described above are categorized as special-purpose drawings. This type of drawing might be required for management control, might aid manufacturing processes, or might be other drawings required by the government. Examples of special-purpose drawings are wire lists that show wiring correlation within an assembly, wiring harness drawings that describe how a wiring harness is built, cable assembly drawings that show how a cable is manufactured, printed wiring master pattern drawings used to produce printed circuit boards, and kit drawings.

Layout Drawings. Layout drawings are design development drawings that provide minimal information. Any of the previously identified types of drawings might start as a layout drawing. The level of detail of a layout drawing does not allow formal item identification. Layout drawings are normally prepared to only the level 1 category of detail.

Configuration Management

As the design of an equipment progresses, the quantity of drawings required for documentation can be enormous. The control of these drawings is known as configuration management. MIL-STD 482A, Configuration Status Accounting Data Elements and Related Features, establishes the procedures for formal control of equipment configuration. During the initial design phase, the drawings change almost daily due to engineering efforts. The point where all documentation becomes subject to configuration control is called *baseline*. At system baseline, the design is normally complete except for changes required as the result of testing and acceptance criteria. The basic system design is complete.

Formal configuration control must be established prior to the beginning of production. After configuration control is established, all changes to the design of the equipment must be approved by the government in accordance with DOD-STD 480A, Configuration Control-Engineering Changes, Deviations, and Waivers.

Contractors submit recommended changes to the government through engineering change proposals (ECP). There are two classes of ECP. A class 1 ECP recommends design changes that affect the form, fit, or function of the equipment. Other changes, such as correction of drawing errors or substitution of parts that do not change form, fit, or function, are class 2 changes.

Configuration management is important to ILS disciplines. It is hard to complete a design and identify required support resources if the design of the equipment continuously changes or if changes occur without being identified. Most contractors have a Configuration Control Board (CCB) that controls and approves all design changes. ILS normally has representation on this board.

The topics discussed in this chapter have a direct relationship to all ILS activities. Understanding the typical government contract that tells what must be accomplished and the methods for describing equipment and its supporting engineering documentation is necessary for a logistician to effectively plan for and manage the resources required to support the equipment when it is fielded.

Chapter 3

Reliability Engineering

Sooner or later all equipment breaks. The more often it breaks, the greater the amount of resources required to support it. The purpose of reliability engineering is to address this problem. Reliability engineering's effort is really twofold: first, participating in the design and development of equipment to make it as failure-free as possible, and second, predicting how the equipment will fail when it is being used. The results of these tasks are used by all ILS disciplines to develop support resource requirements. Reliability engineering accomplishes these tasks using a standard methodology that is established by MIL-STD 785B, Reliability Program for Systems and Equipment Development and Production, and MIL-HDBK 338, Electronic Reliability Design Handbook.

DEFINITIONS

Before starting a lengthy discussion of the Reliability Program the following definitions are necessary.

Reliability. The probability that an item of equipment will perform its intended mission without failing, assuming that the item is used within the conditions for which it was designed, is reliability. Note that there are two key statements in this definition: (1) the word "probability" indicates that reliability deals with statistical calculations and projections of how and when failures may occur, and (2) "within the conditions for which it was designed" sets a boundary on the validity of any predicted failures.

Failure. Failure occurs when an item cannot perform to the requirements for which it was designed. This definition appears to be a simple yes or no. In actuality, there

$$\lambda \text{ (Failure Rate)} = \text{Number of Failures/Total Operating Time}$$

Example: If eight failures occurred in 1000 operating hours

$$\lambda = 8/1000$$
$$\lambda = 0.008$$

Fig. 3-1. Failure rate calculation.

can be gray areas when it is applied. For instance, if the transmission of a truck breaks, there is no doubt that the truck cannot perform its mission of hauling cargo. On the other hand, if an aircraft is capable of flying at a speed of 1000 miles per hour (mph), and for some reason it can only fly 997 mph on a given day, did the aircraft fail? If its mission required a speed of 1000 mph, then it failed. If its intended design mission only required a speed of 950 mph, then it didn't.

Failure Rate. This rate a numeric value that predicts the number of failures of an item, assembly, or piece part that will occur during one hour of operation. Failure rates are developed using tests, field experience, and other significant data. Fig. 3-1 illustrates how a failure rate can be calculated using actual usage data. Dividing the number of failures that occurred over a specific length of time by that length of time results in the failure rate. Note that the Greek letter lambda (λ) is used to represent the failure rate. As shown in Fig. 3-1, failure rates are normally numbers with several decimal places. Many times failure rates are written as failures per million hours of operation.

Mean Time Between Failures (MTBF). This is the reciprocal of a failure rate. It predicts the average number of hours that an item, assembly, or piece part will operate before it fails. Figure 3-2 shows that an item with a failure rate of 0.008 would have an MTBF of 125 hours, which means that, in theory, the item should operate for 125 hours before experiencing a failure. The MTBF of an item is often used in calculating spares requirements, life cycle cost, and other failure-related data. It is easier to use

$$\text{MTBF} = 1/\lambda$$

Example: If an item had a failure rate of 0.008

$$\text{MTBF} = 1/0.008$$
$$\text{MTBF} = 125 \text{ hrs.}$$

Fig. 3-2. MTBF calculation.

$$R(x) = e^{-\lambda t}$$

Where: $R(x)$ = probability of success
e = natural logarithm base
λ = failure rate
t = mission duration

Example: What is the probability of success for an item having a failure rate of 0.008 to complete an eight-hour mission?

$R(x) = e^{-\lambda t}$
$R(x) = (2.718)^{-(0.008)(8)}$
$R(x) = 0.938$ (93% probability of completing the mission)

Fig. 3-3. Mission reliability.

an MTBF than a failure rate when talking about the reliability of an item because people tend to be more comfortable with whole numbers than decimals.

Mission Reliability. One of the predictions that is used to gauge how an item might perform is its predicted mission reliability. This is expressed as the probability of successfully completing a mission of a specified length. Figure 3-3 provides the formula for this prediction. Although the results of this calculation might not look as though they have any usable relationship to the actual performance of the item being developed, they do provide a baseline for evaluation of trade-offs and design changes that affect reliability.

Reliability Engineering. Reliability engineering uses a standard set of mathematical or statistical methods and analyses to predict the reliability of an item and to identify where reliability can be improved by design changes.

RELIABILITY PROGRAM: MIL-STD 785B

DoD contracts for the development of equipment require the contractor to conduct a reliability program. This requirement is imposed to ensure that equipment is designed to meet a desired reliability. The document used to convey to the contractor how the reliability program will be conducted is MIL-STD 785B. This document is formatted by task into three sections:

- Task Section 100—Program Surveillance and Control
- Task Section 200—Design and Evaluation
- Task Section 300—Development and Production Testing

The tasks in MIL-STD 785B provide the guidance necessary for a complete reliability program. Alternatively, the program can be tailored by DoD to include only the tasks that are applicable to the specific equipment being procured. Major weapon system programs normally require that all tasks be accomplished to receive the full benefit of a reliability program.

RELIABILITY PROGRAM SURVEILLANCE AND CONTROL

Task Section 100 of MIL-STD 785B is comprised of tasks that describe how a contractor will establish and monitor the reliability program. These tasks require specific actions to be accomplished to ensure that the contractor considers all aspects of a program that affect equipment reliability.

Reliability Program Plan (Task 101)

The reliability program plan is the foundation of a contractor's reliability program. The plan tells, in great detail, how each objective of the program will be accomplished, managed, and reported. As with all plans, it provides sufficient information to adequately describe how all the required tasks of MIL-STD 785B will be fulfilled. An example of an outline for a reliability program plan is shown in Fig. 3-4. The plan is submitted to the government for approval early in the program and is updated throughout the reliability program to reflect changes as new information becomes available.

1.0 PURPOSE

Describe the intent of the reliability program and tell how the program will fulfill the contractual requirements stated in the Statement of Work. State the desired results of the program.

2.0 EQUIPMENT DESCRIPTION

Provide a brief description of the functional and physical characteristics of the equipment being developed. Equipment reliability requirements established by DoD in the procurement specification, such as MTBF, should also be stated here.

3.0 PROGRAM TASKS

Describe how each MIL-STD 785B task will be performed. Provide examples when possible. Be specific on the methodology to be used. If a task is not required by the Statement of Work, it could possibly still be required in order to perform other tasks. For example, Task 201 may not be required by the Statement of Work, but it must be done if Tasks 202 and 203 are required.

Fig. 3-4. Reliability program plan.

4.0 MANAGEMENT

Describe how the program will be managed. Identify the organizations that will be responsible for accomplishing each task. Include internal procedures for controlling the effort to accomplish each task. Identify the position that has authority for making reliability decisions.

4.1 Organization

Describe the Reliability organization and how it interfaces with other organizations. Provide a diagram of:
- a. the reliability organization,
- b. how the reliability organization relates to other ILS organizations, and
- c. how the above organizations relate to the total program management structure.

4.2 Interfaces

Describe how the reliability program will interface with the total equipment design effort. Specifically tell how reliability data will be integrated into the design and LSA. Flow charts showing these interfaces are very effective.

5.0 SCHEDULE

Provide a milestone schedule showing when each task will be started and completed. Relate reliability milestones to the overall equipment development program milestones. Indicate critical events on the schedule that are dependent on other events, e.g., things that must occur prior to the start or finish of other tasks.

6.0 RELIABILITY PROBLEMS

Describe the known or anticipated reliability problems of the equipment. Examples of these problems could be application of new technology, critical components that have a history of high failure rates, or stringent operating requirements. Provide an assessment of each problem and the proposed or planned method of resolution.

7.0 DATA SOURCES

Describe what sources of data will be used as inputs to the reliability tasks. Data available from previous programs for similar equipment provide the most creditable starting point. Describe how this data will be used and the relative merit of what will be gained through its use. Define what data is expected to be provided by DoD. This is normally historical field data collected by DoD on similar equipment. Identify information expected to be received from suppliers or vendors as a result of their participation in this program. Indicate other anticipated sources of information such as studies, research, etc.

Fig. 3-4. Continued.

Monitor and Control of Subcontractors and Suppliers (Task 102)

It is important that contractors pass along reliability requirements to their subcontractors and suppliers. This assures that final reliability requirements for the equipment will be met. Task 102 provides detailed guidance on the methods of doing this. The task requires that contractors direct subcontractors to establish reliability programs, participate in reliability reviews, and provide reliability data to contractors for inclusion in the overall computation of equipment reliability. This task is extremely important for major weapon system programs where there are normally several tiers of subcontractors and suppliers designing and building portions of the system.

Program Reviews (Task 103)

The purpose of Task 103 is to require formal reliability program reviews at key milestones in the development of equipment. Reviews allow DoD to ensure that the reliability program is proceeding as required by the contract and that program requirements are being met. These reviews allow a free exchange of ideas between the contractor and the government and provide a forum for resolution of problems. Reliability program reviews normally coincide with the Preliminary Design Review (PDR) and Critical Design Review (CDR). Reliability plays a key role in these meetings, because DoD approval of the equipment design is received at PDR and CDR.

Failure Reporting, Analysis, and Corrective Action System (FRACAS) (Task 104)

Task 104 establishes a formal method for reporting failures during engineering model testing or, sometimes, after an item of equipment has been fielded. The task also requires that contractors report their analysis of what caused the failure and a recommended action to correct any problems that were identified. As indicated, there must be a working piece of equipment before this task can be accomplished. It is normally required in conjunction with tasks from Task Series 300, which is discussed later in this chapter. FRACAS is a closed-loop reporting system, which means that each reported failure must be followed by a report of the failure analysis and the corrective action required.

Failure Review Board (Task 105)

Contractors are required to establish a Failure Review Board (FRB) by Task 105. The purpose of the board is to review failure trends and analysis of failures, and to ensure that corrective actions are taken. This task works in tandem with Task 104 (FRACAS) as part of the closed-loop failure reporting and corrective action system.

FRBs are normally comprised of members representing ILS, design engineering, reliability engineering, system safety, maintainability, manufacturing, and quality assurance organizations. The board is required to maintain records of meetings, actions taken, and recommendations. The FRB also reviews failure information received from subcontractors and suppliers that has been provided through Task 102. Figure 3-5 shows where the FRB fits into the overall FRACAS.

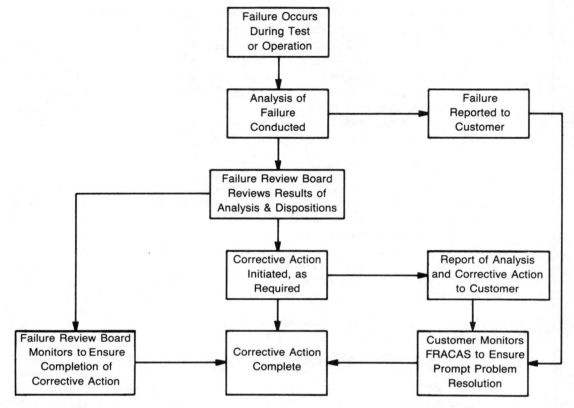

Fig. 3-5. FRACAS flow diagram.

RELIABILITY DESIGN AND EVALUATION

Task Section 200, Design and Evaluation, of MIL-STD 785B contains nine tasks describing the analyses that reliability engineering performs to determine the failure rates of items, assemblies, and piece parts so that equipment designs can be changed to increase overall system reliability. Accurate and successful performance of these tasks is the key to developing a reliable item of equipment.

Reliability Modeling (Task 201)

As previously stated, reliability is a statistical study. Therefore, the first step in initiating the statistical analysis is to develop a creditable method for modeling the reliability of an item. MIL-STD 756B, Reliability Modeling and Prediction, and MIL-HDBK 338 provide detailed procedures for developing models. Reliability mathematical models are developed based on the functional operations of the equipment being analyzed and are tailored to reflect the specific operating requirements of the equipment. The purpose of modeling is to provide an accurate method of predicting failure rates.

In the early stages of an equipment development program, reliability engineering will also develop a reliability block diagram. The purpose of the block diagram is to provide a base for accurate mathematical model preparation. The block diagram is constructed to represent the functional configuration of the equipment. Figure 3-6 is a simplified example of a reliability block diagram. Both the math model and the block diagram are used to accomplish Tasks 202 and 203 of MIL-STD 785B.

Reliability Allocations (Task 202)

During the early design phase of an item of equipment, it is important to have a reliability baseline to guide design engineers as the equipment is developed. Reliability allocations provide the method for establishing this baseline. Allocations are developed using the same functional equipment breakdown as was used to develop the equipment reliability block diagram and math model in Task 201.

The process starts with the overall equipment MTBF or failure rate specified by DoD in the procurement specification. The equipment MTBF is then allocated to the lower functional levels of the equipment. The number of levels addressed is normally identified by DoD in the procurement specification. Allocation is accomplished using past experience of similar equipment or best engineering judgments. Figure 3-7 shows how reliability allocations were developed for the reliability block diagram illustrated by Fig. 3-6. In the example, allocations are expressed in terms of MTBF. Figure 3-7 also shows the conversion of failure rates to MTBF. Note that failure rates are additive, which makes it easier to see how the total equipment failure rate was apportioned.

During the design of equipment, the use of allocated failure rates provides a goal for the reliability of each assembly. As the design develops and the predicted failure rate for each assembly is generated, reliability engineers can compare the predicted failure rates with the allocated failure rates to identify assemblies that will cause the overall equipment to exceed its required MTBF. Using the comparison as a guide,

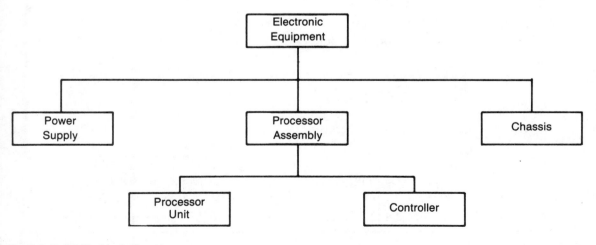

Fig. 3-6. Reliability block diagram.

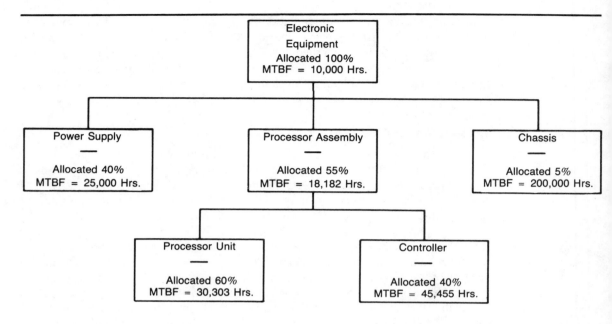

Step 1. Convert the required system MTBF to a failure rate.

1/10,000 (MTBF) = .0001 (system failure rate)

Step 2. Determine the percent of the failure rate to be allocated to each item at the next level down. (Refer to the block diagram)

Item	Allocated % of Failure Rate
Power supply	40
Processor assembly	55
Chassis	5

Step 3. Compute the allocated failure rate.

Item	Allocated Failure Rate
Power supply	.0001 × 40% = .00004
Processor assembly	.0001 × 55% = .000055
Chassis	.0001 × 5% = .000005

Step 4. Convert the allocated failure rates to MTBF.

Item	Allocated MTBF
Power supply	1/.00004 = 25,000 hrs.
Processor assembly	1/.000055 = 18,182 hrs.
Chassis	1/.000005 = 200,000 hrs.

Step 5. Repeat Steps 2 and 3 for each descending level for the complete system.

Fig. 3-7. Computing reliability allocations.

reliability engineers can recommend design changes, such as parts with higher reliability, different technologies, or alternate processes, that will achieve the desired level of reliability. MIL-HDBK 338 provides further details on this process.

Reliability Predictions (Task 203)

The method for conducting reliability predictions varies with the type of equipment under consideration and the status of the design. MIL-STD 756B, Reliability Modeling and Prediction, and MIL-HDBK 338 provide several methods for predicting the reliability of equipment. For the purposes of this discussion, two typical methods are presented to illustrate how reliability predictions are developed.

The first method, parts count reliability prediction, is used in the early stages of design to develop predictions based on the generic failure rates of the anticipated quantity and type of parts to be used in the equipment. The second is the part stress analysis prediction, which uses the actual design configuration to develop detailed predictions. The parts count method is valuable in developing initial predictions for comparison with allocations prepared in Task 202, and in providing information to designers and management on the capability of a proposed design meeting basic reliability requirements. As the design matures, the parts stress analysis provides detailed information on reliability and identification of specific areas where reliability can be improved.

The validity of both methods is dependent on the quality of data used in the predictions and the accuracy of assumptions. The prediction of failure rates for items of equipment, using either method, begins at the piece part level and is summed to the equipment level. MIL-HDBK 217, Reliability Prediction of Electronic Equipment, provides detailed instructions and input data for both of these methods.

Parts Count Reliability Prediction. The parts count method of predicting equipment reliability uses a rather simple formula. The formula, illustrated in Fig. 3-8, shows that, by using generic failure rates from tables in MIL-HDBK 217 and anticipated types and quantities of parts, a baseline equipment failure rate can be derived. The quality factor indicates the level of testing and quality control that the manufacturer used when parts were made. A comparison of the reliability prediction developed for the Processor Assembly in Fig. 3-8 with the reliability allocation in Fig. 3-7 indicates that the initial conceptual design of this item is within the allocated reliability baseline.

Part Stress Analysis Prediction. MIL-HDBK 217 contains a series of complex formulas used to perform a part stress analysis prediction. There is a different formula tailored for each type of part, and each formula considers the electrical stress, heat, environment, and frequency of use of each part type. An example of a stress analysis prediction for a single part is illustrated in Fig. 3-9. This process would be completed for each part for which data did not exist or could not be obtained from the manufacturer. The resulting failure rates are then put into formulas that consider the application of the part and whether it operates in a single, dual, parallel, series, and/or redundant mode. This method, identifies individual parts and assemblies that drive the equipment reliability up or down. Through the use of the part stress analysis prediction method, reliability engineers produce detailed failure rates for all parts, assemblies, and the total

Formula:

$$\lambda_{EQUIP} = \sum_{i=1}^{i=n} N_i \, (\lambda_G \pi_Q)$$

Where:

λ_{EQUIP}	=	Total equipment failure rate (failures/10^6 hr.)
λ_G	=	Generic failure rate for the i th generic part (failures/10^6 hr.)
π_Q	=	Quality factor for the i th generic part.
N_i	=	Quantity of i th generic part
n	=	Number of different generic part categories.

Part Type	Failure Rate[1,2]	Quality[2]	Adjusted	Quantity	Total
Transistors (NPN)	0.86	2	1.72	8	13.76
Resistors (comp)	0.038	1.5	0.057	2	0.114
Resistors (vari)	0.34	1.5	0.51	6	3.06
Capacitors (cer)	0.17	1.5	0.255	10	2.55
Diodes (gen)	0.14	2	0.28	4	1.12
IC (MOS)	0.41	3	1.23	12	14.76
Connector	0.15	2.5	0.375	1	0.375
Printed board	0.01	3	0.03	1	0.03

Total predicted failure rate ($\times 10^{-6}$)	35.769

Predicted MTBF $(1/(35.769)^{10^{-6}}) = 27{,}957.16$

[1]Failure rates expressed as $\times 10^{-6}$.
[2]Failure rates and quality factors from MIL-HDBK 217D.

Fig. 3-8. Parts count reliability prediction.

equipment. These predictions are just that—statistical predictions. Actual failure rates are not developed until the equipment is built and used a sufficient length of time to produce field or test data that is based on actual equipment usage. Figure 3-10 illustrates how failure rates for assemblies are predicted by (1) determining the failure rate for each type of part used in the assembly, (2) multiplying the part failure rate by the quantity of that part used in the assembly, and (3) summing the results. The MTBF was calculated by taking the reciprocal of the assembly failure rate. The failure rate of the equipment is predicted by adding the failure rates of all the assemblies. This example

---

Reliability Worksheet
MIL-HDBK-217D
Monolithic Bipolar and MOS Digital SSI/MSI Devices
(fewer than 100 gates)

Formula:

$$p = Q [C1 \times T \times Y + (C2 + C3) E] L \times 10^{-6}$$

Where:

p = Device failure rate
Q = Quality factor (1)
T = Temperature acceleration factor based on technology (13)
Y = Voltage derating stress factor (1)
E = Application environment factor (4)
C1 = Circuit complexity failure rates based on gate count (0.0077)
C2 . = Circuit complexity failure rates based on gate count (0.0006)
C3 = Package complexity failure rate (0.01)
L = Device learning factor (1)

(Example factors obtained from tables in MIL-HDBK 217D)

Calculation:

$$p = Q [C1 \times T \times Y + (C2 + C3) E] L \times 10 \quad = 0.1425 \times 10^{-6}$$

Predicted failure rate (as computed above): 0.1425×10^{-6}

Fig. 3-9. Part reliability prediction.

has been simplified for ease of understanding. In practice, the theory for developing failure rate predictions remains the same, but other factors such as duty cycle, derating, and environment are applied, resulting in a more accurate prediction.

Failure Modes, Effects, and Criticality Analysis (FMECA) (Task 204)

The identification of all the probable ways that parts, assemblies, and equipment might fail, the causes for each failure, and the effect that the failure will have on the capability of the equipment to perform its mission provides a valuable tool for reliability engineers. A FMECA, developed in accordance with Task 204 and MIL-STD 1629, Failure Modes, Effects, and Criticality Analysis, is a complete analysis of each level of the equipment. Using the FMECA, reliability engineers identify each possible failure mode of the equipment.

A *failure mode* is something that occurs, such as a part failing, that causes the equipment to not function properly. A single part can have several failure modes. When

Item Description	Quantity	Failure Rate	Extended	MTBF
Assembly Level				
Control assembly			16.5924	60268.556
Control card 1	1	1.9956	1.9956	
Control card 2	2	1.4562	2.9124	
Module subassembly	3	3.0582	9.1746	
Multiplexer	1	2.5098	2.5098	
Transceiver assembly			39.1311	25555.121
Power supply	2	6.8835	13.767	
Transmitter module	4	2.9744	11.8976	
Receiver module	4	3.0861	12.3444	
Audio assembly			1.1221	891186.16
Speaker subassembly	1	1.0684	1.0684	
Interconnection	1	0.0537	0.0537	
System Level				
Communications system			95.9767	10419.195
Control assembly	1	16.5924	16.5924	
Transceiver assembly	2	39.1311	78.2622	
Audio assembly	1	1.1221	1.1221	

Summary

Predicted Failure Rate $\quad$ 95.9767 $\times$ 10^{-6}

Predicted MTBF $\quad$ 10419.195 Hours

Fig. 3-10. Reliability prediction worksheet.

considering failure modes, remember that they are actual failures, not symptoms of failures. For example, if you put the key in the ignition of your car, turn it, and nothing happens, that is a failure symptom. The actual failure could be a dead battery, faulty ignition switch, etc. Turning the key and nothing happening is only a *failure indicator*.

The FMECA also provides information on failure indicators or how users know when a failure has occurred. Other information developed by a FMECA includes the following:

- Predictions of the percentage of occurrence of each failure mode for a part
- A description of what caused the failure
- The effect that the failure will have on the capability of the equipment to perform its mission
- Identification of any safety or other type hazard that the failure will cause
- Identification of methods required to fault-isolate the failure
- Corrective action required to correct the failure

MIL-STD 1629 contains detailed tasks that describe the techniques that reliability engineers use to conduct and document a FMECA, and MIL-HDBK 338 contains further information on developing FMECA data. An example of a FMECA worksheet with a description of the information that is recorded on the worksheet in Fig. 3-11. Later in this book, the FMECA is referenced as input information for other ILS activities. These activities include maintainability, maintenance planning, safety, and LSA. The FMECA is an integral part of the process that identifies requirements for logistic support resources.

Sneak Circuit Analysis (Task 205)

Design of electronic equipment is a very complex task, and it is not impossible for design engineers to overlook hidden faults in the extensive circuitry of state-of-the-art items. Hidden faults in design can cause unwanted functions or limit performance. The purpose of a sneak circuit analysis is to identify these hidden faults.

A sneak circuit analysis is accomplished through a time-consuming, circuit-by-circuit analysis of the electronic design. A complete set of engineering drawings and schematics is required. Therefore, the analysis can only be done in the latter stages of design when sufficient detailed documentation exists. The results of the analysis identify undesired circuits, design concerns, or errors in the documentation.

A sneak circuit analysis can be very expensive to perform because of the hours required; and because of the late stages of design when it is performed, the recommended design changes can be very expensive to implement. Because of these constraints, the analysis is normally limited to critical components and circuits. A detailed description of how a sneak circuit analysis is performed is contained in MIL-HDBK 338.

Electronic Parts/Circuits Tolerance Analysis (Task 206)

Changes in operating temperatures will often change the electrical operating characteristics of parts and circuits. The purpose of a tolerance analysis is to identify potential occurrences of this condition that will cause the equipment to exceed its operating specification. This analysis is complex and is most often done using a computer. It is also expensive because of the skill required and the time necessary to develop a computer simulation to do the analysis. So it is normally limited to critical items or items that are historically most susceptible to temperature changes, such as circuits using high power levels and power supplies. The results of the tolerance analysis provide design change recommendations that should increase equipment reliability.

Parts Program (Task 207)

The purpose of the parts program is to standardize, as much as possible, the parts used by a contractor in designing an item of equipment. The procedures for establishing and conducting the parts program are stated in MIL-STD 965A, Parts Control Program. Most DoD acquisition programs require contractors to use this process to ensure that fully qualified parts that meet DoD standards for reliability and quality are produced.

Failure Mode Effects and Critically Analysis
MIL-STD 1629A

SYSTEM ① _____

INDENTURE LEVEL ② _____

REFERENCE DRAWING ③ _____

MISSION ④ _____

DATE _____

SHEET _____ OF _____

COMPILED BY _____

APPROVED BY _____

IDENTIFICATION NUMBER	ITEM/FUNCTIONAL IDENTIFICATION (NOMENCLATURE)	FUNCTION	FAILURE MODES AND CAUSES	MISSION PHASE/ OPERATIONAL MODE	FAILURE EFFECTS			FAILURE DETECTION METHOD	COMPENSATING PROVISIONS	SEVERITY CLASS	REMARKS
					LOCAL EFFECTS	NEXT HIGHER LEVEL	END EFFECTS				
⑤	⑥	⑦	⑧	⑨	⑩	⑪	⑫	⑬	⑭	⑮	⑯

42

1. Identification of the end item using either name or part number.

2. Indenture level of the item being analyzed.

3. Drawing number of the engineering drawing being used as reference for the analysis.

4. Identification of the function of the item being analyzed.

5. Identification number for the function being analyzed using either a functional group code, per MIL-STD 780, or an LSACN, per MIL-STD 1388-2A. For traceability with other ILS activities, it is recommended that the LSACN be used.

6. Identification of the item or function contained in the item identified by note 3, above, that is being analyzed.

7. Statement of the function performed by the item being analyzed.

8. List all possible failures of the item being analyzed. There can be many failures of a single item listed here. Remember that the failures are of the item identified in column 6.

9. Identify the phase of the operation or mission when the failure is predicted to occur, i.e., start-up, peak operation, etc.

10. Identify what will happen to the item being analyzed when the failure occurs.

11. Identify what will happen to the next higher assembly-level item when the failure occurs.

12. Identify what will happen to the end item when the failure occurs.

13. Identify how the operator will know when the failure occurs.

14. Identify design provisions that will permit nullifying the effect of the failure on system operation, i.e., redundancy, reset, etc.

15. Assign the appropriate classification for the severity of the consequences if the failure occurs.
Category I is catastrophic, including loss of equipment or death of personnel.
Category II is critical, including severe injury to personnel or damage to equipment.
Category III is marginal, including minor injury or damage.
Category IV is minor, no serious injury or damage, but still requires unscheduled maintenance.

16. Record additional pertinent information as required.

Fig. 3-11. FMECA worksheet format.

MIL-STD 839, Selection and Use of Parts with Established Reliability Levels, contains guidelines on selection of standard parts.

The basic document produced by the parts program is a Program Parts Selection List (PPSL). The PPSL lists all parts that are approved by DoD for use on a specific contract. This document is extremely valuable on large programs where there might be many subcontractors providing segments of the end-item of equipment. DoD might give the contractor an initial PPSL that is amended, with DoD approval, by the contractor as the design evolves. Nonstandard parts must be qualified by the contractor before they can be added to the PPSL. This list is used by design engineers as they select parts for use in the equipment. Figure 3-12 is an example PPSL.

To assist contractors in selecting nonstandard parts for use in design, DoD has a data repository that is used by government and industry to accumulate historical information about parts. This data is accessible through procedures stated in MIL-STD 1556A, Government/Industry Data Exchange Program (GIDEP). The GIDEP data bank contains four sections; engineering, failure rate, metrology, and failure experience data. This data provides an additional source of information that can prove valuable to a contractor's parts control program.

Reliability Critical Items (Task 208)

The purpose of Task 208 is to provide special emphasis on items (parts or assemblies) that are critical to the equipment's achieving its overall reliability goals. These items have either high predicted failure rate (Task 203) or the greatest potential for increasing the requirements for logistic support resources based on the results of

Generic Part Number	Military Part Number	Part Name	Part Description	Procurement Specification Number
LM741A	M38510/10101B	Microcircuit, linear, monolithic, op amp, bi-polar	Operational amplifier, contains 23 transistors	MIL-M-38510/101
PLT120	03A051C	Fastener, blind, flush	100-degree flush head, A-286, self-locking	NAS1675
RCR05	RCR05G∗∗∗J( )	Resistor, fixed, composition	0.125W/150V, 2.7 ohm to 22 M ohms, 5%	MIL-R-39008/4
1N4148	JANTXV1N4148-1	Semiconductor, diode, silicon, switching	VRWM = 75 V, VF = 1.2 V TA = 25 degree C, IF = 50 MA	MIL-S-19500/116
SE12	SE12XC∗∗( )	Terminal, stud	Stand-off terminal stud non-insulated	MIL-T-55155/12

Fig. 3-12. Preferred parts selection list.

the FMECA. Other items that are classified as reliability critical include potential safety hazards, items requiring special handling or transportation precautions, items difficult to build, or items with a poor performance history. By increasing the reliability of these critical items, reliability engineers will have the greatest impact on the total program.

Effects of Functional Testing, Storage, Handling, Packaging, Transportation and Maintenance (Task 209)

Equipment produced by contractors must be durable enough to withstand the physical movement and repeated handling and testing to which it will be subjected after it is delivered to the government. Task 209 requires contractors to determine the long-term effects that repeated functional testing, storage, handling, packaging, transportation and maintenance will have on the equipment being designed. Specific areas of interest include identification of materials or components that deteriorate with age or when subjected to severe environmental conditions, requirements for testing items that are placed in long-term storage, and special procedures that may be required for maintenance or restoration. The results of this task are used to develop design tradeoffs, field testing requirements, and plans for packaging, handling, storage, and transportation.

RELIABILITY DEVELOPMENT AND PRODUCTION TESTING

Reliability testing starts with testing parts procured for the manufacturing process and continues through the entire equipment production. Task Section 300, MIL-STD 785B, contains four tasks that address methods for testing parts, assemblies, and/or equipment to verify reliability or identify areas where improvements in design, materials, or procedures will provide increased equipment reliability. The use of these tests validates reliability of the final equipment design and manufacturing process. These tests are normally part of the integrated test plan for the equipment being procured by the government.

Environmental Stress Screening (ESS) (Task 301)

By subjecting parts to test conditions that simulate the anticipated field conditions in which they will be used, contractors can identify weak parts, parts that will not perform in the predicted field environment, or design deficiencies that appear in extreme environmental conditions. ESS, also known as "shake and bake," applies stress to items to weed out reliability risk parts. The term "shake and bake" refers to subjecting the parts to vibration and temperature variations while they are being used in realistic situations. Similar tests are performed on assemblies and, finally, total equipment. MIL-STD 810D, Environmental Test Methods and Engineering Guidelines, provides detailed information on environmental testing. MIL-STD 785B recommends that 100% ESS be performed on critical parts, assemblies, and equipment. This testing is nondestructive, will identify weak items and workmanship defects, and should eliminate early field failures. ESS can be expensive due to the time required to perform the testing, but it is well worth the expense because it identifies potential reliability problems early, and reduces field failures.

Reliability Development/Growth Test Program (RD/GT)(Task 302)

Contractors are normally required to perform RD/GT on the first full-scale engineering development model of the equipment being designed. The purpose of RD/GT is unique in that it is conducted to cause the equipment to fail. Using the test-analyze-fix process, RD/GT tests equipment under the actual, simulated, or even accelerated environment in which the equipment will be fielded to identify design deficiencies and defects. The iterative RD/GT process provides for early incorporation of corrective measures that will enhance reliability. The intended results of this process are to increase operational effectiveness of the equipment when it is actually fielded and to reduce the failures that require higher maintenance and logistics support costs. MIL-STD 1635, Reliability Growth Testing, and MIL-STD 2068, Reliability Development Tests, provide a detailed philosophy and methodology for performing RD/GT.

Reliability Qualification Test (RQT) Program (Task 303)

As the equipment design matures and nears time for starting production, RQT is used to verify that the contractor's equipment design meets the performance requirements established by the procurement contract. The preproduction testing is done on one or more samples of equipment representative of the approved production configuration. RQT differs from RD/GT in two ways: (1) it is intended to prove the equipment design, not make it fail; and (2) it is normally performed by an independent testing agency, not the contractor. MIL-STD 781, Reliability Design Qualification and Production Acceptance Tests: Exponential Distribution, provides a detailed description of how RQT will be conducted.

By using an independent test agency, DoD receives an unbiased evaluation of the equipment's achieved reliability. The results of RQT can be used by DoD to authorize the contractor to start full production of the equipment.

Production Reliability Acceptance Test Program (PRAT)(Task 304)

After production has started, PRAT is used by DoD to verify that equipment being produced continues to meet the contractual reliability performance requirements. This type of testing is described in MIL-STD 781 and is very similar to RQT except that it is done on full production items. As with RQT, PRAT is done by an independent testing agency. The quantities of items tested and the frequency of testing are dependent on the desires of DoD. PRAT can be very expensive due to the comprehensive test facilities and time required. The results of this testing are supplied to the contractor for incorporation in design changes if needed to meet reliability requirements.

The purpose of reliability engineering is to aid in the design of equipment, ensuring that it works as long as possible without failing. Additionally, it predicts statistically how and when the equipment will fail. Reliability engineers work with design engineers throughout the design of equipment. The reliability program, per MIL-STD 785B, consists of 18 interrelated tasks that provide the framework for all reliability-related activities. Figure 3-13 shows when these tasks are normally accomplished during the acquisition phases of a program.

PROGRAM TASK		ACQUISITION PHASE			
		Concept	Dem/Val	FSD	Prod
	100 Series				
101	Reliability program plan	X	X	X	X
102	Monitor/control of subcontractors and suppliers	X	X	X	X
103	Program reviews	X	X	X	X
104	FRACAS		X	X	X
105	Failure review board		X	X	X
	200 Series				
201	Reliability model	X	X	X	X
202	Reliability allocation	X	X	X	
203	Reliability prediction	X	X	X	X
204	FMECA	X	X	X	
205	Sneak circuit analysis			X	
206	Tolerance analysis		X	X	
207	Parts program	X	X	X	X
208	Reliability critical items	X	X	X	X
209	Effects of functional testing, maintenance, and PHS&T		X	X	
	300 Series				
301	Environmental stress screening			X	X
302	Reliability development/growth testing		X	X	
303	Reliability qualification test			X	
304	Production reliability acceptance test				X

Fig. 3-13. Reliability program tasks by acquisition phase.

The results of the reliability program are used extensively by other ILS disciplines in determining the logistic support resources that will be required to sustain the equipment once it is fielded. As illustrated in subsequent chapters, the accuracy of reliability predictions and the FMECA determine requirements for resources, which, in turn, determine the projected life cycle costs of the equipment.

Chapter 4

Maintainability Engineering

Chapter 3 described how reliability engineering participates in the design process to produce equipment that is as failure-free as possible. Maintainability engineering also participates in the design process, but its effort is focused on making equipment as easy and inexpensive to fix as possible when it fails. The goals of maintainability engineering are to provide input into the design process that results in an equipment design that has easily identifiable faults, requires minimum labor and other logistic support resources to perform maintenance, and has the lowest life cycle cost possible. The concepts used by maintainability engineering to accomplish the goals stated above, such as modeling, allocations, predictions, and testing, are similar to those used by reliability engineering. Many of the results of reliability tasks are used by maintainability engineers, which becomes evident later in this chapter.

The initial task of maintainability engineering is to develop predictions of how long it will take to repair equipment when it fails. Using these predictions, the design can be analyzed to identify possible changes that would reduce the time required to perform maintenance. As the design matures, the maintainability aspects of the equipment can be determined through actual testing and demonstration of maintenance actions. The combination of equipment reliability—how often it will fail—and equipment maintainability—how long it takes to repair a failure—are directly correlated to the amount of time that the equipment will be capable of performing its mission. Fewer failures require less time for maintenance, so the more time the equipment is operational. Because of this fact, it is not uncommon to see these disciplines referred to as Reliability and Maintainability (R&M).

DEFINITIONS

An understanding of the following terms used in this chapter will aid in understanding maintainability engineering.

- **Maintainability.** The probability that a failed item can be repaired in a specified amount of time using a specified set of resources is known as maintainability. Note that this is a statistical prediction, which means that, like reliability, maintainability can be greatly influenced by variables such as availability of resources and environmental conditions where maintenance is performed.
- **Mean Time To Repair (MTTR).** This is the average time required to perform maintenance over a specified operating period. Initially, the MTTR is developed using predicted times to perform maintenance tasks. When the design is complete, the MTTR can be refined by actually measuring the time to perform tasks. The accuracy of the MTTR is then dependent on the correctness of reliability predictions.
- **Mean Manhours per Maintenance Action (MMH/MA).** This is the average time required to perform a maintenance action. The MMH/MA is used to develop a prediction of the total quantity of labor that will be required to perform maintenance. Multiplying the MMH/MA of an item of equipment by the predicted number of failures of that equipment during a specific amount of time results in the anticipated number of manhours that will be required to perform maintenance.
- **Mean Manhours per Operating Hour (MMH/OH).** The ratio of manhours required to perform maintenance to one hour of equipment operation is known as the MMH/OH. This number is used as a gauge to develop tradeoff comparisons between different ways of doing maintenance.
- **Fault Isolation.** Fault isolation is the act of identifying a failure to the level that will enable corrective maintenance to begin. In other words, it is doing some kind of testing or evaluation to get down to the level where something can be replaced or repaired in order to fix the problem.
- **Maintenance Task.** This is any action that is taken to fix a failed item or prolong the serviceability of an item. Types of maintenance tasks are described in detail in Chapter 5, Maintenance Planning.

THE MAINTAINABILITY PROGRAM

The goal of the maintainability program is to improve the operational readiness of equipment while reducing the requirements for manpower and other logistics resources. The methods used by maintainability engineers to achieve this goal are described in MIL-STD 470, Maintainability Program for Systems and Equipment. This document provides a standard framework for contractors to use when conducting maintainability-related activities. MIL-STD 470 is similar in format and methodology to MIL-STD 785. It contains three sections that divide the maintainability activities into categories of efforts required to accomplish a complete program. These sections are Program Surveillance and Control, Design and Analysis, and Evaluation and Test. Each section contains tasks that are directly related to that phase of the program. As with MIL-STD 785, DoD normally tailors the maintainability program requirements by selecting those MIL-STD 470 tasks that are applicable to the equipment being designed.

Program Surveillance and Control (Task Section 100)

The purpose of Task Section 100 is to provide actions necessary to plan and control the maintainability program. This section has four tasks. They are similar in purpose to the corresponding tasks in MIL-STD 785, and are sometimes conducted simultaneously to eliminate duplication of effort.

Maintainability Program Plan (Task 101). The maintainability program plan describes and controls a contractor's maintainability efforts. The plan provides an overall description of how a contractor will fulfill contractual maintainability requirements, as stated in the Statement of Work, and the methods that will be used to ensure that the equipment being designed meets the maintainability requirements of the procurement specification. It states the goals of the maintainability program. Included in the plan is the identification of each MIL-STD 470 task that will be accomplished by the contractor with a description of how the tasks will be performed. Procedures to be used to evaluate and control the progress toward task completion will be identified. The plan provides a description of the maintainability organization and how it interfaces with other ILS disciplines such as reliability, safety, logistic support analysis, and support equipment. A schedule of when maintainability tasks will start and when they will be completed will also be included. The schedule should show a cross-reference to the efforts of other disciplines, such as reliability and logistic support analysis, because many tasks are interrelated and dependent on outputs from, or provide inputs to, other ILS activities. Figure 4-1 is an outline for a typical maintainability program plan.

1.0 PURPOSE

Describe the purpose of the Maintainability Program and tell how the program will fulfill the contractual requirements stated in the Statement of Work. State the desired results of the program.

2.0 EQUIPMENT DESCRIPTION

Provide a brief description of the functional and physical characteristics of the equipment being developed. Equipment maintainability requirements established by the government in the procurement specification, such as MTTR or MMH/OH, should also be stated here.

3.0 PROGRAM TASKS

Describe how each MIL-STD 470A task will be performed. Provide examples when possible. Be specific on the methodology to be used. If a task is not required by the Statement of Work, it could possibly still be required in order to perform other tasks.

Fig. 4-1. Maintainability program plan.

4.0 MANAGEMENT

Describe how the program will be managed. Identify the organizations that will be responsible for accomplishing each task. Include internal procedures for controlling the effort to accomplish each task. Identify the position that has authority for making maintainability decisions.

4.1 Organization

Describe the maintainability organization and how it interfaces with other organizations. Provide a diagram of
a. the maintainability organization,
b. how the maintainability organization relates to other ILS organizations, and
c. how the above organizations relate to the total program management structure.

4.2 Interfaces

Describe how the maintainability program will interface with the total equipment design effort. Specifically tell how maintainability data will be integrated into the design and LSA. Flow charts showing these interfaces are very effective.

5.0 SCHEDULE

Provide a milestone schedule showing when each task will be started and completed. Relate maintainability milestones to the overall equipment development program milestones. Indicate critical events on the schedule that are dependent on other events, e.g., things that must occur prior to the start or finish of other tasks.

6.0 MAINTENANCE CONCEPT

Describe the methods to be used to develop maintainability criteria that achieve the desired support for the maintenance concept of the system. The maintenance concept tie-in is critical to determine if the equipment design can be maintained through the concept, or if the concept must be modified to achieve the overall supportability goals.

7.0 DATA SOURCES

Describe what sources of data will be used as inputs to the maintainability tasks. Data available from previous programs for similar equipment provide the most creditable starting point. Describe how these data will be used and the relative merit of what will be gained through their use. Define what data are expected to be provided by DoD. This is normally historical field data collected by DoD on similar equipment. Identify information expected to be received from suppliers or vendors as a result of their participation in this program. Indicate other anticipated sources of information such as studies, research, etc. Describe how the data gathered during the maintainability demonstration will be used to improve the system design.

Fig. 4-1. Continued.

Monitor/Control of Subcontractors and Vendors (Task 102). It is important that contractors require their subcontractors to comply with the maintainability program requirements identified by DoD. Task 102 establishes the requirement for contractors to pass maintainability requirements to subcontractors and monitor their progress toward meeting the program goals. The first step in this process is to ensure that maintainability requirements are imposed on subcontractors through the same contract that buys whatever hardware is being obtained from the subcontractor. It must state the following requirements that the program must meet:

- Maintainability constraints and requirements
- Maintenance and support concepts and requirements
- Standardization and interchangeability requirements
- Maintainability and fault detection and isolation demonstration requirements

Contractors must require subcontractors to implement a maintainability program that will fully support the contractor's program. Subcontractors should be required to participate in design reviews that relate to their hardware. Contractors should review the subcontractor's maintainability predictions and analysis techniques and results for accuracy and correctness of approach. Failure on the part of the contractor to ensure that subcontractors are completely involved in the maintainability program will result in continued failure of equipment to meet overall program goals.

Program Reviews (Task 103). Maintainability program reviews are normally conducted in conjunction with other regularly scheduled reviews, such as reliability reviews. The purpose of these reviews is to evaluate the progress, consistency, and technical accuracy of the maintainability program. Maintainability plays an important role in both the Preliminary Design Review (PDR) and Critical Design Review (CDR). At these reviews, the design of the equipment being developed is reviewed and approved by the government. The influence of maintainability engineering on the proposed design is reviewed to determine if changes are needed in either the approach or philosophy being used. Maintainability predictions are normally prepared for presentation at these reviews. The predictions are used by the government to determine how well the design will meet specified maintainability goals.

Data Collection, Analysis, and Corrective Action System (Task 104). Every development program generates more data than could ever be digested and used to its fullest extent unless specific guidance is provided about what should be collected and what to do with it after it is collected. Task 104 delineates the kinds of maintainability data that should be collected, and it tells how the data should be used. Relevant maintainability data is derived from maintainability analyses, engineering tests, demonstration tests, and user tests. These data are compiled and analyzed to develop trends and identify critical areas for further investigation to resolve design problems or enhance overall performance. They are also used as input to the LSA process to assess the adequacy of technical manuals, test equipment, and training; to determine personnel requirements; and to develop a historical data base for comparison with maintainability predictions. The format and method for compiling data are normally the contractor's preference; however, they should contain sufficient information to identify maintenance actions, time required, number of personnel required, adequacy of support and test equipment, methods for fault detection and isolation, and any special or unusual occurrences.

Design and Analysis (Task Section 200)

The design and analysis section contains seven tasks that establish the basis for the methods used by maintainability to review, evaluate, and influence the equipment design. The implementation of each task is dependent on the type of equipment being developed, the phase of the acquisition cycle, and the needs of the government.

Maintainability Modeling (Task 201). The use of models to determine the maintainability of an item of equipment enables those involved to predict the ease with which maintenance will be performed on the equipment. Models should be developed early in the concept phase. Using an appropriate model, maintainability engineers can develop an initial prediction of equipment maintainability. Then, they can determine the effect of changes on the total system and evaluate the need for redesign to achieve maintainability goals. The selection or tailoring of the appropriate model is the first step in predicting the number of hours that the equipment will be unavailable for use.

MIL-HDBK 472 contains five basic procedures that can be used to predict the maintainability of equipment. The models that are provided in the handbook must be tailored to reflect the appropriate configuration and complexity of the equipment. Figure 4-2 illustrates the model used to predict the MTTR of an item. The model for MTTR prediction uses the common tasks required to repair a failed item. Each of the elements of the model must likewise be predicted using models, so a submodel for each sub-element must be developed. The complexity of the equipment being modeled dictates how detailed a model should be. Small items of equipment might require only selected

$$ \text{MTTR} = \overline{T}_p + \overline{T}_{FI} + \overline{T}_{FC} + \overline{T}_A + \overline{T}_{CO} + \overline{T}_{ST} = \sum_{M=1}^{m} \overline{T}_M $$

Where:

$\overline{T}_p$ = Average preparation time

$\overline{T}_{FI}$ = Average fault isolation time

$\overline{T}_{FC}$ = $\overline{T}_D + \overline{T}_I + \overline{T}_R$

$\overline{T}_D$ = Average disassembly time

$\overline{T}_I$ = Average interchange time

$\overline{T}_R$ = Average reassembly time

$\overline{T}_A$ = Average alignment time

$\overline{T}_{CO}$ = Average checkout time

$\overline{T}_{ST}$ = Average startup time

$\overline{T}_M$ = Average time of the Mth element of MTTR

Fig. 4-2. MTTR prediction model.

portions of the model shown at Fig. 4-2, while major weapon systems require extensive submodels to determine the input for the overall system model. For equipment such as a tank, which has several distinct levels, some assemblies are primarily mechanical, others are electronic, and some are a combination. The model for the tank might be composed of submodels that are tailored for each type of assembly. The accuracy and appropriateness of the model will determine the validity of the resulting predictions. The model must represent the task functions required to maintain the equipment. As with all models, the resulting predictions form a baseline for comparison when determining the effect of recommended changes to the system.

Maintainability Allocations (Task 202). The maintainability design process begins with development of an allocation of the overall equipment maintainability goals. The maintainability goals of an equipment are normally found in the procurement specification. These goals may be expressed as MTTR figures for each level of maintenance, e.g., "MTTR at the organizational level will not exceed 0.25 hours; and at the intermediate level, 0.75 hours." Another method for expressing maintainability goals is as a ratio of equipment operating hours to maintenance hours, e.g., "MMH/OH will not exceed 0.005." For design planning, these goals represent the maximum mean time allowable for maintenance at a given maintenance level.

The contractor begins the allocation process using the goals allocated for each lower equipment level. The maintenance concept for the equipment plays an important part in the allocation process. For example, if the maintenance concept calls for only two levels of maintenance, then the maintainability allocation should consider only those levels. Additionally, the maintenance concept may establish criteria for the maintenance actions that will be accomplished at each level. If the maintenance concept states that organizational maintenance will consist of removal and replacement of modules, then maintainability considerations at the organizational level should be centered on improving the accessibility and ease of removing and replacing modules. The allocation of maintenance time should reflect this maintenance concept. Figure 4-3 illustrates

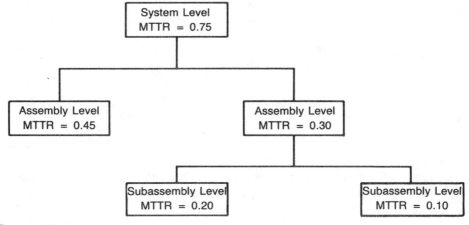

Fig. 4-3. MTTR allocation.

allocation of an MTTR from system level down two indentures. The design engineers responsible for incorporating the maintainability of the equipment must be given these allocations so that they understand the requirements for maintenance that have been levied on their area of responsibility. The allocation process also provides a bookkeeping procedure for establishing a baseline for achieving the equipment maintainability goals. As the design matures, problems involving these goals can be evaluated to determine their effect on the overall equipment. This allows continual evaluation of the equipment maintainability in relation to the goals. It also enables tradeoff analyses to be conducted, determining where changes will be most beneficial.

The allocation process also helps contractors achieve equipment goals when dealing with subcontractors. By allocating the overall equipment goals down to levels that are supplied by subcontractors, the contractor can manage the total maintainability program rather than having to rely on chance to achieve the contractually stated maintainability goals. The initial allocation process should be based on historical information about the maintainability of similar equipment, initial predictions, or best engineering judgments. The allocations might change as the design matures, but the goals cannot be changed unless the government concurs.

Maintainability Predictions (Task 203). The purpose of a maintainability prediction is to determine, using mathematical calculations, if an equipment design will meet the established maintainability goals. The prediction process is also used to identify designs that will not meet the goals. During the concept phase, there is insufficient information to perform detailed prediction calculations, so predictions may be based on the performance of previous equipment, using design changes to emulate the new design. In order to perform detailed calculations, information, such as the maintenance concept, functional block diagrams, identification of replaceable units, and reliability estimates, is necessary. Using this information, maintainability engineers can develop predictions for the equipment. As the design matures, the quantity and quality of information increases, allowing refinement of early predictions.

MIL-HDBK 472, Maintainability Prediction, contains instructions on how predictions are to be accomplished. The handbook contains methods for conducting predictions early in the acquisition process when detailed information is not available, and later in the FSD phase when detailed information is available. Figure 4-4 shows the results of a MTTR prediction for an electronic assembly. Note that the reliability prediction for each item in the assembly is used to predict the frequency of the anticipated maintenance actions. The maintenance actions represent those normally associated with the repair of a failed item. (The terminology used might differ slightly among branches of the military, but the basic maintenance functions are the same.)

The actual prediction method chosen must correspond to the equipment and parameters to be predicted. Figure 4-5 shows a simple calculation of MMH/OH using the results of the above MTTR prediction. This prediction process is updated throughout the design process to continually assess the maintainability status of the equipment and to provide information for the decision-making process. The results of this task become input to other analyses and the activities of other logistics disciplines.

ITEM: Electronic assembly   **PART NUMBER:** 12345-6
METHOD OF REPAIR: Remove/replace Assy at "O" Level - Remove/replace CCAs at "I" Level

Drawing No.	Nomenclature	Qty (N)	Fail Rate (λ)	($N\lambda$)	Maintenance Task Times (hours)						Rp	($N\lambda$)(R_p)
					Locat	Isola	Disay	Intch	Reasy	Ckout		
					Organizational Level							
12345-6	ASSEMBLY	1	48.0712	48.0712	0.01	0.04	0.08	0.06	0.08	0.01	0.28	13.459936
					Intermediate Level							
12345-6	ASSEMBLY	1	48.0712	48.0712	0.05	0.25	0.05	0	0.05	0.25	0.65	31.24628
4685-4	CCA #1	4	7.5437	30.1748	0.01	0.15	0.02	0.02	0.02	0.15	0.37	11.164676
2694-5	CCA #2	2	3.9874	7.9748	0.01	0.15	0.02	0.02	0.02	0.15	0.37	2.950676
6379-4	CCA #3	3	4.3869	13.1607	0.01	0.15	0.02	0.02	0.02	0.15	0.37	4.869459
1693-5	CAA #4	1	2.3476	2.3476	0.01	0.15	0.02	0.02	0.02	0.15	0.37	0.868612
8258-4	CCA #5	2	5.3968	10.7936	0.01	0.15	0.02	0.02	0.02	0.15	0.37	3.993632
8326-5	CCA #6	3	3.6524	10.9572	0.01	0.15	0.02	0.02	0.02	0.15	0.37	4.054164
5274-6	CCA #7	3	1.8643	5.5929	0.01	0.15	0.02	0.02	0.02	0.15	0.37	2.069373

Predicted Mean Time to Repair (MTTR)   0.2800000 hours at organizational level
0.4742817 hours at intermediate level

0.7542817 hours total assembly MTTR

Fig. 4-4. MTTR prediction worksheet.

Failure Modes, Effects and Criticality Analysis (FMECA) Maintainability Information (Task 204). As stated in Chapter 3, the purpose of the FMECA is to identify the predicted ways that an item will fail and what will happen when the failure occurs. The maintainability portion of the FMECA relates to documenting information about the failure that can be used to plan how to fix the failure when it occurs. The key information relates to how maintenance personnel will know when the failure occurs and what actions should be taken. Figure 4-6 illustrates the maintainability information portion of the FMECA. This figure is basically the same as Fig. 3-11, with the exception of blocks 13 and 15. Block 13 contains information that can be used to predict when a failure is likely to occur, and block 15 states the actions that must be taken to correct the failure. This information is used as input to both the reliability-centered maintenance (RCM) analysis and the maintenance task analysis. MIL-STD 1629A contains detailed instructions for determining and documenting maintainability information. The contractor must closely monitor the development of this information, because the results of this task form the basis for determining many other logistics-related requirements for the equipment.

Maintainability Analysis (Task 205). The maintainability analysis is one of the key tasks of the maintainability program. The purposes of this task are to identify equipment maintainability design features that will enable the equipment to meet maintainability goals, evaluate design alternatives, provide detailed input to the

Formula: MMH/OH = (MTTR × C × F)/MTBF

Where:

MTTR = Mean time to repair

C = Crew size

F = Operation service ratio

MTBF = Mean time between failures

Item: Name: Radar set
Number: 12345-1

Data:
MTTR = 0.25
C = 2
F = 1.5
MTBF = 1500

Results: MMH/OH = 0.0005

Fig. 4-5. Mean manhours per operation hour calculation.

maintenance planning process, and continually evaluate the design to ensure that goals are met. This task has a significant overlap with analyses being conducted simultaneously by other disciplines, so close coordination is required to avoid duplication of effort. To accomplish this task, maintainability engineers require detailed information on the equipment design and other information such as the maintenance concept and plan, anticipated test and fault isolation capabilities, maintenance skills, operational information, and reliability predictions.

The maintainability analysis process develops detailed design criteria that are necessary to meet established goals. These criteria can consist of requirements for accessibility, tool usage, test points, standardization of tools and procedures, connectors and fasteners, and modularization. These criteria serve as guidelines for engineers as the equipment is designed. As the design evolves, alternatives are identified and evaluated to determine the approach that provides the best maintenance capabilities. These alternatives must be evaluated as early as possible to allow implementation without unnecessary redesign. The maintainability analysis is a key element in selecting the most desirable methods for testing the equipment when a failure occurs. The adequacy and efficiency of the equipment testability has a large impact on the maintainability of an equipment. The easier it is to find a failure, the quicker it can

Failure Mode Effects and Criticality Analysis
Maintainability Information
MIL-STD 1629A

SYSTEM ①
INDENTURE LEVEL ②
REFERENCE DRAWING ③
MISSION ④

DATE _____
SHEET _____ OF _____
COMPILED BY _____
APPROVED BY _____

IDENTIFICATION NUMBER	ITEM/FUNCTIONAL IDENTIFICATION (NOMENCLATURE)	FUNCTION	FAILURE MODES AND CAUSES	FAILURE EFFECTS			SEVERITY CLASS.	FAILURE PREDICT-ABILITY	FAILURE DETECTION MEANS	BASIC MAINTENANCE ACTIONS	REMARKS
				LOCAL EFFECTS	NEXT HIGHER LEVEL	END EFFECTS					
⑤	⑥	⑦	⑧	⑨	⑩	⑪	⑫	⑬	⑭	⑮	⑯

Fig. 4-6. FMECA-maintainability information worksheet. (Continued on page 60.)

59

NOTE: Several items are duplications of like entries of the FMECA worksheet shown in Fig. 3-11. The use of a computerized data base is recommended for compiling all FMECA data to reduce errors and eliminate duplicated effort.

1. Identification of the end item using either name or part number.

2. Indenture level of the item being analyzed.

3. Drawing number of the engineering drawing being used as reference for the analysis.

4. Identification of the function of the item being analyzed.

5. Identification number for the function being analyzed using either a functional group code, per MIL-STD 780, or an LSACN, per MIL-STD 1388-2A. For traceability with other ILS activities, it is recommended that the LSACN be used.

6. Identification of the item or function contained in the item identified by note 3, above, that is being analyzed.

7. Statement of the function performed by the item being analyzed.

8. List all possible failures of the item being analyzed. There can be many failures of a single item listed here. Remember that the failures are of the item identified in column 6.

9. Identify what will happen to the item being analyzed when the failure occurs.

10. Identify what will happen to the next higher assembly level item when the failure occurs.

11. Identify what will happen to the end item when the failure occurs.

12. Assign the appropriate classification for the severity of the consequences if the failure occurs.
    Category I is catastrophic, including loss of equipment or death of personnel.
    Category II is critical, including severe injury to personnel or damage to equipment.
    Category III is marginal, including minor injury or damage.
    Category IV is minor, no serious injury or damage, but still requires unscheduled maintenance.

13. Identify any methods or data collection that can be used to predict when a failure is likely to occur, i.e., number of landings, miles, hours, etc.

14. Identify how the operator will know when the failure occurs.

15. Describe the anticipated maintenance actions that will be required to correct the failure.

16. Record additional pertinent information as required.

be fixed. The clock starts for down-time when the failure occurs and does not stop until the equipment is repaired. Excessive time requirements for testing increase the overall equipment down-time and degrade the equipment maintainability. Inputs to the maintainability analysis come from reliability analyses and predictions, information on the recommended skill levels and number of maintenance personnel, system safety analyses, and maintenance planning. Outputs from the maintainability analysis are used as inputs to other maintainability tasks and by other ILS disciplines for determining detailed resources required to support the equipment when it is fielded.

Maintainability Design Criteria (Task 206). The purpose of Task 206 is to quantify the criteria for equipment design to achieve established maintainability goals. This task documents the results of previous maintainability tasks and translates requirements into guidelines that can be used by design engineers during equipment design. Figure 4-7 provides examples of maintainability design criteria.

Preparation of Inputs to the Detailed Maintenance Plan and Logistic Support Analysis (LSA)(Task 207). The maintainability program provides outputs that are used by other ILS disciplines. The purpose of this task is to coordinate the availability and its dissemination of information. Information from the maintainability program applies directly to maintenance planning and logistic support analysis, as discussed in later chapters. The maintainability information that results from the tasks described above will vary according to specific contractual and program requirements. Examples of information that is provided to other activities is shown in Fig. 4-8.

Evaluation and Test (Task Section 300)

The proof of the effectiveness of the maintainability program is determined through maintainability evaluation and testing. The requirements for this effort are in Task 301. This demonstration is necessary to verify that the maintainability design criteria have been incorporated into the equipment design, resulting in the achievement of contractually stated maintainability goals.

Maintainability Demonstration (Task 301). MIL-STD 471, Maintainability Verification/Demonstration/Evaluation, provides detailed instructions on how to conduct a maintainability demonstration. The purpose of this demonstration is to physically show that the equipment is capable of being maintained. The process for conducting the demonstration is relatively simple. It is conducted using the actual equipment late in

---

- Assemblies and repair parts having the same part numbers will be functionally and physically identical.
- Access to maintenance-significant items will be provided through entries that do not require removal of other components.
- Scheduled maintenance, alignments, and calibration requirements will be avoided.
- Special tools or test equipment will not be required to perform maintenance at organizational- or intermediate-maintenance levels.
- Captive, quick-release fasteners will be used to secure maintenance access panels or covers.
- All screws and bolts will be of standard dimensions to reduce tool requirements.

Fig. 4-7. Maintainability design criteria.

- Results of maintainability predictions.
- Unscheduled and scheduled maintenance requirements.
- Tool and test equipment requirements.
- Skill-level requirements for each maintenance level.
- Fault identification requirements.
- Types of maintenance tasks required at each level.
- Test requirements for fault isolation.

Fig. 4-8. Maintainability information.

PROGRAM TASK		ACQUISITION PHASE			
		Concept	Dem/Val	FSD	Prod
	100 Series				
101	Reliability program plan		X	X	X
102	Monitor/control of subcontractors and vendors		X	X	X
103	Program reviews	X	X	X	X
104	Data collection		X	X	X
	200 Series				
201	Maintainability model	X	X	X	X
202	Maintainability allocations	X	X	X	X
203	Maintainability predictions		X	X	
204	FMECA		X	X	
205	Maintainability analysis	X	X	X	
206	Design criteria		X	X	
207	Inputs to maintenance plan and LSA		X	X	
	300 Series				
301	Maintainability demonstration			X	

Fig. 4-9. Maintainability program tasks by acquisition phase.

the FSD phase. Also required are the technical manuals, required tools, and other support equipment necessary to accomplish maintenance. This demonstration obviously affects not only maintainability, but virtually all ILS disciplines.

The object of the demonstration is to take an operational system, induce failures into it, and use only the technical manuals and support equipment that will be available to maintenance personnel when the equipment is fielded to find and fix the failure. The government might require that military technicians perform the maintenance rather than contractor personnel to get a true picture of the maintenance of the equipment. In other words, this demonstration is a test of how well the contractor has designed the equipment for maintenance and how usable the logistics support package is in actually performing maintenance. The maintainability demonstration is a significant program milestone that occurs close to the end of FSD. Failure to successfully pass the demonstration might result in equipment redesign, technical manual changes, and delays in starting full production of the equipment.

TASK SELECTION

The contract statement of work for equipment procurement will normally state the maintainability tasks to be performed for a specific program. Task selection is based on the complexity of the equipment, the acquisition phase of the program, and the desired maintainability participation in the design process. Figure 4-9 shows when maintainability program tasks are typically accomplished by acquisition phase.

The maintainability program provides a unique input to the design process. While the reliability program seeks to make the equipment as failure-free as possible, the maintainability program takes the opposite approach. Assuming that the equipment will fail, it addresses how to best design the equipment so that it can be fixed when it does fail. The two programs must be coordinated to receive maximum benefit from both viewpoints.

Chapter 5

Maintenance Planning

Chapters 3 and 4 describe how ILS participates in the design process so that the final equipment design is as reliable and maintainable as possible. This does not alter the fact that eventually all equipment will require maintenance. *Maintenance* should not be confused with *maintainability*. Maintainability engineering is accomplished during the design of equipment, so that it can be fixed when it fails. Maintenance is any action taken to keep equipment in a serviceable condition or to fix it when it fails. Maintenance planning is the process that develops the anticipated maintenance requirements for the equipment and proposes who will do required maintenance tasks and where they will be done. This chapter deals with the general concepts of maintenance planning, types of maintenance actions, and typical organizations that perform maintenance.

The majority of the activities that have been and will be discussed in the text deal with DoD as a generic user of equipment, and make the broad assumption that all branches of the military operate in the same manner. The current goal of DoD is to make this true. In reality, however, each service is organized differently for maintenance. This is because each branch of the military has different missions, different support systems for accomplishing maintenance, and varying levels of capability within their own organizational structure. The maintenance plan for an item of equipment must be tailored to fit the maintenance structure of its ultimate user.

MAINTENANCE CONCEPT

Maintenance of equipment does not just happen. It is the result of extensive planning and preparation that start during the preconcept phase and continue through full-scale development. The maintenance planning process begins with the maintenance concept, a statement of general guidelines to be used in developing the maintenance plan for an item of equipment. The guidelines established by the maintenance concept are the foundation for maintenance planning. Areas addressed by the maintenance concept include (1) a strategy for allocating maintenance tasks to the different levels of maintenance; (2) the repair policy regarding similar types of items contained in the equipment; (3) the criteria for scheduling maintenance tasks; and (4) the anticipated availability of resources, in gross terms, to support maintenance. An example of a typical maintenance concept statement is in Fig. 5-1.

Development

The maintenance concept is initially developed by the government during the concept phase. The initial concept is the result of an analysis of pertinent information concerning maintenance of equipment. The maintenance concept provides a description of the anticipated environment, both operationally and logistically, where the equipment will operate. This description might include the organizational structure of both the using and supporting units that will perform maintenance. An assessment of existing and emerging maintenance technologies is used to determine opportunities that could significantly affect the maintenance program for the new equipment. These technological opportunities could include new ways of performing maintenance tasks, improved methods for testing and identification of faults, design configurations that enhance maintenance, and changes in materials that reduce failures.

Historical logistics and maintenance data of similar equipment provide invaluable sources of information in developing the maintenance concept. The historical data form a measurement base against which the logistics and maintenance support of the new equipment can be gauged. These data provide lessons learned on previous equipment that can be applied to the new equipment.

Refinement

The proposed maintenance concept is refined by comparing the concept with the projected resources and constraints that will exist when the equipment is fielded. The intent of this refinement is to cause as few changes as possible to the existing maintenance system when the new equipment becomes operational. The proposed

---

Maintenance tasks at the organizational level will be limited to unscheduled removal and replacement of failed modules or components and scheduled maintenance that can be accomplished without the aid of special tools or support equipment. Intermediate-level maintenance will have the capability of repairing electronic, electromechanical, and hydraulic assemblies, including the replacement of failed components. Overhaul, refurbishment, and fabrication of structural parts will be accomplished at the depot level. Maximum use will be made of existing tools, support equipment, test equipment, and associated support resources.

Fig. 5-1. Maintenance concept.

operational and maintenance operations are compared with existing operations to determine the optimum maintenance concept. The resulting maintenance concept considers the complexity of the new equipment, its mobility requirements, permissible time that can be spent in maintenance, critical maintenance skills that must be considered by maintenance planning, and minimum maintenance procedures that must be accomplished to ensure that the equipment can perform its assigned mission.

LEVELS OF MAINTENANCE

The term "levels of maintenance" is commonly used to describe the different capabilities for performing maintenance that are inherent in the organizational structure of the military services. There are basically three levels of maintenance common to all services: organizational, intermediate, and depot. The type of maintenance that each level is capable of performing depends on the tools, test equipment, and training of personnel available. The goal of maintenance planning is to maintain equipment at the lowest level possible, or closest to the user of the equipment. This limits the amount of time until equipment is maintained when it fails, and it limits the amount of resources that must be available to support maintenance.

Organizational Maintenance

Maintenance performed by the owner or user of the equipment is categorized as organizational ("O" level) maintenance. The capabilities of "O" level are normally limited to periodic servicing of equipment, troubleshooting to identify failures, and removing/replacing major components. Again, the factors limiting what maintenance can be done at "O" level are the tools, test equipment, and training of personnel. Using a piece of electronic equipment as an example, "O" level might be limited to daily testing of the equipment to determine its capability to perform its mission, removing the item when it fails, and replacing the entire failed item with one that works. Remember, at "O" level, the equipment user's mission is to fulfill a requirement using the equipment; therefore, maintenance planning must consider possible short-term actions to keep the equipment working. It is not uncommon for "O" level maintenance actions to be limited to an MTTR of less than one hour.

Intermediate Maintenance

Maintenance actions that are not within the capabilities of "O" level are passed to the next higher level—intermediate ("I" level) maintenance. "I" level is more capable of performing maintenance because that is its primary mission, providing maintenance support to subordinate units. Because the complexity of maintenance tasks increases at "I" level, it has a greater range of tools and test equipment available, and personnel are trained to perform the required maintenance tasks. "I" level maintenance on electronic equipment might consist of testing items removed by "O" level for repair, and replacing failed modules. In some instances, "I" level also repairs circuit card assemblies by replacing failed components. This can only be done if the required test equipment is available to isolate the failed component and to test repaired units to verify that the repair action corrected that problem. Personnel must be trained in accomplishing this task.

Depot Maintenance

Maintenance actions that cannot be accomplished at "O" or "I" levels are passed to depot ("D" level) maintenance. "D" level maintenance has the capability to do anything necessary to repair failed equipment. Normally, "D" level maintenance facilities have the widest range of tools, test equipment, and knowledgeable maintenance personnel. Fabrication of structural parts, major overhauls and refurbishment, and complete rebuilding of equipment can be done at "D" level. Referring to the example of electronic equipment, "D" level would be capable of relocating all components on circuit card assemblies, fabricating replacements for damaged chassis parts, or rebuilding the entire equipment.

Figure 5-2 illustrates the relationships among the levels of maintenance. A single depot maintenance facility might support several intermediate maintenance facilities, as might an intermediate maintenance facility support many organizational facilities. This type of maintenance structure consolidates maintenance actions requiring more sophisticated and expensive tools and test equipment, and training at higher levels, which reduces overall cost for performing maintenance.

Actual Maintenance Levels

"O", "I", and "D" levels of maintenance are common to all branches of the military. However, each service is different, due to the nature of their missions and how maintenance must adapt to support them. For example, the Air Force operates from fixed installations; aircraft must have landing strips. So, the maintenance levels of the Air Force are centered around performing maintenance on aircraft from fixed installations. The Army, on the other hand, is mobile. Its elements are capable of moving to any location on the globe; therefore, its maintenance levels must be able to support it anywhere. The Navy is really two distinct services—naval aviation and naval ships. In both cases, however, the Navy must be more self-sufficient than the Air Force or

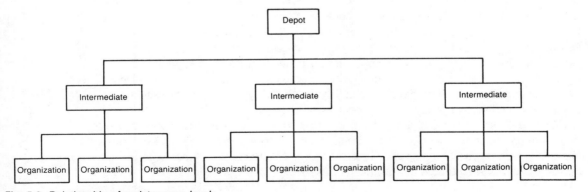

Fig. 5-2. Relationship of maintenance levels.

the Army, because its missions require deployment by sea, which limits the size of maintenance facilities that can follow and the lines of supply required to sustain operations.

U.S. Air Force. The actual maintenance levels of the Air Force closely match the generic levels described earlier. "O" level maintenance performed by aircraft crews is referred to as "flight-line" maintenance. It is also performed by the unit owning the aircraft. "I" level maintenance is, in many cases, performed in the same location as "O" level by other units whose mission is to provide this level of maintenance support. It might also be provided by mobile crews or centrally located facilities. "D" level is normally the responsibility of a single facility for each type of equipment or major assembly. For instance, one depot might be responsible for the repair and overhaul of jet engines used on the F-16 aircraft. All depot-level work on these engines would be done at this one location. Each location that provides depot maintenance normally is responsible for the management of the entire support resources of the items it maintains. These installations are referred to as air logistics centers (ALC). Typical ALCs are located at San Antonio, Texas; Sacramento, California; and Ogden, Utah. Overall responsibility for maintenance and materiel management rests with the Air Force Logistics Command, Wright-Patterson Air Force Base, Dayton, Ohio.

U.S. Army. Because the Army is much larger than the Air Force, its maintenance-level structure is expanded. For many years, the Army has had five levels of maintenance: (1) crew, (2) organizational, (3) direct support, (4) general support, and (5) depot. Crew and organizational levels correspond to "O" level as described above, and direct and general support correspond to "I" level. Crew-level maintenance consists of the maintenance that can be performed by the crew assigned to the equipment using the tools provided with the equipment. These tools are discussed in Chapter 7. Organizational maintenance is performed by the maintenance personnel assigned to the unit that owns the equipment, as in the Air Force. The biggest distinction between direct support and general support is the mobility of the units that provide this service. Direct support maintenance units are normally assigned to combat divisions, which means that these units must be very mobile. This limits the number of tools and other maintenance equipment assigned to direct support units; therefore, direct support maintenance could be described as "I" level maintenance that can be accomplished in minimal time using mostly standard tools and test equipment. General support maintenance is "I" level maintenance that normally requires more sophisticated tools and test equipment and a greater length of time to accomplish. (The Army is moving toward combining direct and general support levels into a composite direct-support/general-support level that more closely matches the generic "I" level.) Depot maintenance is performed by fixed installations in the Army; however, the delineation of sole responsibility for a given equipment being delegated to a single facility is not as clearly defined as in the Air Force. The Army's top-level management structure for maintenance and support of equipment is similar to the Air Force's. Individual commands are responsible for specific types of equipment. Examples of these commands are the Army Tank-Automotive Command (TACOM), which is responsible for tracked and wheeled vehicles; Communications Electronic Command (CECOM), which manages radios and other communications equipment; and Missile Command (MICOM) which manages rockets and missiles. The Army Materiel Command (AMC), in Alexandria, Virginia, has overall responsibility for managing the acquisition, maintenance, and support of Army equipment.

U.S. Navy. The Navy's philosophy for performing maintenance is close to the generic description of "O", "I", and "D." The problem with clearly relating these levels to the Navy is, again, due to the missions it is assigned. As stated, there are really two distinct parts to the Navy: naval ships and naval air. As long as naval aircraft operate from fixed, dry-land locations, the supporting maintenance organization is much like the Air Force's "O" and "I" levels. Depot maintenance is performed by facilities located mainly on the East and West Coasts. However, each of these facilities is capable of total depot overhaul and repair. When naval aircraft are based on carriers, the division of maintenance is not as clear. Aircraft carrier task forces are capable of performing virtually all levels of maintenance if combat aircraft can't wait for maintenance until the aircraft carrier returns to port.

Naval ships are more difficult to categorize, because they must be self-sufficient or supported by *tenders*, repair ships that accompany a task force at sea. "O" level maintenance is performed by the crew of the ship. Many of the "I" level maintenance actions relating to ships are divided into two categories: afloat and ashore. Both afloat and ashore maintenance actions are normally "I" level, although they can be "D" level depending on the type of action and how it relates to the ability of the equipment to perform its mission. "D" level maintenance can be performed by tenders at sea or at shipyards. Obviously the major overhaul of ships must be done by facilities that have the capability. Naval Air Systems Command (NAVAIR) is responsible for aircraft support and maintenance, and Naval Sea Systems Command (NAVSEA) is responsible for naval ships.

When planning for the maintenance of an item of equipment, it is important to know who the ultimate user will be. As shown above, the maintenance structure of each service is different and must be taken into consideration when developing the equipment maintenance plan.

MAINTENANCE TASKS

After determining who does maintenance, you must determine what maintenance is to be done. The FMECA, as described in Chapter 3, analyzes an item of equipment to determine all the possible failures that could occur. Maintenance planning determines the maintenance tasks that will be required to fix or prevent the possible failures identified by the FMECA. Maintenance tasks are simply actions that can be combined into step-by-step procedures to perform maintenance. The most common maintenance tasks, as defined by MIL-STD 1388-2A, DoD Requirements for a Logistic Support Analysis Record, are:

- **Access**—to perform operations necessary to gain access to an item of the next lower level of indenture or an item blocking accessibility to the item under analysis.
- **Adjust**—to maintain or regulate, within prescribed limits, by bringing into proper

or exact position, or by setting the operating characteristics to specified parameters.

- **Align**—to adjust specified variable elements of an item to bring about optimum or desired performance.
- **Calibrate**—to determine accuracy, deviation, or variation by special measurement or by comparison with a standard.
- **Disassemble/Assemble**—to take to pieces; to take apart to the level of the next smaller unit or down to all removable parts.
- **Fault Location**—the process of investigating and detecting the cause of equipment malfunctioning; the act of isolating a fault within an item of equipment.
- **Inspect**—to determine the serviceability or detect incipient failures by comparing an item's physical, mechanical, and/or electrical characteristics with established standards through examination.
- **Install**—to perform operations necessary to properly fit a spare/repair part into the next higher assembly.
- **Lubricate**—to apply a substance (e.g., oil, grease, graphite) to reduce friction.
- **Operate**—to control equipment in order to accomplish a specific purpose.
- **Overhaul**—maintenance effort to restore an item to completely serviceable and operational condition. Overhaul is normally the highest degree of maintenance performed.
- **Package/Unpackage**—the action required to prepare equipment for storage and transportation. Also includes the action required to unpack.
- **Preserve**—the action required to treat equipment whether installed or stored to keep it in a satisfactory condition.
- **Remove**—to perform operations necessary to take a spare/repair part out of the next higher assembly.
- **Repair**—utilized as a corrective maintenance action or task function to restore to a serviceable condition an end item, assembly, subassembly, module, or component. Also to be utilized as maintenance action or task function to restore an item removed from the end item through replacement of lower-order nonrepairable items and through rework, such as welding, grinding, or resurfacing, to correct a specific fault. Repair actions include discrete sequences for locating faults, correcting faults or malfunctions, and verifying that the faults have been corrected.
- **Replace**—to substitute serviceable items for malfunctioned, damaged, or worn-out items.
- **Service**—operations required periodically to keep an item in proper operating condition, i.e., clean, drain, paint, or replenish fuel, lubricant, fluids, or gases.
- **Test**—to verify serviceability by measuring mechanical, pneumatic, hydraulic, or electrical characteristics of an item and comparing those characteristics with prescribed standards.

Rather than elaborating on each of these maintenance tasks, an example is in order. The following example addresses the repair of a failure using typical steps and maintenance tasks.

The controller circuit card of an item of electronic equipment fails to operate properly. Assume that the using service has the three generic levels of maintenance and that the electronic equipment is a component in a major system. The maintenance planning process to fix this failure might be as follows:

Step 1. The failure occurs.

Step 2. The system operator observes that the failure occurred because the system quit. Using a test capability (discussed in Chapter 7), the operator isolates the fault to the electronic equipment.

Step 3. The failed unit is removed and replaced with a spare. This is the quickest way to get the system operational again.

Step 4. When the new unit is in place, the operator tests the system to ensure that replacing the unit corrected the system failure. (At this point, the system failure has been corrected, but the failed electronic equipment has not been fixed. The operator sends the failed unit to a supporting maintenance facility.)

Step 5. Upon receiving the failed unit, the supporting maintenance facility tests the unit to isolate the fault to an assembly within the unit. In this case, the test indicates that the controller circuit card is the failed item.

Step 6. Maintenance personnel open the unit.

Step 7. The failed item is removed and replaced with a spare.

Step 8. The unit is closed.

Step 9. The unit is tested to ensure that the failure was corrected by replacing the failed item. (Now the electronic equipment has been fixed, but the failed controller circuit card is still bad. Because the support maintenance facility does not have the capability to fix it, the item is forwarded to the next higher support maintenance organization.)

Step 10. The failed item is tested upon receipt to determine which component on the circuit card failed. The test identifies a failed transistor.

Step 11. The failed item is repaired by removing the failed transistor and replacing it with a spare.

Step 12. The repaired item is tested to ensure that it is now functional.

Through the execution of these 12 steps, the failure was corrected. It might seem that this is an overly complicated way of doing maintenance, but the concept of limiting the capabilities of different levels of maintenance and passing failed units and items to higher support units is extremely cost effective and results in the highest possible system availability. It would have been possible to allow the operator to replace the failed transistor. But, the amount of resources (test equipment, maintenance manual, spare parts, and operator training) required to permit such an action would be unrealistic. This method enabled the system to be operational in a short period of time. If the operator had been allowed to find and replace the failed transistor, the system could have been nonoperational for many hours, or even days.

Figure 5-3 shows how these 12 steps would be documented using the levels of maintenance and the maintenance tasks described above. Note that the maintenance tasks provide a step-by-step process for completing the required actions to fix the failure.

MAINTENANCE TASK ANALYSIS

Maintenance actions cannot be performed without the necessary resources, such as manpower, tools, test equipment, spare parts, appropriate facilities, training, and time. The purpose of maintenance task analysis is to identify these resources using a step-by-step analysis of each task. As each step of a task is analyzed, the resources required to support that step are identified.

Steps 5 through 9 of the repair scenario can be combined into a single maintenance action. Figure 5-4 is a maintenance task analysis worksheet that illustrates a sequential analysis of this maintenance action. The results of this analysis indicate the time required to perform the tasks, the skill level of the maintenance personnel, tools and test equipment, recommended training, replacement parts, and facility. This process is repeated for each maintenance action identified by the FMECA as a requirement to fix a potential failure. The analysis generates an extensive data base that provides input to other ILS disciplines for analysis.

The summation of the results of a complete maintenance task analysis for all maintenance tasks for an item of equipment should identify all the support resources, in terms of both type and quantity, required to perform maintenance. By summing the task times by maintenance level, the labor required to support maintenance at each level is identified. This is the tried and proven method for developing the total logistics support package for an item of equipment. Further discussion of maintenance task analysis is provided in Chapter 16 and Chapter 17.

Step	Maintenance Task	Maintenance Level		
		O	I	D
1	Failure occurs in system			
2	Fault location of system	X		
3	Remove/replace unit	X		
4	Test system	X		
5	Fault location of unit		X	
6	Disassembly of unit		X	
7	Remove/replace circuit card		X	
8	Assembly of unit		X	
9	Test unit		X	
10	Fault location of circuit card			X
11	Remove/replace component			X
12	Test circuit card			X

Fig. 5-3. Maintenance tasks to complete a repair action.

MAINTENANCE TASK ANALYSIS

ITEM NAME: Unit   PART NUMBER: 2345-6   NHA: System   NHA P/N: 12345-6

Maintenance Level: ORG (INT) DEP   TASK DESCRIPTION: Repair unit by removal/ replacement
MAINTENANCE TYPE: (UNSCHED) SCHED   of failed circuit card

Task No.	Step No.	Task/Step Description	Maint Time	No. Pers	Elapsed Time	Total Time	Support Equipment Required
046	001	Fault Location of Unit	0.05	1	0.05	0.05	Test Station P/N 1664-5
	002	Disassemble Unit	0.02	1	0.02	0.02	Screwdriver P/N AC1471
	003	Remove/Replace Circuit Card	0.02	1	0.02	0.02	Extractor P/N 883-A6
	004	Assemble Unit	0.02	1	0.02	0.02	Screwdriver P/N AC1421
	005	Test Unit	0.05	1	0.05	0.05	Test Station P/N 1664-5
	Totals for Time and Personnel				0.16	0.16	

PREPARED BY J. Jones   DATE 10/16/86

Fig. 5-4. Maintenance task analysis.

TYPES OF MAINTENANCE

Maintenance planning divides required maintenance actions into two types: scheduled and unscheduled. The distinction between these two is self-evident. Scheduled maintenance actions are those performed at specific intervals. The intervals could be in calendar measurement (days, weeks, months), or they could be based on the usage of the equipment (hours, miles, sorties). Unscheduled maintenance is performed on an as-required basis to fix the equipment when it breaks.

As maintenance tasks required to support an item of equipment are identified, determining which should be scheduled and which should be unscheduled tasks would appear to be easy in most cases. Common sense would dictate that such tasks as adding oil or checking tires would be scheduled at intervals required by equipment operation. It might prove cost effective to perform other tasks on a scheduled basis in order to prevent equipment failures rather than to wait until a failure occurs and fix it. These scheduled maintenance tasks are called *preventative maintenance* tasks. The method to determine which tasks should be performed as preventative maintenance is defined in MIL-STD 2173, Reliability-Centered Maintenance Requirements for Naval Aircraft,

MAINTENANCE PLANNING BY ACQUISITION PHASE

PRECONCEPT

- Development of preliminary maintenance concept.
- Identification of existing maintenance capabilities.
- Preparation of preliminary maintenance plan.

CONCEPT

- Update maintenance concept based on equipment concept.
- Develop preliminary maintenance requirements.
- Update preliminary maintenance plan.

DEMONSTRATION/VALIDATION

- Update maintenance plan.
- Initial identification of maintenance support requirements.
- Validation of maintenance concept based on equipment design.

FULL-SCALE DEVELOPMENT

- Conduct detailed maintenance task analysis.
- Complete detailed identification of maintenance support resources.
- Update maintenance plan.
- Verification of technical manual adequacy to support maintenance.
- Identification of support equipment required for maintenance.
- Identification of personnel required to support maintenance.
- Identification of facilities required to support maintenance.
- Identification of maintenance training requirements.

PRODUCTION

- Update maintenance plan.
- Participate in design change process to improve maintenance.
- Use operational data to verify maintenance capabilities.

Fig. 5-5. Maintenance planning by acquisition phase.

Weapon Systems and Support Equipment. Reliability-centered maintenance (RCM) is the process of evaluating required maintenance tasks to determine which tasks should be accomplished as part of a preventative maintenance program. A detailed discussion of RCM is provided in Chapter 15.

Maintenance planning starts in the preconcept phase and continues through full-scale development, as illustrated in Fig. 5-5. Detailed maintenance planning is necessary to establish the overall requirements for sustained support of the equipment throughout its useful life. Failure to accomplish effective maintenance planning will result in insufficient support resources being available for maintenance and will therefore degrade the ability of the equipment to perform assigned missions.

Chapter 6

Provisioning

When equipment fails, spare parts are required to fix or replace the failed items. Provisioning is the process for identifying and obtaining the initial stock of spare parts required to support fielded equipment. The method for accomplishing provisioning has been developed over the years through trial and error. Until 1974, each branch of the military had its own way of documenting requirements for spare parts due to the unique needs of the different branch. This caused many problems. First, there was no uniform way of correlating the parts used by one service with those of another. Second, the data bases that were developed by the services were incompatible.

In 1974, MIL-STD 1552, Uniform Department of Defense Requirements for Provisioning Technical Documentation, and MIL-STD 1561, Uniform Department of Defense Provisioning Procedures, were issued. These documents established a standard method for accomplishing and documenting provisioning requirements. MIL-STD 1552 has since been superseded by MIL-STD 1388-2A, DoD Requirements for a Logistic Support Analysis Record; however, the basic guidance of this document has not changed.

Provisioning is one of the few ILS disciplines that uses data from all other disciplines. Therefore, it represents one of the final outputs of the ILS effort. As stated previously, one of the major goals of ILS is to identify support resources required to sustain operations. The final provisioning documentation identifies all material required to support maintenance.

THE PROVISIONING PROCESS

Provisioning is accomplished by generating data lists, called provisioning lists, that identify the parts of an equipment item. Logically, this should not be done until the

75

design is finalized. Therefore, provisioning activities occur as late in the FSD phase as possible, but early enough to allow procurement and delivery of spares required to support the first equipment that is fielded. Initial provisioning starts with the award of a government contract for the provisioning effort and ends with the delivery of spares to support the equipment. Major milestones in the provisioning process are conferences between the contractor and the government. At these conferences, discussions are held about provisioning activities, or the government reviews and approves or disapproves the contractor's spare parts recommendations. Figure 6-1 illustrates provisioning milestones.

Contract Award

Provisioning planning begins on, or before, contract award depending on the phase of acquisition of the equipment. Contracts that require provisioning contain two stan-

Contract Award

Guidance Conference

Long Lead Time Items
Provisioning Conference

Provisioning Preparedness
Review Conference

Submission of DLSC Screening

Submission of PTD

Provisioning Conference

Submission of Post
Conference List

Receipt of Provisioned
Item Order

Submission of Priced
Provisioned Item Order

Delivery of Spares

Fig. 6-1. Provisioning milestones.

dard documents: DD form 1949-1, LSAR Data Selection Sheet, and DD Form 1949-2, Provisioning Requirements Statement (PRS). These two documents are the foundation for all provisioning activities. (Note: DD form 1949-1 is found in MIL-STD 1388-2A. Easier identification of purely provisioning data is found in MIL-STD 1552 (obsolete). (DD form 1949-2 is found in MIL-STD 1561.) DD form 1949-1 identifies all the information that must be submitted on each part provisioned as a spare part. This form is discussed in Chapter 17 and illustrated in Fig. 17-17. DD form 1949-2, shown at Fig. 6-2, provides specific directions about when provisioning will occur, conferences, and other relevant information. In addition to the forms stated above, a Provisioning Performance Schedule (PPS) may also be included in the contract. The PPS identifies key contractual provisioning milestones. Sample PPS is in Fig. 6-3.

Provisioning Guidance Conference

Provisioning is one of the few activities that a contractor agrees to accomplish before the total extent of the requirement is known. A provisioning guidance conference is normally scheduled within 60 to 90 days after the contract is awarded. The guidance conference is held to enable the provisioning specialists from the government to meet the contractor and agree on what provisioning will be accomplished. (Even though MIL-STD 1552 was supposed to make provisioning a uniform process, each service has addendums to the standard that tailor provisioning requirements to meet their own needs. Therefore, there is a lot of room for interpretation that must be resolved.) The results of the guidance conference are published as minutes of the conference and are signed by representatives of the government and the contractor. These minutes become part of the contract.

Long Lead-Time Item-Provisioning Conference

The first conference at which the government reviews a contractor's recommendations for spare parts is the long lead-time item-provisioning conference. Long lead items cannot be bought or manufactured within the normal provisioning cycle and, therefore, must be obtained as early as possible to meet operational needs. These items are identified on a Long Lead Time Items List (LLTIL). The contract delivery requirements dictate what length of delivery time makes an item long lead. Normally they are items that require more than 9 to 12 months for delivery.

Interim Support Items Provisioning Conference

Sometimes the provisioning process is just too slow to ensure that spares will be available to support operational equipment. The interim support items provisioning conference is much like the long lead-time item conference, except that every item discussed is probably a candidate for sparing because the operational need date must be met. An Interim Support Items List (ISIL) is prepared, identifying interim support items required to support equipment maintenance until full provisioning can be accomplished. Some programs have an interim support period in which the contractor supports limited numbers of fielded units. This is where an ISIL is most applicable.

MIL–STD–1561B

PROVISIONING REQUIREMENTS STATEMENT	FORM APPROVED OMB NO. 0704-0188 EXP. JUN 30 1986

EQUIPMENT NOMENCLATURE

MODEL TYPE NUMBER

CONTRACT AND ITEM NUMBER	DATE (YYMMDD)

PR MIPR NUMBER	DATE (YYMMDD)

SOLICITATION NUMBER	DATE (YYMMDD)

PROVISIONING ACTIVITY (Address and Zip Code)	CONTRACTOR NAME (Address and Zip Code)

A This Provisioning Requirements Statement (PRS) is furnished in accordance with MIL-STD-1561B Deliverable Provisioning Technical Documentation (PTD) and Supplementary Provisioning Technical Documentation (SPTD) Requirements, will be specified on DD Form 1423, Contract Data Requirements List

B When the PRS is furnished after contract award the Contractor shall submit a priced proposal within 30 days after receipt of the PRS This PRS may be modified or changed by a supplemental agreement to the contract

C A Statement of Prior Submission (SPS) submitted in accordance with paragraph 5 4 MIL STD 1561B may result in reduction or elimination of PTD and SPTD requirements specified on DD Form 1423 and conference requirements of the PRS

PROVISIONING REQUIREMENTS

1 GUIDANCE CONFERENCE   (Check one) ☐ IS REQUIRED   ☐ IS NOT REQUIRED IF REQUIRED THE CONFERENCE WILL BE HELD AT (Paragraph 5 1 1)

A PLACE	B DATE (YYMMDD)

C TIME	D ESTIMATED NUMBER OF DAYS

2 PROVISIONING CONFERENCE   (Check one) ☐ IS REQUIRED   ☐ IS NOT REQUIRED IF REQUIRED THE CONFERENCE WILL BE HELD AT (Paragraph 5 1 4)

A PLACE	B DATE (YYMMDD)

C TIME	D ESTIMATED NUMBER OF DAYS

E THE CONTRACTOR - (Check one) ☐ IS REQUIRED   ☐ IS NOT REQUIRED TO HAVE A SAMPLE ARTICLE OF THE COMPONENT/END ITEM AT THE CONFERENCE (Paragraph 5 1 4 a)
SAMPLE ARTICLE (Check one) ☐ WILL BE VIEWED   ☐ WILL BE DISASSEMBLED AT CONFERENCE

3 PROVISIONING PREPAREDNESS REVIEW CONFERENCE - (Check one)
☐ IS REQUIRED   ☐ IS NOT REQUIRED (Paragraph 5 1 2)

4 LONG LEADTIME ITEMS PROVISIONING CONFERENCE - (Check one)
☐ IS REQUIRED   ☐ IS NOT REQUIRED (Paragraph 5 1 3)

5 INTERIM SUPPORT ITEMS CONFERENCE - (Check one)
☐ IS REQUIRED   ☐ IS NOT REQUIRED (Paragraph 5 1 5)

6 MANUFACTURERS OR COMMERCIAL MANUALS – (Check one)
☐ ARE REQUIRED   ☐ ARE NOT REQUIRED (Paragraph 5 3 15)

7 INCREMENTAL SUBMISSION OF PTD – (Check one)
☐ IS AUTHORIZED   ☐ IS NOT AUTHORIZED (Paragraph 5 5)

8 PROVISIONING SCREENING   (Check one) ☐ IS REQUIRED   ☐ IS NOT REQUIRED
RESULTS (Check one) ☐ ARE REQUIRED   ☐ ARE NOT REQUIRED TO BE ENTERED ON THE PL (Paragraph 5 6)

9 DELIVERY FOR SUPPORT ITEMS WILL BE - (Check one)
☐ CONCURRENT (Paragraph 5 8 1)   ☐ SCHEDULED (Paragraph 5 8 2 1)   ☐ SCHEDULED (Paragraph 5 8 2 2)

DD FORM 1949-2, 84 OCT   PREVIOUS EDITION IS OBSOLETE

Fig. 6-2. Provisioning requirements statement.

MIL-STD-1561B

10. RESIDENT PROVISIONING TEAM (RPT) (Check one)
☐ WILL BE ESTABLISHED   ☐ WILL NOT BE ESTABLISHED (Paragraph 5.2.1)

11. INTERIM RELEASE (Check one) ☐ IS AUTHORIZED   ☐ IS NOT AUTHORIZED (Paragraph 5.7.5)

12. SPTD SPECIFICATIONS DRAWINGS WILL BE FURNISHED ON
☐ MICROFILM   ☐ HARD COPY   ☐ APERTURE CARDS (Paragraph 5.3.13.2)

13. SPTD WILL BE SEQUENCED BY (Check one) ☐ PLISN   ☐ REFERENCE NUMBER
☐ REFERENCE DESIGNATION   ☐ OTHER _____ (Paragraph 5.3.13.2).

14. THE INITIAL PIO WILL BE SUBMITTED WITHIN _____ DAYS AFTER APPROVAL OF PTD/SPTD OR WITHIN _____ DAYS AFTER COMPLETION OF PROVISIONING CONFERENCE OR WITHIN _____ DAYS AFTER ACCEPTANCE OF THE PCL (Paragraph 5.7.1)

15. TOOLS AND TEST EQUIPMENT (Check one)
☐ WILL BE   ☐ WILL NOT BE INCLUDED IN PPL (Paragraph 5.3.1)

16. PPS (Check one) ☐ IS   ☐ IS NOT REQUIRED (Paragraph 4.2)

17. REPAIR KITS AND REPAIR PART SETS (Check one)
☐ WILL BE   ☐ WILL NOT BE INCLUDED IN THE PPL (Paragraph 5.3.1)

18. COMMON AND BULK ITEMS LIST (Check one) ☐ OPTION 1   ☐ OPTION 2
☐ OPTION 3   ☐ OPTION 4 (Paragraph 5.3.3)

19. MILITARY SERVICE AGENCY ADDENDUM ☐ IS ATTACHED

DD FORM 1949-3, 84 OCT (Reverse)

Fig. 6-2. Continued.

MIL-STD-1561B
PROVISIONING PERFORMANCE SCHEDULE (EXAMPLE)

END ITEM:	CONTRACTOR
END-ITEM DELIVERY DATE:	SOLICITATION OR CONTRACT NO.:

Events	Not Later Than	Calendar Date of Events
Contract Award		
Provisioning Requirements Statement		
Guidance Conference		
Interim Support Items Conference		
Provisioning Preparedness Review Conference		
Long Lead Time Item—Provisioning Conference		
Submission of PTD		
Provisioning Conference		
Post Conference List		
Provisioned Item Order		
Priced Provisioned Item Order		
Delivery of Support Items		

Fig. 6-3. Provisioning performance schedule.

Provisioning Preparedness Review Conference

The purpose of this conference is to allow the government to preview provisioning documentation for a provisioning conference. Because provisioning conferences are attended by representatives of many organizations, the review ensures that the contractor is ready for a provisioning conference and that funds will not be wasted by convening a useless conference.

Provisioning Conference

The formal provisioning conference, or source coding conference, is the forum for contractor presentation of recommended spares to the government. The conference consists of government review of the contractor's provisioning lists to approve recommended selection of spares. One of the key activities is government approval of source, maintenance, and recoverability (SMR) codes recommended by the contractor.

The duration of a provisioning conference depends on the complexity of the item being provisioned. Small items require only two or three days, while the provisioning conference for a major weapon system might take several months. Conference attendees normally include provisioning analysts, cataloging specialists, and maintenance engineers representing both DoD and the contractor.

PROVISIONING TECHNICAL DOCUMENTATION

Provisioning Technical Documentation (PTD) is any data submitted to the government as part of the provisioning process. As indicated above, there are several different types of provisioning lists that a contractor might be required to prepare. In addition to the LLTIL and the ISIL, a contract might require a Provisioning Parts List, Tool and Test Equipment List, Common Bulk Items List, and Repairable Items List. The data reflected on these lists are basically the same. The difference in these lists is the types of items for which they are prepared and the intended use of the lists.

Provisioning Parts List

The Provisioning Parts List (PPL) is a complete listing of all parts that make up an item of equipment. This list is prepared by the contractor and submitted to the government at, or before, the provisioning conference. A complete PPL not only identifies everything required to build the equipment, it also provides information about each part. This information is divided into cataloging data, which describes the part, and application data, which tells how it is used and maintained.

Tools and Test Equipment List

A Tools and Test Equipment List (TTEL) contains information about unique tools and test equipment required to perform maintenance on the equipment. The format of the list is the same as for a PPL. Submission of the TTEL is dependent on the contract delivery requirements, but it is normally submitted with the PPL.

Common Bulk Items List

It is easier to document the usage of bulk items such as wire, adhesives, solder, etc., using the Common Bulk Items List (CBIL). The CBIL is a companion document to the PPL. In a large item of equipment, a common bulk item could possibly be used several hundred times, which might require an identical number of entries on the PPL. The item would require only one entry on the CBIL because it consolidates all usages into a single total quantity required for the equipment.

Repairable Items List

The Repairable Items List (RIL) is an abbreviated version of a PPL. It contains provisioning data on the repairable items contained in the equipment. Submission of the RIL normally precedes the PPL. In many cases the RIL is used to cross-check the maintenance task list to ensure that all items identified by Provisioning as repairable have been addressed by Maintenance Engineering.

Data Preparation

Logistics engineers use the engineering drawings as the basis for generating provisioning lists. The parts lists that describe the equipment design provide the starting point for all provisioning activity. The drawing package can be supplemented by other documentation such as manufacturer's catalogs, MIL-SPECs, etc. All the parts included on a provisioning list must be fully described using these source data, which are termed Supplementary Provisioning Technical Documentation (SPTD). The preparation of provisioning lists is accomplished as illustrated in Fig. 6-4.

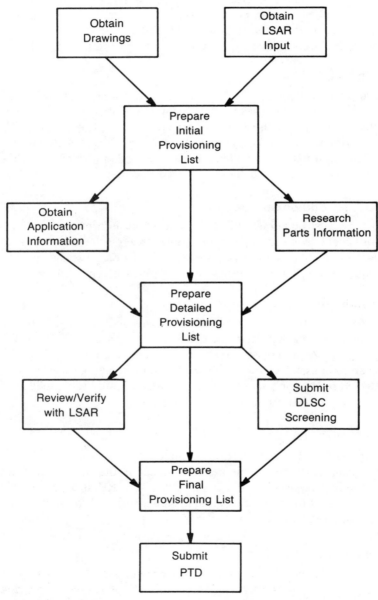

Fig. 6-4. Provisioning process.

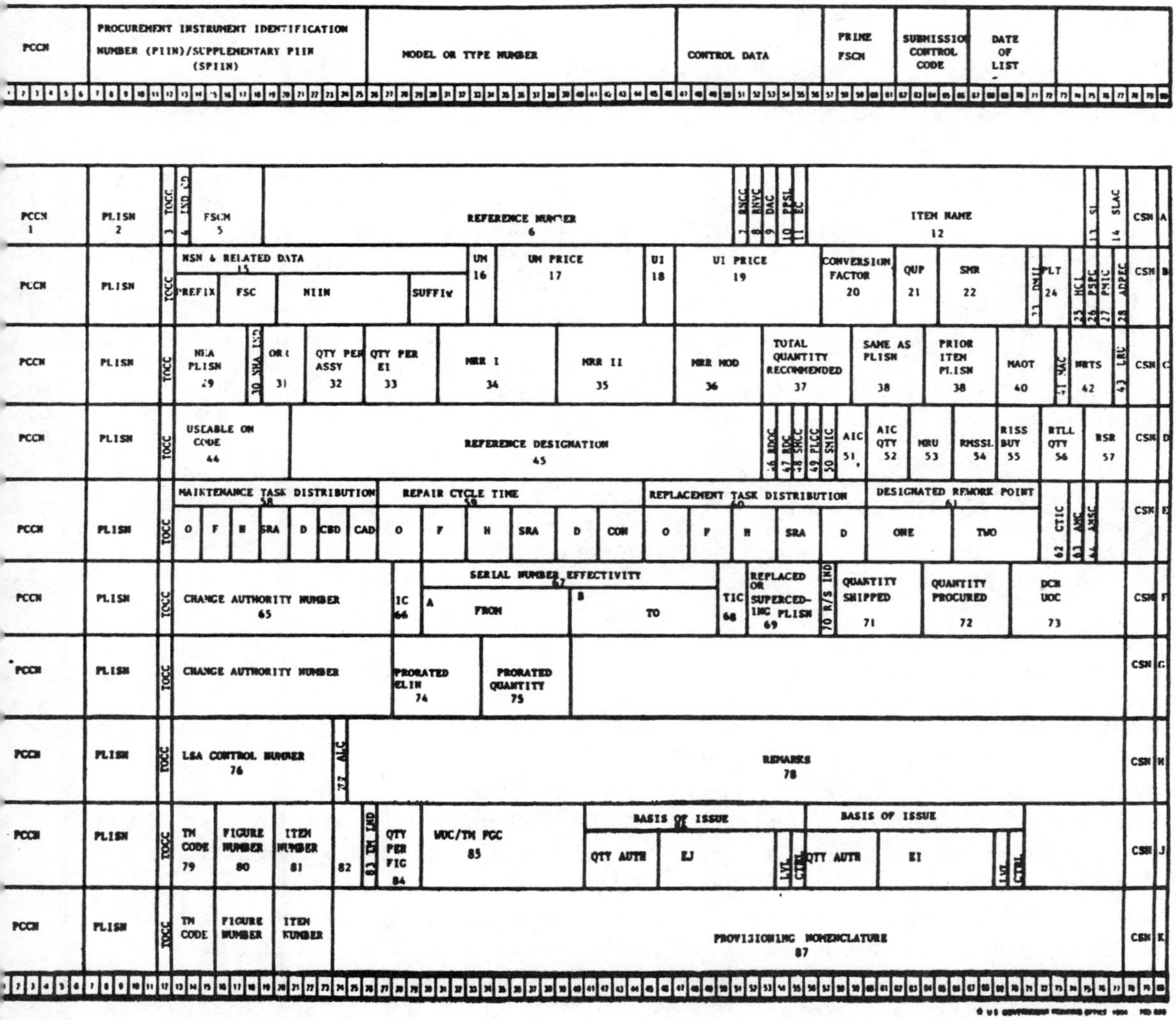

Fig. 6-5. Provisioning format.

The equipment PPL is normally sequenced by starting at the top level and documenting each level of the equipment down to the lowest level, showing the indenture relationship between each part on the list. This is called a *topdown breakdown sequence*. Other lists are sequenced in part number order. All data reflected on provisioning lists are formatted in the standard provisioning format established by MIL-STD 1552 (obsolete) so the format is commonly referred to as "1552 format." An example of this format is shown at Fig. 6-5. The format is a series of data lines that have dedicated spaces for each data element. Each item appearing on a provisioning list must have at least four data lines. Figure 6-6 illustrates completed data lines for four parts. Notice the indenture relationship between each part.

```
ABF666A001   A9999912345-65                              2     RADAR SYSTEM           01A
ABF666A001        6580015438746     EA0051753800EA0051753800000001001PAOODX 48        01B
ABF666A001               0010000010002134000042680                                    01C
ABF666A001   UF6                                                                       01D
ABF666A001   A                                                                         01H
ABF666A001   0010000100001                                                             01K


ABF666A002   B99999387209-1                              2     RADAR SUBSYSTEM #1      01A
ABF666A002        6580017364962     EA0018574600EA0018574600000001001PAOFFX 48        01B
ABF666A002   A001      0010000010002134000042680              000008                   01C
ABF666A002   UF6      A1                                                               01D
ABF666A002   AA                                                                        01H
ABF666A002   0010000100002                                                             01K


ABF666A003   C99999287451-1                              2     ASSEMBLY #1             01A
ABF666A003        6580011524375     EA0003456000EA0003456000000001001PAFDDX 24        01B
ABF666A003   A002      0010000010002134000042680              000016                   01C
ABF666A003   UF6      A1A1                                                             01D
ABF666A003   AAAA                                                                      01H
ABF666A003   0010000100003                                                             01K


ABF666A004   B99999598451-1                              2     RADAR SUBSYSTEM #2      01A
ABF666A004        6580018576306     EA0012670000EA0012670000000001001PAOFFX 48        01B
ABF666A004   A001      0010000010002134000042680              000008                   01C
ABF666A004   UF6      A2                                                               01D
ABF666A004   AB                                                                        01H
ABF666A004   0010000100023                                                             01K
```

Fig. 6-6. Provisioning parts list.

Data Elements

The data elements on provisioning lists can be divided into two categories: cataloging data and application data. Cataloging data describe the part without regard to its usage. This type of data—item name, shelf life, NSN, unit of issue, etc.—applies to the part itself, and can be saved by the contractor, hopefully by computer, and reused each time the part is provisioned. Application data relate to the specific application of the part being reported on the provisioning list. Data elements such as quantity per assembly, maintenance task distribution, technical manual references, etc., tell how the part is used. Application data can be different for each use of a part in a system.

The data elements on provisioning lists are identified by the government on contract DD form 1949-1. A complete description of each data element listed on DD form 1949-2 is provided in MIL-STD 1388-2A. The data requirements of each provisioning contract are tailored to fit the equipment. However, several data elements are commonly required on all contracts.

Provisioning List Item Sequence Number (PLISN). The PLISN is a sequentially assigned number that appears at the beginning of each data line on a provisioning list. It is used to sequence the list and provide a cross-reference to subsequent listed items. The Next Higher Assembly (NHA) PLISN relates a part to its next indenture level, and the Same As PLISN identifies subsequent appearances of the same items.

Manufacturer's Part Number. The part number of each item on the provisioning list must be the true number assigned by the manufacturer or agency that rols the specification or drawing to which the part conforms. It is extremely important that this number be transferred correctly from the source document. Errors in transfer will cause duplication or erroneous provisioning. For example, part numbers 123-4 and 123/4 are read by computers as being different numbers when, in fact, they might be the same part. Part numbers that are reflected on assembly drawing parts lists should be challenged when inconsistencies arise.

Federal Supply Code for Manufacturers (FSCM). Each manufacturer that provides equipment or parts to the government is assigned a unique FSCM. The FSCM is a five-digit code that is used to relate a part number to a manufacturer. Notice on Fig. 6-5 that the FSCM appears just before the part number. Once a PPL is submitted to the government, the combination of FSCM and part number are used to identify each part. For example, one manufacturer builds a model of truck and assigns that model the part number 12345, and another manufacturer who makes light bulbs has a type of bulb with part number 12345. If the government issues a contract for 100 items of part number 12345, would they get trucks or light bulbs? The answer to this problem is the use of the correct FSCM in conjunction with the part number in order to get the desired items. A complete listing of current FSCMs is found in the *Cataloging Handbook H4/H8 Series*, microfiche available from the Government Printing Office. This code may also be referred to as a Commercial and Government Entity (CAGE) code.

Source, Maintenance, and Recoverability (SMR) Code. The SMR is probably the most important data element on any provisioning list. It is directly related to the maintenance task analysis and is also reflected in technical manuals. The SMR is a six-digit code that is divided into three sections that tell where a part comes from, who does maintenance on it, and what is done with the part when it can't be repaired. Figure 6-7 is an SMR code matrix. The first and second positions identify the source of the part. The source can be through procurement, as a component or a kit; be manufactured from bulk material; or be assembled from a set of components. The "X" source code is used to signify an item that cannot be obtained. The third and fourth positions describe the maintenance of the item. The third position tells the lowest level of maintenance ("O" is organizational maintenance level, "F" is intermediate, and "D" is depot) that is authorized to remove and replace the item. Here is the direct tie-in to the maintenance task analysis discussed in Chapter 5. Chapter 9 *Technical Manuals* discusses how the maintenance task analysis is used to generate technical manuals.

If the technical manual for organizational maintenance indicates that an item is to be removed and replaced at "O" level, but the SMR indicates that "F" (intermediate) is the lowest level authorized to remove and replace the item, the item will not be available at organizational level to support the maintenance task. This is an all too frequent error. The fourth position identifies the lowest level of maintenance authorized to perform a repair on the item. The situation described above for the third position also applies to the fourth. The fifth position tells what level of maintenance is authorized to determine if the item is beyond repair and should be discarded. The first through the fifth positions are standardized codes for all services. The sixth position is reserved for the unique requirements of each branch of the service. AFR 66-45, Joint Regulation

SOURCE		MAINTENANCE		RECOVERABILITY	RESERVED
		USE	REPAIR		
1st Position	2nd Position	3rd Position	4th Position	5th Position	6th Position
P — Procurable	A Stocked			Z — Nonreparable Condemn by 3rd Position Level	Reserved for Individual Service Applications
	B Insurance				
	C Deteriorative				
	E Support Equipment, Stocked	O — Remove/Replace by Organizational level	Z — No Repair		
	F Support Equipment, Nonstocked		B — No Repair Recondition	O — Reparable Condemn by Organizational (or Field or Depot)	
	G Life of System Support				
K — Component of a Repair Kit	F Intermediate Kit		O — Repair by Organizational		
	D Depot Kit	F — Remove/Replace by Intermediate Level		F — Reparable Condemn by Intermediate (or Depot)	
	B In Both Kits		F — Repair by Intermediate		
M — Manufacture	O Organization				
	F Intermediate				
	D Depot		D — Limited Repair by O or F Level		
A — Assemble	O Organization	D — Remove/Replace by Depot Level		D — Reparable Condemn by Depot Only	
	F Intermediate		D — Overhaul by Depot		
	D Depot				
X — Nonprocured	A Requisition NHA				
	B Reclamation or Requisition by Part Number		L — Repair by Depot	A — Special Handling	
	C Mfg Drawings				

Fig. 6-7. Joint military services uniform SMR coding matrix.

Governing the Use and Application of Uniform Source Maintenance and Recoverability Codes, provides detailed information about the SMR code.

National Stock Number (NSN). All standard items procured by the government are assigned an NSN that uniquely identifies the item. Everything from beans to bullets to aircraft has an NSN. An NSN is a 20-digit number, although the three-digit prefix and four-digit suffix are not commonly used when referring to this number. Figure 6-8 shows how an NSN is divided into seven different parts. The Federal Stock Classification (FSC) identifies the type of item. The North Atlantic Treaty Organization (NATO) code identifies the country that is cognizant for procurement. For example, the United States is represented by NATO codes 00 and 01. The National Item Identification Number (NIIN) is made up of a two- digit NATO code and a seven-digit, randomly assigned, identification number. The Master Cross Reference List (MCRL), microfiche available from the U.S. Government Printing Office, provides a cross reference between NSNs and corresponding manufacturer's FSCM and part numbers. The identification of parts listed on provisioning lists that have been assigned an NSN normally relieves the contractor of having to provide complete identification data for that part.

Quantity per End Item (QPEI). The total number of times a part is used in an item of equipment is known as the QPEI. The QPEI is essential when computing the number of spares required and in planning maintenance for like items.

Recommended Quantity. Contractors might be required to recommend initial quantities of items to be procured by the government as spares. There are many different methods for computing the recommended quantity. The selection of the method to be used is important, and should be a priority topic of discussion at the provisioning guidance conference. Each service has a preferred method for spare quantity computation.

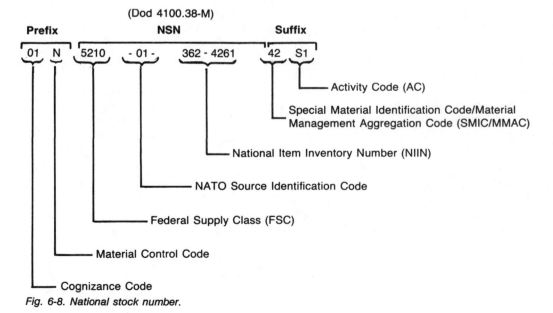

Fig. 6-8. National stock number.

Most formulas require basic information such as MTBF, annual operating hours, mission duration, QPEI, sparing confidence level, and number of items being supported. The MTBF is obtained from Reliability Engineering. QPEI is generated in the preparation of the PPL. Most of the remaining data must be obtained from the government. Poisson distribution is sometimes used in computing sparing levels. Other data are unique to the services. Figure 6-9 illustrates the basic philosophy used in computing the recommended quantity for initial spares.

Production Lead Time (PLT). The production lead time indicates the time between placement of an order for an item and delivery of the first unit. PLT is the key for identification of long lead items that are provided to the government on the LLTIL.

Unit Price. The unit price is one of the most difficult data elements for a contractor to provide on a provisioning list. The problem is related to the credibility and useability of the price quoted. At the time items are initially provisioned, procurement of material for production of the equipment might not have started. Therefore, unit prices for spares are often best guesses or inflated budgetary prices, because they can vary considerably with the quantity produced or procured at a given point. This problem is discussed further in Chapter 19.

Item Name. The government has a standard method for expressing the name of an item. Item names are comprised of the root noun followed by modifiers. For example, the item name for an M-1 tank is tank, combat, full- tracked, M-1. *The Federal Item Name for Supply Cataloging, H6-1*, published by the U.S. Government Printing Office, contains standard item names.

Provisioning Screening

Contractors are required to screen provisioning lists to identify parts on the list that have already been assigned national stock numbers (NSNs). Screening is accomplished by sending a list of the part numbers and corresponding FSCMs contained on the provisioning list to the Defense Logistics Service Center (DLSC). DLSC, located in Battle Creek, Michigan, is responsible for assignment of NSNs and maintenance of the records of all parts that have been assigned NSNs. It is also responsible for the assignment of FSCMs. The list submitted by the contractor is compared by computer with the DLSC master files. The comparison identifies parts that have previously been procured by the government. Then, DLSC generates a report that includes cataloging data about the part, such as item name, unit of issue, shelf life code, etc., and, most important, the NSN assigned to the part. The results of this screening are returned to the contractor for inclusion in the provisioning list.

DLSC information can make the provisioning task much easier for the contractor. Items that have already been assigned an NSN normally require fewer entries on provisioning lists, because the NSN allows full identification by the government provisioning team. To allow sufficient time for DLSC to process a screening request

$$\text{Recommended Quantity} = \frac{\text{QPEI} \times \text{Number of Systems} \times \text{Operating Hours}}{\text{MTBF}}$$

Fig. 6-9. Basic spares requirement computation philosophy.

and for the contractor to receive and record the results, screening is normally accomplished 45 to 60 days prior to a provisioning conference.

Provisioning Submittal

Provisioning lists are submitted to the government for review prior to each conference. Lists may be submitted in printed copy or by computer magnetic data tape. The government should provide direction at the guidance conference as to when and how the lists will be submitted. The important point to be considered is that submittal requirements must be consistent.

The SPTD to support the provisioning conference is normally held by the contractor if the conference is to be held in the contractor's facility. If the conference is to be held at a government facility, the SPTD is usually submitted with the provisioning list. The government normally requires two copies of each drawing submitted. History has proven that it is best to assemble one of these drawing packages in part number sequence and one in PLISN sequence. This aids in researching the parts at the actual provisioning conference.

Provisioning Schedule

The scheduling of provisioning activities is always somewhat questionable. It would be great to be able to wait until the design was complete before starting to provision. If this were done, spares needed to support the equipment when fielded would not be available because of the lead time required for procurement. Starting provisioning activity too early in a program is wasteful because of changes that will be made to the equipment before the design is finalized. Therefore, the reasonable solution is to use backward planning to determine the optimum start date. The key to this process is determining the date when the equipment will be fielded. Starting from this date and working backwards, allowances must be made for the following:

- Processing time for the receipt of spares at the final destination
- Transportation of spares from the contractor
- Acceptance from the contractor and administrative processing time
- Production and testing by the contractor
- Receipt of contract and ordering of material
- Administrative time required to issue a contract
- Submittal of the final provisioning list that has changes made at the provisioning conference
- The actual provisioning conference.

This process, illustrated in Fig. 6-10, identifies the latest date that the provisioning conference can occur and still support the equipment when it is fielded. This process also identifies how long lead items are determined. If an item has a PLT that exceeds this length of time, it must be ordered early on the LLTIL to be available when the equipment is fielded.

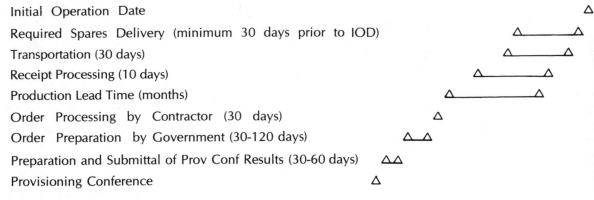

Initial Operation Date

Required Spares Delivery (minimum 30 days prior to IOD)

Transportation (30 days)

Receipt Processing (10 days)

Production Lead Time (months)

Order Processing by Contractor (30 days)

Order Preparation by Government (30-120 days)

Preparation and Submittal of Prov Conf Results (30-60 days)

Provisioning Conference

Fig. 6-10. Provisioning planning.

This schedule must then be matched with the design schedule to see if the equipment will be far enough along in the process to support full provisioning. Remember, it might take the contractor a year to prepare for the provisioning conference if the equipment is complex. Major weapon systems sometimes require several years to complete provisioning. Use of this type of planning process ensures that spares will be available when needed.

POST PROVISIONING

After the contractor has submitted the final provisioning list, the government does several things. First, the list is submitted by the government to DLSC. When the contractor screened the list prior to the provisioning conference, the results were for information only. Remember, provisioning occurs close to the end of FSD, so the design might not be final in all cases when screening was done. DLSC takes the results of the provisioning conference and rescreens each item. Items that do not have an NSN are cataloged.

Cataloging is the process of detailed identification, according to military specification, that culminates in the assignment of an NSN. That is how the government's master parts file at DLSC is kept up to date. Normally, no orders for spares are placed until DLSC has completed cataloging and NSN assignment.

After cataloging, the government uses the latest data on projected equipment utilization and MTBF, or failure rate, to recalculate quantities of spares. Each service has its own method of doing this. The USAF D220 computerized provisioning system has proven to be very effective in accomplishing this task in a timely manner, because it automates virtually all functions required to start spares procurement. The remaining government actions that occur after provisioning are discussed further in Chapter 11.

DESIGN CHANGE NOTICE

After the final provisioning list has been submitted to the government, the contractor is normally required to update the list to reflect design changes that occur through the duration of the contract. These changes are submitted to the government in a design change notice (DCN). A DCN consists of a new or replacement set of data lines in the same format as the original list. Each DCN must show the number of the ECP, or change order, that the government approved to cause the change. A copy of the drawing or other SPTD that changed must also accompany the DCN.

The provisioning process, when accomplished properly, provides information to the government about the type and quantity of spare and repair parts required to support maintenance. It also identifies the configuration of an item of equipment. When submitted to the government, the list becomes part of the DLSC master data file that contains information on parts procured by the government.

Chapter 7

Support Equipment
and Testing

Weapon systems or other items of equipment normally require the use of additional equipment to support operations or maintenance. Any item of equipment required to support operation or maintenance is categorized as *support equipment*. The support equipment can be a special item designed for only one specific use, or it can be items that have multiple uses. The discussion of support equipment in this chapter focuses on items required to support maintenance. Maintenance support equipment consists of items used to perform maintenance tasks and items that are used to perform testing. Definition of the range of different types of test equipment used by DoD is contained in MIL-STD 1309C, Definition of Terms for Test, Measurement, and Diagnostic Equipment.

It can be argued that testing is a maintenance task; however, the test-fix-test method of performing maintenance provides a clear distinction between support equipment types. Test, measurement, and diagnostic equipment (TMDE) is used to perform testing to *identify* what is wrong with an item, and maintenance support equipment is used to *fix* it. The two are dependent; they must be developed and used together.

SUPPORT EQUIPMENT

There are several different ways to classify support equipment (SE). As stated above, it can be classified by its use for testing or actual maintenance. Other classifications include its availability, application, complexity, or cost. The most frequent method used to classify support equipment is to refer to the items as either *common* or *special support* equipment. Each of these broad categories is open to interpretation, depending on the specific application.

Common Support Equipment

Generally, any item of support equipment that is currently used by the military and has multiple applications is classified as common support equipment. This includes hand tools (such as pliers, screwdrivers, hammers, and wrenches), compressors, hydraulic lifts, oscilloscopes, volt meters, battery testers, etc. All of these items are commonly used to perform maintenance and have a broad range of applications. Items that have been adopted into the DoD inventory and have been assigned an NSN are classified as standard items. An item such as a new kind of wrench or test equipment, that could be used to replace an existing standard item, but has not been adopted into the DoD inventory, is classified as nonstandard common support equipment.

It is reasonable to assume that a hand tool exists within the DoD inventory for every maintenance task requirement. The process of identifying the proper tool for a task is accomplished by first finding the correct size needed to perform the function and then cross-referencing the requirement to the Master Cross-Reference List (MCRL) to determine the correct NSN for the item. A General Services Administration (GSA) catalog is extremely useful in finding the proper standard hand tool to perform a maintenance task. The GSA catalog contains sufficient information to verify that the identified tool will perform the task, and the catalog also contains the applicable NSN for the item.

Identification of standard test equipment can be a little more difficult; however, references are available that provide sufficient information to accomplish proper identification. MIL-STD 1364, Standard General Purpose Electronic Test Equipment, is an extensive reference document that identifies standard test equipment that can be used to perform testing. MIL-STD 1387, Preparation and Submission of Data for Approval of Nonstandard General Purpose Electronic Test Equipment, describes the requirements for adopting standard test equipment.

Special Support Equipment

Items of support equipment that have limited application, or that have been developed to perform a specific support function for a single weapon system, are categorized as special support equipment. An example of special support equipment is a sophisticated electronic test station designed to test components of a weapon system that cannot be used for any other purpose. This type of support equipment is normally procured concurrently with the weapon system it is to support.

Special support equipment tends to be expensive in both the short and long term. Up-front engineering and development followed by small-quantity procurement makes special support equipment expensive. The follow-on support of this equipment, with standard documentation and training requirements and limited demand for spares, increases the cost of maintaining the support equipment. Special support equipment should be avoided when possible to reduce the life cycle cost of a weapon system.

Requirement Identification

The identification of requirements for support equipment must begin as early in the acquisition process as possible. Early identification of anticipated requirements in the concept phase is necessary to plan for the funding of SE. This budget planning may be developed using projections based on similar systems or the equipment being replaced. It is important to recognize that a requirement for SE exists, and that funds must be set aside to support the requirement. Lack of sufficient funding for SE will adversely affect maintenance's planning.

The maintenance concept for the system plays a large role in determining the requirements for support equipment. The number of maintenance levels to be used and the anticipated maintenance tasks to be performed at each level will determine the gross requirements for SE. In theory, the organizational level requires less SE than the intermediate or depot levels. Also, the cost for SE at the organizational level is normally less than at the intermediate, which, in turn, is normally less than at the depot.

Another significant factor in identifying SE requirements is the anticipated built-in test capability of the equipment. For example, if an item of equipment has the capability to fault-isolate a failure to a small module within the system, and the operator can remove and replace the defective module using common tools, the need for sophisticated fault-isolation SE at both the organizational and intermediate levels may be avoided. The intermediate level can then concentrate on the repair of modules, rather than having to fault-isolate the entire system, remove and replace modules, and repair failed modules.

As the design matures, specific types and quantities of SE can be identified based on the maintenance task analysis. The maintenance task analysis, as described in Chapter 5, is a detailed identification of all resources required to support the equipment. As each maintenance task is analyzed, the type and quantity of SE required to support the task is documented. The results of this analysis can then be used to quantify the total requirements for SE to support the system at each level of maintenance. Figure 7-1 shows how annual requirements for an item of SE can be calculated. The calculation must be repeated for each item at each maintenance level. This process should not be considered the final decision on SE requirements, because the data base does not account for concurrent task completion; however, it does provide a prediction for the amount of time that each item of SE will be required to support maintenance.

The prediction can be used to determine if the acquisition of an item is justified based on its anticipated use. Items that do not appear to warrant procurement should be analyzed further to determine if the maintenance task can be accomplished using

$$SE = Q(\Sigma \ TF \times ETT)$$

Where: Q = Number of systems supported
TF = Task frequency
ETT = Elapsed task time

Fig. 7-1. Annual support equipment requirements.

other resources. This type of trade-off analysis should result in a rational decision about which items of SE will be used enough to justify procurement to support maintenance. For example, if an item of SE is predicted to be used only 40 hours per year and costs $100,000, the justification for buying the item must be very strong.

Remember, maintenance tasks are written assuming that the required SE will be available. If the SE is not going to be available, the maintenance task must be rewritten. This causes a waterfall effect, because the change to the maintenance task will also affect the technical manuals, training, and provisioning of the system. Therefore, it is important to solidify detailed SE requirements as early as possible in the acquisition process to avoid rework and unnecessary costs.

Provisioning

Support equipment planning and acquisition have a direct impact on provisioning activities in two areas. First, provisioning documentation is normally required for SE, just as it is required for the weapon system being acquired. PTD for SE might consist of complete documentation of the equipment to identify spares and repair parts, or it might be limited to a tools and test equipment list (TTEL). The DD forms 1949-1 and 1949-2 (contained in the SOW) establish the requirements for provisioning of SE.

The second area where SE has an impact on provisioning is in the assignment of SMR codes. If an SMR code is assigned to a spare that indicates the item is to be removed and replaced at the organizational level, then the SE required to perform that maintenance task must be available at that maintenance level. Inconsistencies in SE availability and SMR code assignment will impair maintenance capabilities. In either case, provisioning efforts must be coordinated with support equipment planning to ensure that the PTD accurately reflects the appropriate SE requirements.

Technical Manuals

Maintenance tasks documented in technical manuals (TMs) are written based on the availability of appropriate SE. As the manuals are being generated, the requirements for SE are identified and justified as stated above. Changes to maintenance tasks or changes to the availability of SE have a wide-ranging effect on both TMs and SE. The development of technical manuals is a lengthy process, and changes to SE availability can cause extensive rework to TMs, especially in the later stages of development. It is extremely important that SE availability be verified as early as possible in the FSD phase in order to avoid the unnecessary expense of having to change maintenance procedures in the equipment TM.

A second consideration for impact of support equipment on TMs is that a manual might be required for the support equipment. Complex items of support equipment that will be operated and maintained by maintenance personnel require documentation. MIL-M-81919A, Preparation of Support Equipment Technical Manuals, provides detailed requirements for these documents.

Training

Contractors normally develop training requirements and training courses as a part of the overall acquisition process. This training consists of how to operate and maintain the equipment. Remember that SE should also be included in this training. Maintenance training, especially, uses the SE to accomplish maintenance tasks. Personnel might need to be trained on how to operate and maintain the SE as well as the weapon system. Such training is included as part of the maintenance training program. If SE is not operated and maintained properly, then maintenance tasks cannot be performed as required to fix the equipment when it fails.

Facilities

SE planning must be coordinated with facility planning. Many items of SE have specific power, space, cooling, and other environmental requirements that must be considered during the facility planning process. SE usefulness will be degraded or nonexistent if proper operating conditions are not present where the equipment is to be used. Specific SE requirements must be identified as early as possible in the acquisition cycle due to the long length of time required to acquire adequate facilities to support maintenance.

Planning

The completion of detailed and justified SE requirements is accomplished in the latter stages of FSD. The orderly process of attaining this goal begins during the concept phase and evolves as the equipment design matures. Information on SE requirements is needed by other ILS disciplines throughout the development of the equipment, so the baseline for SE established during the concept phase must not be haphazard. A clear definition of the SE that is anticipated on the program must be documented. The logical source for this process is to determine which SE supports the current equipment that is to be replaced, and what SE is currently available to maintenance personnel who will maintain the new equipment.

Maintenance personnel are normally issued a standard tool kit, which they use to perform maintenance. Supplementary tool kits are also available for general use by those assigned to a maintenance activity. The composite of these two sources should form the baseline for SE planning and should be used to support maintenance on the new equipment if possible.

The documenting of an initial SE list is extremely useful to the maintenance task analysis process. As tasks are identified and documented, the necessary SE to support the task can be selected from the SE list rather than having the analyst find an item that can be used. On a large equipment development program, there might be as many as 50 people writing maintenance tasks. If each person identifies his or her own preferred tools to support maintenance rather than having a list to select from, the result is a duplication of information and erroneous reporting of tool requirements.

The SE list should include identification of the SE part number, FSCM, item name, and sufficient information, such as size or capacity, to allow proper selection. Use of a standard SE list has proven to be cost effective, because it significantly reduces the amount of research that must be done to identify SE required to support a maintenance task.

TESTING

Repair of a failure can begin only when the part that failed has been identified. Testing is the process of identifying failures. The inherent testability of an item has a direct impact on its maintainability. The easier it is to identify a failure, the less time and resulting expense are required to perform maintenance. A detailed explanation of the process of planning and conducting tests would require more space than is available in this book. Therefore, this section is limited to providing a basic description of the terminology and procedures associated with testing. The logistician must understand basic testing concepts in order to plan for resources required to support maintenance.

Test Philosophy

The broad guideline for development of test requirements for an item of equipment is the test philosophy. The test philosophy is established during the concept phase, and guides design engineers throughout equipment development. It is normally a statement of the desired test characteristics of the equipment. The philosophy must be compatible with the maintenance concept for the equipment. For example, if the maintenance concept is based on removal and replacement of modules at the organizational level, then the test philosophy must address how maintenance personnel will be able to test to determine which modules should be removed and replaced in order to fix a failure. The same holds true for all levels of maintenance where testing must support the maintenance concept.

Test Requirements

Test requirements are definitive goals established contractually by the procurement specification. The test philosophy must be compatible with these requirements. For example, the specification might require testing to identify 80 percent of equipment failures for repair at the organizational level. The requirements might also state that fault identification will be accomplished within a specified amount of time.

Test Terminology

The logistician must understand the following test terminology to identify maintenance test requirements.

- **Built-in Test (BIT)**. BIT is the inherent capability of an item of equipment to detect, diagnose, or isolate failures using a self-contained automated test. BIT can be either a continuously executed test or a test that is initiated by the equipment operator. The key is that BIT is performed while the equipment is installed or operational.
- **Built-in Test Equipment (BITE)**. BITE is hardware or software designed into an item of equipment to provide BIT capability.
- **Unit Under Test (UUT)**. UUT is any equipment or assembly undergoing testing. This term is most commonly used to describe items that are being tested at intermediate or depot levels.

- **Automatic Test Equipment (ATE)**. ATE is equipment designed to perform testing of a UUT with little or no operator intervention. ATE is normally controlled by a computer.
- **Factory Test Equipment (FTE)**. FTE is equipment used during the manufacturing process to test assemblies or units prior to delivery to the government. FTE can be used as depot test equipment after production ceases; however, the intended use might require modification, because the FTE is designed to prove that an item is failure free, and depot test equipment is designed to find failures. FTE is normally a type of ATE.
- **Special Test Equipment (STE)**. STE is equipment designed to test specific functions or parameters. STE is normally ATE and can be used as FTE.
- **General Purpose Test Equipment (GPTE)**. GPTE is equipment that can be used to test a range of parameters common to two or more systems whose basic design is different.
- **General Purpose Electronic Test Equipment (GPETE)**. GPETE is electronic test equipment that can be used to test more than one system without modification. Each service has an approved listing of GPETE that should be used to select test equipment for a system being developed when at all possible.
- **Special Purpose Test Equipment (SPTE)**. SPTE is equipment that can be used to test only one system or parameter group of a UUT.
- **Special Purpose Electronic Test Equipment (SPETE)**. SPETE is electronic test equipment that can only be used to test a specific system or UUT. This type of test equipment should be avoided if at all possible to reduce the cost of supporting maintenance.
- **Test Fixture**. A test fixture is equipment that is required to connect a UUT with test equipment. The fixture may be electrical, mechanical, or both. The test fixture may also be called an interface device.
- **Test Point**. A test point is an electrical contact designed into an electrical or electronic circuit that provides a capability to test the circuit. Test points are used to enhance the testability of a UUT.
- **Manual Troubleshooting**. Manual troubleshooting is the process of detecting or isolating a failure using methods that rely on the operator or maintenance personnel to make decisions as to the cause of the failure. This type of testing should be limited to simple processes whenever possible to reduce the time required to perform maintenance.

Requirements Identification

The process for identification of maintenance test requirements begins with the FMECA. The FMECA identifies how the equipment will fail. This process creates test requirements because maintenance cannot be performed until the failure is identified.

As a potential failure is identified by the FMECA, an appropriate method of knowing that the failure has occurred is documented. This forms the basis for planning the testing of equipment. Testing can be accomplished by a combination of BIT and manual troubleshooting on equipment, or through the use of ATE on equipment. The results

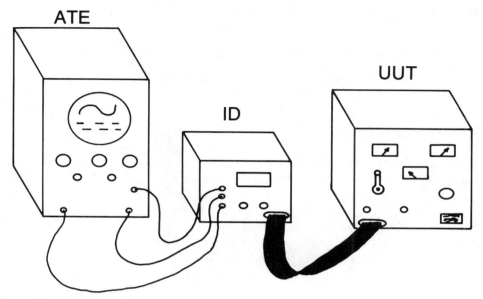

Fig. 7-2. Test set-up.

of this process is a complete identification of the requirements of testing to support maintenance.

Test Setup

A typical test setup is illustrated in Fig. 7-2. The setup consists of the UUT, a test fixture, the appropriate ATE, and test software. A single piece of ATE can be used to test many UUTs by changing test fixtures and test software. This method of testing reduces the number of different items of test equipment required to test an item of equipment.

Test Program Set

The method of documenting a test setup is contained in a test program set (TPS). A TPS is composed of those items that are necessary to test a UUT on an ATE. This includes the actual computer test program (TP), the necessary interface device (ID), a test program instruction (TPI), and supplementary data. The TP contains a coded sequence of computer commands that provide the ATE a set of instructions to automatically test the UUT. The ID is a fixture that electrically or mechanically connects the UUT to the ATE. A TPI provides information needed to conduct the test, such as installation instructions or instructions the test operator must follow to complete the test. Supplementary data may consist of any information necessary to aid in the testing process, and may include logic diagrams, schematics, or method for interpreting test data. MIL-STD 2077, General Requirements for Test Program Sets, contains guidance on the preparation of a TPS.

SUPPORT EQUIPMENT AUTHORIZATION

The support equipment required to perform maintenance does not magically appear at the appropriate maintenance facilities when the system is delivered. Authorization for ownership of support equipment is accomplished in two ways: (1) support equipment that is issued with the item it is to support, and (2) documents that authorize maintenance organizations to requisition and use equipment.

Basic Issue Items

Support equipment that is provided with the weapon system to support crew or organizational level maintenance is termed basic issue items (BII). For example, when a tank is delivered to its unit, it is accompanied by a set of tools to be used by the crew in maintenance of the tank. The tool kit includes wrenches, screwdrivers, a hammer, fixtures to change the vehicle tracks, and other items necessary to perform crew-level maintenance. BII is considered part of the tank system, and is accounted for as part of the tank on appropriate property accountability documentation. The technical manual for the tank contains a list of authorized BII.

Authorization Documents

Authorization for a maintenance organization to have support equipment other than BII is established through documents that address the equipment requirements for the unit. Each service publishes authorization documents that dictate the equipment that a unit is authorized to own. This includes support equipment.

Documentation for units that will own the weapon system and perform organizational-level maintenance provide a clear relationship and requirement for ownership of required support equipment. The authorization for support equipment at intermediate and depot levels is not as clear cut, because a determination must be made as to which units will be performing maintenance for what weapon systems. Close coordination between the contractor and the government is required to ensure that the appropriate authorization documents reflect authority to own the necessary support equipment.

Supply Catalogs

There is one other type of authorization document that must be considered when discussing support equipment: supply catalogs. In most cases, the authorization documents establish the fact that a unit is authorized to own a tool kit; however, the document does not define the components that make up the kit. This is accomplished through the use of supply catalogs.

The catalogs are designed to accommodate changes to the components of a kit without changing authorization documents. If the components of a kit are to be changed as the result of introduction of new maintenance requirements to support a new item of equipment, then the applicable supply catalog must be changed. Again, close coordination between the contractor and the government is required to ensure that the changes are accomplished accurately and efficiently, providing the necessary support to the new equipment when it is fielded.

1. Identification of the unit under test configuration.
2. General data of the physical and functional characteristics of the unit under test.
3. Interface requirements between the unit under test and the automatic test equipment.
4. Performance characteristics of the unit under test.
5. Detailed test information.
6. Outline installation drawings.
7. Assembly drawings.
8. Wiring diagrams.
9. Functional block diagrams.
10. Test flowcharts and diagrams.

Fig. 7-3. Test requirements document.

DOCUMENTATION

Support equipment planning and identification results in the preparation of documentation that provides detailed information to the government concerning requirements for support equipment and testing. This documentation consists of support equipment recommendations data (SERD)and test requirements documents (TRD). A contractor will normally prepare these documents during the FSD acquisition phase.

Support Equipment Recommendations Data

A separate support equipment recommendations data (SERD) form is prepared for each item of support equipment that is required to support maintenance except common hand tools, which are normally excluded from this requirement. The purpose of a SERD is to provide detailed justification for using an item of support equipment to perform maintenance or testing. Information presented by a SERD includes identification of maintenance tasks that the SE will support, the total quantity required, and recommended sources for procurement.

Test Requirements Document

The purpose of a test requirements document (TRD) is to provide the data necessary to select or design the manual or automated test equipment to test a UUT, design test fixtures or interface devices needed to connect the UUT to the test equipment, and develop the test procedures required to actually perform the test. Procedures for preparation of a TRD vary among services. Detailed procedures for preparation are provided in MIL-STD 1345B(Navy), Preparation of Test Requirements Document; MIL-STD 1519(USAF), Preparation of Test Requirements Document; and MIL-STD 2076(AS), General Requirements for Unit Under Test Compatibility with Automatic Test Equipment.

Typical contents of a TRD are shown in Fig. 7-3. The information contained in the TRD should be sufficient to stand alone as the single reference for developing all tests required to support maintenance of an item of equipment.

Chapter 8

Personnel

Regardless of the weapon system being designed, personnel will be required to operate and maintain the equipment when it is delivered to the government. Therefore, the logistician planning for operation and maintenance of the system must have a basic understanding of how the military services classify personnel, how personnel requirements are determined, and how a weapon system acquisition program might affect personnel resources.

CLASSIFICATION

Personnel in the military services are classified in two ways: grade (rank) and skill specialty. These two methods allow a match of the requisite experience and technical capabilities with tasks to be accomplished. Grade can normally be equated, to some extent, by the length of time that a person has spent in the military. Skill specialty is dependent on the training that he or she has received. Every person in the military has a grade and a skill specialty.

Grade Structure

There are three basic categories of grades in the military: commissioned officer, warrant officer, and enlisted personnel. Commissioned officers are charged with the responsibility of commanding or assisting the commanders of the organizations to which they are assigned. The ultimate responsibility for the success or failure of a military unit in accomplishing its assigned missions rests with these individuals.

Warrant officers are technical experts in their designated fields, which might be maintenance, supply, administration, etc. Warrant officers do not have command responsibilities.

Enlisted personnel are further divided into noncommissioned officers (NCOs), who are senior enlisted personnel, and lower-grade enlisted personnel. The noncommissioned officer (NCO) is called the backbone of the military. An NCO is the first-line supervisor who ensures that all tasks necessary to accomplish unit missions are completed.

In each category, individuals start at the bottom grade and advance to higher grades through a promotion process that is based on training, experience, past performance, and length of time in grade. When planning for the operation and maintenance of equipment, it is important to remember that the persons who will be doing most of the actual work are the lower-grade enlisted personnel. Therefore, the requirements for accomplishing the tasks must be commensurate with their capabilities. A young, lower-grade enlisted person, just out of high school, should not be expected to perform a complex maintenance task in the same manner as a seasoned technician with 20 years of experience.

Skill Specialty

Every type of job in the military is uniquely identified as a skill specialty. The purpose of the skill specialty system is to ensure that the person assigned to a job has the training and experience necessary to accomplish its task requirements. Each skill specialty, or job type, in the military has its own unique identification number, a skill specialty code (SSC). The numbering system differs among the military services, but the concept for using this method for skill-specialty identification is the same. The Army and Air Force systems use five digits, and the Navy and Marine Corps use four digits. For example, SSC 63B10 in the Army is for a light-wheeled vehicle mechanic, and SSC 30454 in the Air Force is for a heavy ground radio maintenance technician.

SSCs for the Army and Air Force include provisions for both a skill specialty identifier and proficiency level. The proficiency level can normally be equated to the grade of the individual who holds the SSC. The higher the proficiency number, the higher the grade of the individual. The formal titles for skill specialty codes are Military Occupational Specialty (MOS) for the Army, Navy, and Marine Corps, and Air Force Specialty Code (AFSC) for the Air Force.

Each specialty has a definitive job description that tells which duties the individual assigned the SSC is expected to perform and how the SSC is earned. Descriptions are maintained for every skill specialty in the military in documents prepared by each service. Most SSCs are awarded upon completion of a training course designed for that SSC. Some can be awarded through on-the-job training (OJT).

PLANNING

As a weapon system is being developed, the personnel needed to operate and maintain the system must be identified in sufficient time to allow acquisition and training prior to the equipment being used by the government. If a contractor delivers the equipment and there are no trained operators or maintenance personnel, the equipment will probably either sit idle or be misused or damaged by untrained personnel. Therefore, as a part of the overall ILS effort, the contractor is normally required to develop an initial recommendation for personnel requirements to support the equipment. To accomplish this task, the contractor must identify the requirements for personnel; develop recommendations for SSCs that should be used as is, changed or modified, or created; and provide a formal report to the government.

Requirements Identification

The first step in developing requirements for personnel is to identify exactly what tasks must be accomplished and the predicted frequency that the tasks must be performed. There are several approaches to identifying personnel requirements, but the best way is to use the information generated by other ILS disciplines as a data base and expand on this information. Requirements for personnel to actually operate the equipment are usually pretty obvious. Because the operators are assigned this task as a primary responsibility, this part of the identification process is easy. For example, if a contractor is building tanks and each tank requires a four-man crew of tank commander, gunner, loader, and driver, then it is a simple matter of multiplication to determine the number of personnel required to operate the number of tanks that the contractor is building.

The identification of maintenance personnel is somewhat more difficult. There is not a one-to-one relationship of maintenance personnel to a single item of equipment, so the process of determining the number and types of maintenance personnel required to support the equipment must be based on other information. The best way is to use the FMECA, maintenance task analysis, and RCM analysis as the source information for determining the labor required to maintain an item of equipment. Remember, the FMECA identifies all the ways that an item of equipment is predicted to fail; the maintenance task analysis identifies the tasks required to fix these failures along with the predicted hours for each task and the predicted frequency; the RCM analysis is used to select scheduled maintenance tasks. By extracting the appropriate information from these sources, a contractor can identify all maintenance tasks to be performed.

The second step in identifying personnel requirements is to predict the annual hours of labor required to support each task. The driving factor in this process is the maintenance task frequency and the hours required to accomplish each task. The task frequency is normally based on a predetermined number of operating hours, or a standard per "x" hours of operation. (Note: The process for calculating task frequency often differs slightly among services and specific contracts, so ensure that a clear understanding of this process exists each time personnel requirements are being developed.) Figure 8-1 shows how to develop a formula that will predict the annual hours of labor required to support each maintenance task. The results of this calculation are then used to develop detailed personnel requirements. By repeating this calculation for each maintenance task and summing the results, the total predicted maintenance hours to support one item of equipment can be determined. If multiplied by the number of items to be supported, the total hours of labor can also be predicted. These figures

$$MH/MT = TF \times TT \times Q$$

Where: MH/MT = Manhours per maintenance task
TF = Task frequency
TT = Task time
Q = Number of systems supported

Fig. 8-1. Annual manhours per maintenance task.

are applicable not only to personnel planning, but also to life cycle cost predictions discussed in Chapter 15.

Skill Specialty Development

The requirements identification process produces a list of tasks that must be accomplished and the predicted time required to perform them. This information is used to develop actual requirements for specific skill specialties. This list is normally extensive, so the most logical approach to sort out who should do what tasks is to try to standardize maintenance tasks with the existing service skill specialty structure. The best way to accomplish this is to determine the existing personnel resources that can be available to support the new equipment. If the new equipment will replace an item that is currently in use, it is logical to assume that the personnel who were supporting the old system will be used to support the new one. If the new equipment is not a replacement for an existing item, the job becomes more difficult. The best way to approach this situation is to find a similar item of equipment and develop personnel requirements using the similar equipment as a guide.

In either situation, after a base framework of SSCs has been developed (this information is often provided by the government), the next step is to categorize the maintenance tasks in two groups: (1) tasks comparable to those required by existing SSCs and (2) tasks not covered by existing SSCs. The category for a specific task might be dictated by the technology required, or the use of new support equipment.

This process of matching tasks to SSCs should result in information that can be used to justify using existing SSCs, modifying SSCs, or creating new SSCs to support maintenance. Those tasks that fall into the first category justify the use of existing SSCs. Tasks in the second category are used to recommend changes to existing SSCs or, if necessary due to the number of distinctly new tasks, to create a totally new SSC. In practice, the creation of a new SSC should be avoided if at all possible, due to the lengthy administrative process required on the part of the government.

After the tasks have been matched to an SSC, the annual hours of labor per SSC can be calculated by totaling the task times per SSC and multiplying by the number of items to be supported.

The final step in this process is to predict the annual hours of labor per SSC by maintenance level. Using the results of a repair-level analysis and other analyses that determine which tasks are accomplished at what level of maintenance, the tasks are divided into levels of maintenance and SSC at each level, summed for each level, and multiplied by the number of items supported. This final step identifies the specific SSCs and total hours of labor required at each level of maintenance to support equipment maintenance. The personnel requirements can then be developed based on this information.

Reporting

The results of the personnel planning process are provided to the government by several different methods. One method is through submission of a report prepared in accordance with MIL-D-26239A, Qualitative and Quantitative Personnel Requirements Information (QQPRI) Data. A QQPRI report provides the government with a complete description of the information developed during the personnel planning process. Figure 8-2 is an outline of a typical QQPRI report. This information is used to develop personnel staffing in support of the new equipment. It is also used for planning the modification of existing training courses or the development of new training courses

---

Qualitative and quantitative personnel requirements information

Part I.	QQPRI for operation and organizational and intermediate-level maintenance
Section 1.	Introduction
	A brief description of the purpose of the report and the equipment.
Section 2.	System Description
	A detailed description of the equipment to include its purpose, operational characteristics, maintenance, and operational concepts.
Section 3.	Maintenance and Operations Summary
	A summary of the skill specialties required to perform operation and maintenance, predictions of utilization times, and support equipment used.
Section 4.	Skill Specialty Descriptions
	A detailed description of each skill specialty required to perform operation and maintenance, including a complete task listing, support equipment the individual is required to operate, task times and frequencies, annual manhour requirements, and task proficiency level.
Section 5.	Preliminary Staffing Estimates
	A preliminary recommended staffing for the new equipment, including staffing data, organizational diagrams, and other pertinent information.
Section 6.	Special Problem Areas
	A detailed description of any problems or concerns that affect personnel planning.
Part II.	QQPRI for depot-level support
	Part II is essentially a repeat of Sections 4 through 6 of Part I, but addresses depot-level requirements. In some cases, the skill specialty descriptions will be more descriptive if the tasks are to be performed by civilian personnel employed by an installation maintenance facility.

Fig. 8-2. QQPRI data (report outline per MIL-D-26239A).

for the skill specialties required to operate and maintain the equipment. Another method of reporting personnel requirements is by submitting LSA summary reports, discussed in Chapter 17.

STAFFING

The method for staffing in the military differs significantly from that in the civilian sector. In the civilian sector, a business maintains a staff of people based on work load and other economic factors, and the number of people employed by a company varies from year to year. Additionally, the number of employees within divisions or groups in a company might vary even if the total number of employees does not change. The military is quite different.

Authorization

Congress establishes a ceiling for military personnel strength in each service. Based on this number, the services prepare internal authorization documents that tell the number, grade, and skill specialty of personnel to be assigned to each organization. Once issued, these documents normally change only when an organization's assigned equipment or missions change. For example, the Army uses two types of authorization documents: Table of Organization and Equipment (TOE) for units that are assigned a combat mission, and Table of Distribution and Allowances (TDA) for noncombat units. The TOE for a typical armor battalion would tell the number of personnel by grade and MOS that are authorized for the unit, and the types and quantities of equipment that the unit must have to perform its missions. This would include tank crews, mechanics, cooks, supply personnel, tanks, machine guns, radios, trucks, rifles, and anything else that the unit is authorized to have in order to perform its mission.

Changes

The branch of the military that will receive the new equipment being developed must issue the necessary changes to personnel authorization documents so that the required staffing is available when the equipment is fielded. The process of issuing a change might require several years due to the administrative process that must be completed.

Remember that the resources must be obtained from existing allocations authorized by Congress, so it is much easier to use personnel allocations that previously supported the equipment being replaced. The QQPRI report submitted by the contractor plays an important role in this process; and therefore, it must be as accurate and timely as possible.

Planning for personnel requirements must be accomplished as early as possible in the acquisition process to ensure that sufficient personnel with the proper skills are available to support the equipment when it is fielded. Limited personnel resources might be one of the key factors that determines the level and quality of support provided to a weapon system. ILS disciplines can provide a positive impact on the availability of personnel resources through accurate documentation of detailed, justifiable personnel requirements.

Chapter 9

Technical Manuals

The technical manual (TM) is the only documentation that is received by the ultimate equipment user. Therefore, it is one of the most important documents that a contractor prepares. The generic term "technical manual," or technical order (TO), refers to a series of documents that provides instructions to the user on how to operate and maintain the equipment. Included in this series can be an operator's manual, a maintenance manual for each level of maintenance, lubrication orders (LO), depot maintenance work requirements (DMWR), repair parts and special tools list (RPSTL), and illustrated parts breakdown (IPB).

The purpose of a TM is to provide users or maintenance personnel with all the information and instructions required to operate and maintain the equipment. TMs contain detailed narrative and pictorial descriptions of operation and maintenance procedures, necessary support and test equipment, reference information, and identification of spare and repair parts. The lack of proper technical manuals will significantly impair the ability of the user to operate and maintain the equipment, which, in turn, renders the equipment ineffective in performing assigned missions.

TYPES OF MANUALS

The history of the development of technical manuals differs with each branch of the military. As the need for manuals to support increasingly complex equipment evolved, each service created its own methods for preparing manuals to fulfill the specific requirements of the time. This evolutionary process led to each service having different requirements for the content and format of their manuals. This makes it difficult for the logistician to have a detailed understanding of all the options available for preparing the manuals.

There are several different methods for formatting, organization, and printing. Therefore, each contract awarded that requires the preparation of TMs must be read thoroughly to determine the exact requirements that are to be met. For the purposes of this discussion, the topics addressed will be those most common to technical manuals and their preparation. TMs can be divided into three basic groups: operator manuals, maintenance manuals, and parts manuals. Each manual type has a specific purpose and scope. Most acquisition programs require the preparation of one or more of these manuals.

OPERATOR'S MANUAL

An operator's manual, in general terms, is a manual that contains all the information that the operator, or crew, requires to operate and maintain an item of equipment. MIL-M-63036A, Preparation of Operator's Technical Manual, contains detailed requirements for the preparation of an operator technical manual and includes instructions for operation and maintenance, lists of support equipment, and other reference material. Figure 9-1 is an outline for a typical Army operator's manual prepared in accordance with MIL-M-63036A, which illustrates the information that can be contained in an operator's manual. Operator's manuals for other services can have different formats; however, their content, tailored for particular applications, is basically the same.

Front Matter. The term "front matter" refers to the first few pages of a TM that provide the operator with basic information about the manual. Included in the front matter are the front cover, safety warnings, table of contents, and instructions on how to use the TM. Figure 9-2 shows the cover for an Army TM. Standard items found on all TM covers are the TM number in the upper right corner, identification of the equipment, and type of manual. The government agency procuring the TM provides the TM number to the contractor.

Note that this manual is actually for two different items of equipment, an infantry fighting vehicle and a cavalry fighting vehicle. These items are very similar in design; therefore, the Army chose to procure a single manual that covers both.

Figure 9-3 illustrates a warning page that identifies safety hazards that might occur during operation or maintenance of the equipment. A TM might contain several such pages. Identification of potential safety hazards is accomplished by system safety engineering, which is discussed in Chapter 13. The "How to Use This Manual" section is a brief discussion on the TM that instructs the user, especially a novice, on how the TM should be used as a tool and guide for operation or performance of maintenance.

Chapter 1 - Introduction. The purpose of the Introduction is to familiarize the user with the equipment. Contained in this chapter is narrative and pictorial information that addresses the general aspects of the equipment. One of the first things that normally appears in this section is a picture of the equipment, as shown in Fig. 9-4.

There is a complete description of the characteristics and capabilities along with a technical discussion of the principles of functional operation of the equipment. The Introduction should contain enough information to familiarize the operator with physical

Front Matter
  Cover
  Warning Page
  Table of Contents
  How to Use This Manual

Chapter 1 - Introduction
  Section 1 - General Information
  Section 2 - Equipment Description
    Equipment Characteristics, Capabilities and Features
    Location and Description of Major Components
    Differences Between Models
    Equipment Data
  Section 3 - Technical Principles of Operation

Chapter 2 - Operating Instructions
  Section 1 - Description of Operator's Controls and Indicators
  Section 2 - Preventative Maintenance Checks and Services
  Section 3 - Operation Under Usual Conditions
  Section 4 - Operation Under Unusual Conditions

Chapter 3 - Maintenance Instructions
  Section 1 - Lubrication Instructions
  Section 2 - Troubleshooting Procedures
  Section 3 - Maintenance Procedures

Chapter 4 - Maintenance of Auxilliary Equipment

Chapter 5 - Ammunition

Appendix A - References

Appendix B - Components of End Item and Basic Issue Items

Appendix C - Additional Authorized List Items

Appendix D - Expendable Supplies and Materials List

Index

Fig. 9-1. Operator's manual (outline per MIL-M-63036A).

TM 9-2350-252-10-2

OPERATOR'S MANUAL

FIGHTING VEHICLE, INFANTRY, M2
(2350-01-048-5920)

AND

FIGHTING VEHICLE, CAVALRY, M3
(2350-01-049-2695)

TURRET

**HEADQUARTERS,
DEPARTMENT OF THE ARMY**

JANUARY 1985

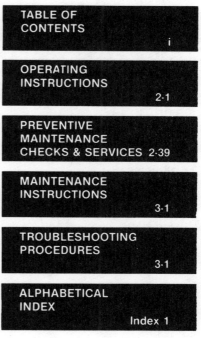

Fig. 9-2. Technical manual cover.

TM 9-2350-252-10-2

 **WARNING**

Noise from vehicle or weapons can damage hearing of soldiers in vehicle. All personnel in vehicle MUST WEAR DOUBLE HEARING PROTECTION when gun or vehicle is operated. Hearing protection devices must be properly worn to provide effective protection.

If DOUBLE HEARING PROTECTION is not worn, the safe level of noise exposure will be exceeded in a short time. Hearing loss occurs gradually. Each noise exposure that exceeds the ear protection guidelines below will cause a temporary hearing loss. Over time, the loss in hearing will become permanent. Plan each day's operation, and be sure all crew and riders have the required ear protectors. Spare earplugs must be available.

DEFINITIONS:

DH-132 — The "tankers helmet," also called "CVC" helmet. Must be in good condition, with liner and earcups fitted tightly, and chin strap worn at all times.

EARPLUGS — Only standard issue earplugs are acceptable. All of the dismounted squad soldiers must be trained in how to use them. Since they may be removed and lost, spares must be carried.

H-251 HEADSET — The listen-only headset provided for the dismounted squad while in the vehicle.

DOUBLE HEARING PROTECTION — Use of two hearing protection devices at the same time. For this vehicle, use earplugs with either the DH-132 helmet or the H-251 headset.

Fig. 9-3. Technical manual warning page.

CHAPTER 1

INTRODUCTION

Section I. GENERAL INFORMATION

TURRET FOR INFANTRY FIGHTING VEHICLE (IFV), M2 / CAVALRY FIGHTING VEHICLE (CFV), M3

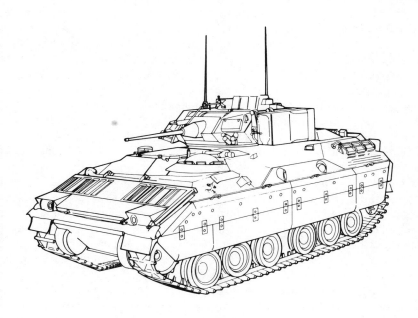

Left Front View

Fig. 9-4. Full view equipment illustration.

and functional aspects of the equipment encountered during operation and operator-performed maintenance. The Introduction can be divided into three sections; general information, equipment description, and technical principles of operation.

Section 1 - General Information. The general information section provides an equipment orientation, identification of applicable regulations and forms that should be used in conjunction with the operation or maintenance of the equipment, and reporting requirements. A glossary and list of abbreviations used in the TM may be included in this section. Any information pertaining to equipment warranty is also identified.

Section 2 - Equipment Description. The purpose of the equipment description section is to provide detailed information on the specifics of the equipment. This section includes subsections addressing equipment characteristics, capabilities, and features; location and description of major components; differences among models; and equipment data.

Figure 9-5 is a typical characteristics, capabilities, and features subsection that explains what the equipment is supposed to do. Note that the terminology used is as nontechnical as possible. It is important that the operator be familiar with the names and locations of the major components of the equipment, because the TM uses standard names for each component throughout the manual. Figure 9-6 is an example depicting the location and description of major components of an item of equipment.

In some cases, where a TM is used for more than one model, the TM contains a table listing the differences between them. This is essential information for the operator because operation and maintenance can vary depending on the specific model. The information also addresses various optional equipment or configurations that the operator might encounter. Also contained in the equipment description are specific tabular equipment data that apply to the operator's functions. Figure 9-7 shows typical equipment data found in TMs. These data provide the operator with detailed information on the specific capabilities and limitations of the equipment.

Section 3 - Technical Principles of Operation. This section contains a functional description of the equipment operation. Only information that pertains to the operator's responsibilities should be included. The purpose of the technical principles of operation section is to tell how the operator's controls and indicators interface with the equipment. The information should be directed toward the operator's actions and should avoid highly technical functions outside of the operator's scope. Figure 9-8 illustrates information from a typical technical principles of operation section.

Chapter 2 - Operating Instructions. The purpose of Chapter 2 is to provide the operator with instructions to effectively and efficiently operate the equipment. The information must be completely accurate because the operator uses this information for actual operation of the equipment. The level of detail and technical content should reflect the skills of the operator. Overly technical instructions for simple equipment should be avoided. Likewise, simplistic instructions for highly technical equipment might lack the proper detail to allow safe and efficient operation. Operating instructions are typically divided into four sections: description and use of operator's controls and indicators, preventative maintenance checks and services, operation under usual conditions, and operation under unusual conditions.

Section 1 - Description and Use of Operator's Controls and Indicators. This section

Section II. EQUIPMENT DESCRIPTION AND DATA

EQUIPMENT PURPOSE, CAPABILITIES, AND FEATURES

The purpose of the turret is to provide a two-man weapon station for the IFV and the CFV. These vehicles are able to defeat enemy armored vehicles because of the turret's fire power and assault features. Its components can also be used against low-flying aircraft, gun emplacements, and other targets.

The turret can traverse in either direction at a high rate of speed. Either the commander or the gunner can select, arm, and fire the following weapons: a 25mm gun (the main gun), coax machine gun, or TOW launcher. In addition, two smoke grenade launchers are provided. Crew members are assigned individual weapons which are stowed in the squad area.

An integrated sight unit (ISU) is in front of the gunner. The ISU allows day and night viewing from inside the turret. A relay sight extends from the ISU to the commander's station.

Fig. 9-5. Equipment description section.

TM 9-2350-252-10-2

LOCATION AND DESCRIPTION OF MAJOR COMPONENTS

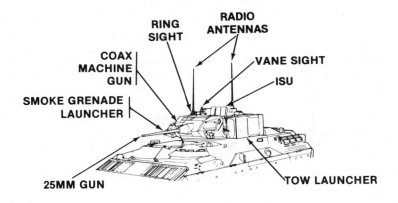

Turret Exterior

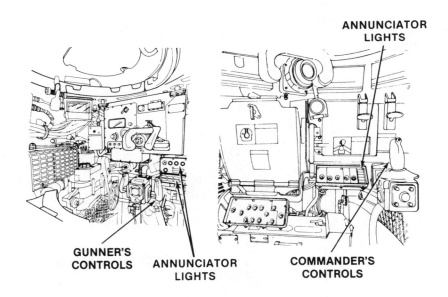

Turret Interior

Fig. 9-6. Location and description of major components.

EQUIPMENT DATA

TURRET (TWO-MAN)

Armament . . . . 25mm gun
TOW missile launcher
Coax machine gun
Traverse. . . . . 360° continuous
Elevation
   25mm gun and coax machine gun . . . . +59° to –9°
   TOW missile launcher . . . . +29° to –19°
Slew rate, maximum elevation and traverse . . . . 60°/sec
Slew rate, TOW . . . . 15°/sec
Tracking rate, minimum . . . . 0.05 mil/sec
Stabilization system . . . . Electric
Ring gear, pitch diameter . . . . .60 in (152.4 cm)

COMMUNICATIONS (COMMANDER'S VEHICLE)

Radio, IFV (standard) . . . . AN/VRC-46,1 set
AN/GRC-160, 1 set
Radio, CFV (standard) . . . . AN/VRC-12, 1 set
AN/PRC-77, 1 set

Other configurations exist

NIGHT VISION EQUIPMENT

Sight, gunner. . . . . Thermal imagery
Sight, commander. . . . . Optical relay from gunner's sight

BACKUP SIGHT

Depth . . . . 11.62 in (29.5 cm)
Width . . . . 5.25 in (13.3 cm)
Height . . . . 16.75 in (42.5 cm)
Weight . . . . 49.5 lb (22.5 kg)
Line of sight, elevation . . . . −10° to +60°
True field of magnification . . . . 10°
Magnification 5X
Focus . . . . –4° diopters to +4° diopters

25MM GUN

Caliber. . . . . .25mm dual feed
Weight
   Receiver Assembly . . . . 95 lb (43.1 kg)
   Barrel Assembly . . . . 90 lb (40.9 kg)

1-17

Fig. 9-7. Equipment data.

Section III. TECHNICAL PRINCIPLES OF OPERATION

TURRET

Hatches

Two torsion bar-assisted hatch covers serve the commander and gunner. Both hatch covers can be latched in OPEN or CLOSED positions. The commander's hatch cover can also be latched in POP-UP position. From this position, the commander can see 360°. both hatch covers have inside locks which prevent entry from the outside.

The commander's hatch cover has an emergency cable release in the squad area. In the event that the commander and gunner are injured, the hatch cover can be popped by pulling the cable release. This permits entry to the turret from the outside when both hatch covers are locked inside.

Turret Drive System

The turret drive system consists of azimuth, gun elevation, and TOW elevation drives. It also has the TOW launcher mechanism, electronic control assembly, gunner's and commander's handstations, gun gyro, and hull/turret gyro.

An electric motor drives a gearbox which connects with the azimuth drive, TOW and rotor elevation, or TOW launcher lift mechanism. Three gyro assemblies signal vehicle rate of movement to the electronic control assembly. Turret drives can be activated from handstations. To prevent accidental control action, palm switches on the hand controls must be pressed while the handstation is moved. All drives can be manually operated by the rotation of handwheels at the gunner's station.

Turret Travel Lock

A turret travel lock is located above the turret entrance. The travel lock's linkage and gear mechanism can lock the turret in a stationary position. It can also prevent turret rotation when the azimuth drive is in action.

Fig. 9-8. Technical principles of operation.

describes the functions of all the controls and indicators that the operator must use to operate and maintain the equipment. It is important to describe functions from the operator's point of view, to avoid confusion and make the information as understandable as possible. Figure 9-9 shows a typical description and use of operator's controls. Note that the example contains three specific items of information: (1) the name of the control, (2) sufficient pictures to identify the control and its location in the equipment, and (3) a description of the function of the control.

Section 2 - Preventative Maintenance Checks and Services. Periodic scheduled maintenance is the responsibility of the operator in the normal course of equipment operation. Preventative maintenance checks and services (PMCS) is the structured method for accomplishing scheduled maintenance. PMCS actions are divided into before operation, during operation, after operation, and monthly checks and services. Normally the PMCS section contains a checklist to be used by the operator to ensure that each preventative maintenance action is accomplished when required. Figure 9-10 shows a section of a PMCS checklist. Note that the checklist contains a numerical sequence of accomplishment, the item to be checked or serviced, procedure to be used, references to operations required to perform the procedure, and criteria for evaluating discrepancies to determine if the equipment is not ready or not available for operation. The source information for PMCS is developed through the reliability-centered maintenance analysis process discussed in Chapter 15.

Section 3 - Operation Under Usual Conditions. Operating the equipment under normal or usual conditions is described in Section 3. This section contains step-by-step instructions to the operator on how each function of the equipment is accomplished. Figure 9-11 shows instructions on how to adjust crew seats. Note that the instructions include the personnel required to accomplish the task, the condition of the equipment when the task is performed, and detailed instructions and illustrations for the task. Similar operation tasks are provided for each function that the operator is required to perform under normal operating conditions. Source information for this section comes from the logistic support analysis process discussed in Chapter 16.

Section 4 - Operation Under Unusual Conditions. During the operation of equipment, there might be times when the normal operating instructions are not applicable due to either unusual environmental conditions or combat situations. This section provides the instructions for operating the equipment under these unusual conditions. For vehicle operation, the instructions might include driving in snow or on icy roads. Figure 9-12 shows an operation task that might be required under combat conditions. Note that the task is identical in format and composition to a task for usual operational conditions.

Chapter 3 - Maintenance Instructions. The purpose of Chapter 3 is to provide the operator with all the instructions required to perform scheduled and unscheduled maintenance tasks. The source information for this chapter is the maintenance task identification and maintenance task analysis process discussed in Chapter 16 and the logistic support analysis data discussed in Chapter 17.

Section 1 - Lubrication Instructions. Specific emphasis is placed on lubrication of mechanical equipment as a necessity for continued equipment operation. Procedures for lubricating the equipment might be contained in this section, or might be referenced

TM 9-2350-252-10-2

COMMANDER'S HATCH CONTROLS

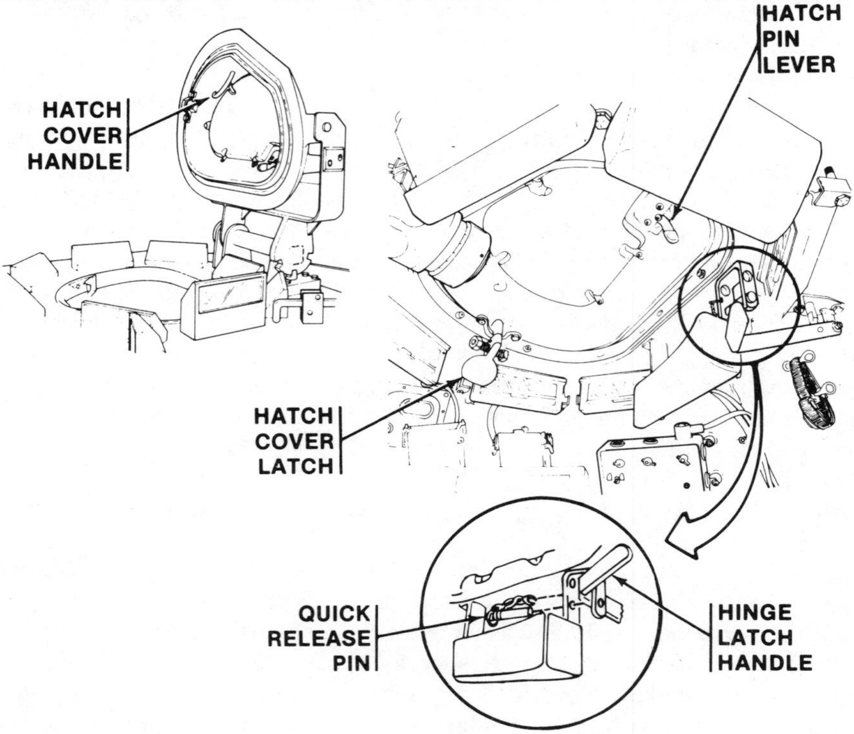

HATCH COVER HANDLE	Hand grip raises and lowers commander's hatch cover.
HATCH COVER LATCH	Locks or releases commander's hatch cover.
HATCH PIN LEVER	Allows commander's hatch cover to be moved to LEVEL position.
HINGE LATCH HANDLE	Allows commander's hatch cover to be moved to POP-UP, UPRIGHT, or FULL OPEN position.
QUICK RELEASE PIN	Locks and unlocks hinge latch handle.

Fig. 9-9. Description and use of operator's controls.

ITEM NO.	ITEM TO BE INSPECTED Procedure	Equipment will be reported NOT READY/AVAILABLE if:

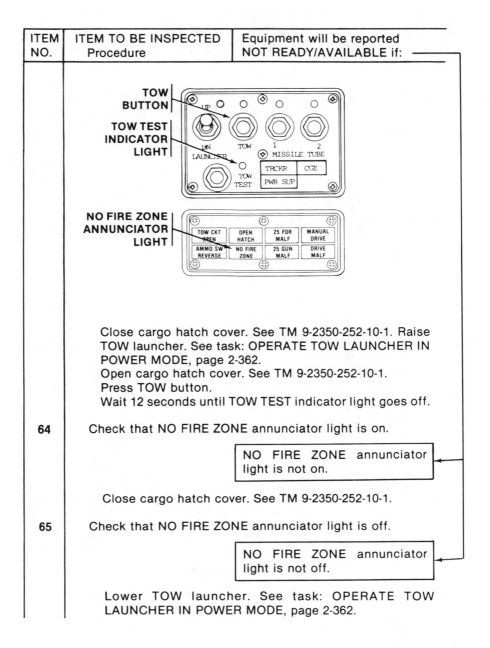

Close cargo hatch cover. See TM 9-2350-252-10-1. Raise TOW launcher. See task: OPERATE TOW LAUNCHER IN POWER MODE, page 2-362.
Open cargo hatch cover. See TM 9-2350-252-10-1.
Press TOW button.
Wait 12 seconds until TOW TEST indicator light goes off.

64 Check that NO FIRE ZONE annunciator light is on.

> NO FIRE ZONE annunciator light is not on.

Close cargo hatch cover. See TM 9-2350-252-10-1.

65 Check that NO FIRE ZONE annunciator light is off.

> NO FIRE ZONE annunciator light is not off.

Lower TOW launcher. See task: OPERATE TOW LAUNCHER IN POWER MODE, page 2-362.

Fig. 9-10. PMCS checklist.

TM 9-2350-252-10-2

ADJUST GUNNER'S/COMMANDER'S SEATS

INITIAL SETUP

Personnel Required:

Gunner
Commander

Equipment Conditions:

Gunner or commander seated
in turret

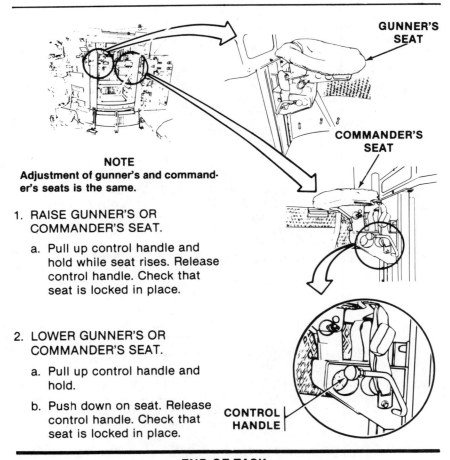

GUNNER'S SEAT

COMMANDER'S SEAT

NOTE
Adjustment of gunner's and command-er's seats is the same.

1. RAISE GUNNER'S OR COMMANDER'S SEAT.

   a. Pull up control handle and hold while seat rises. Release control handle. Check that seat is locked in place.

2. LOWER GUNNER'S OR COMMANDER'S SEAT.

   a. Pull up control handle and hold.

   b. Push down on seat. Release control handle. Check that seat is locked in place.

CONTROL HANDLE

END OF TASK

Fig. 9-11. Operator's task.

TM 9-2350-252-10-2

OPERATE VANE SIGHT

| INITIAL SETUP |

Personnel Required:

Commander

Equipment Conditions:

Turret operating in power mode (page 2-152)
Commander's hatch cover open (page 2-145)

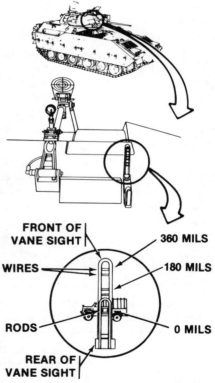

NOTE
Commander uses vane sight to quickly aline turret with target.

1. TRAVERSE TURRET TO ALINE TARGET WITH VANE SIGHT. See task: OPERATE TURRET IN POWER MODE, page 2-152.

2. DETERMINE ELEVATION OF TARGET.

   a. Look through vane sight. Aline rods of rear sight between two wires of front sight and with center of target.

   b. If rods aline between bottom wires and with center of target, elevation of target is 0 mils.

   c. If rods aline between middle wires and with center of target, elevation is 180 mils.

   d. If rods aline between top wires and with center of target, elevation is 360 mils.

FRONT OF VANE SIGHT

WIRES

RODS

REAR OF VANE SIGHT

360 MILS

180 MILS

0 MILS

END OF TASK

Fig. 9-12. Operator's task under unusual conditions.

as a separate lubrication order (LO). Figure 9-13 shows a typical LO, which indicates the places where an item of equipment is lubricated, the intervals when lubrication is performed, and the lubricant that is required.

Section 2 - Troubleshooting Procedures. This section of the TM provides the operator with a logical sequence of actions for troubleshooting equipment malfunctions. The actions prescribed are based on the symptoms exhibited by the equipment. Only those actions authorized for operator accomplishment are included in the procedures. Figure 9-14 illustrates a troubleshooting table. Note that the table has three basic steps, plus additional information as necessary to assist the operator. The malfunction being investigated is a failure of the 25mm gun to fire. The first instructions are to render the gun safe for maintenance. To perform the first step, the operator must remove a portion of the gun assembly. The troubleshooting table provides a reference to the maintenance task for removing the assembly. Next, the table tells the operator what to check and how the check is done. Finally, there is a decision logic for the operator to follow based on the results of the check. In the illustration, decision "a" tells the operator that if a lockpin does not function, the operator must refer to organizational maintenance to fix the failure. Decision "b" tells the operator to go to the next step, because the lockpin was not the cause of the malfunction. This process of elimination is followed by the operator until the cause of the malfunction is identified.

Section 3 - Maintenance Procedures. The procedures for accomplishing all maintenance tasks necessary to maintain the equipment are contained in this section. The tasks are divided into step-by-step procedures that identify all actions that must occur to complete the task. Typical tasks include inspections, checks, adjustments, alignments, removal, installation, disassembly, assembly, repair, cleaning, and testing. Figure 9-15 shows a single maintenance task. Note that the task identifies the conditions necessary to accomplish the task, personnel requirements, and step-by-step procedures. The source information for maintenance procedures is developed through the maintenance task analysis process described in Chapters 5 and 16. The LSAR data base discussed in Chapter 17 serves as the actual source for writing these procedures.

Chapter 4 - Maintenance of Auxiliary Equipment. Many times an item of equipment actually consists of a prime item plus supporting or auxiliary equipment. When this occurs, the TM for the prime equipment provides references for where the operator should go to find instructions for maintenance of the auxiliary equipment. Figure 9-16 shows typical references for auxiliary equipment maintenance.

Chapter 5 - Ammunition. When the equipment requires ammunition, the TM contains a description and illustrations of the types of ammunition that the operator must use. The information is sufficient to allow the operator to readily identify ammunition that should and, more importantly, should not be used. Figure 9-17 shows a typical example of how ammunition is illustrated in a TM. Notice the references to color coding of ammunition by type. Color coding is standardized for all ordnance used by U.S. military forces.

Appendixes. Technical manuals also provide identification of additional equipment and supplies necessary to operate and maintain the equipment. These items are divided into four basic categories: components of end item (COEI), basic issue items (BII),

MIL-M-63004B (TM)

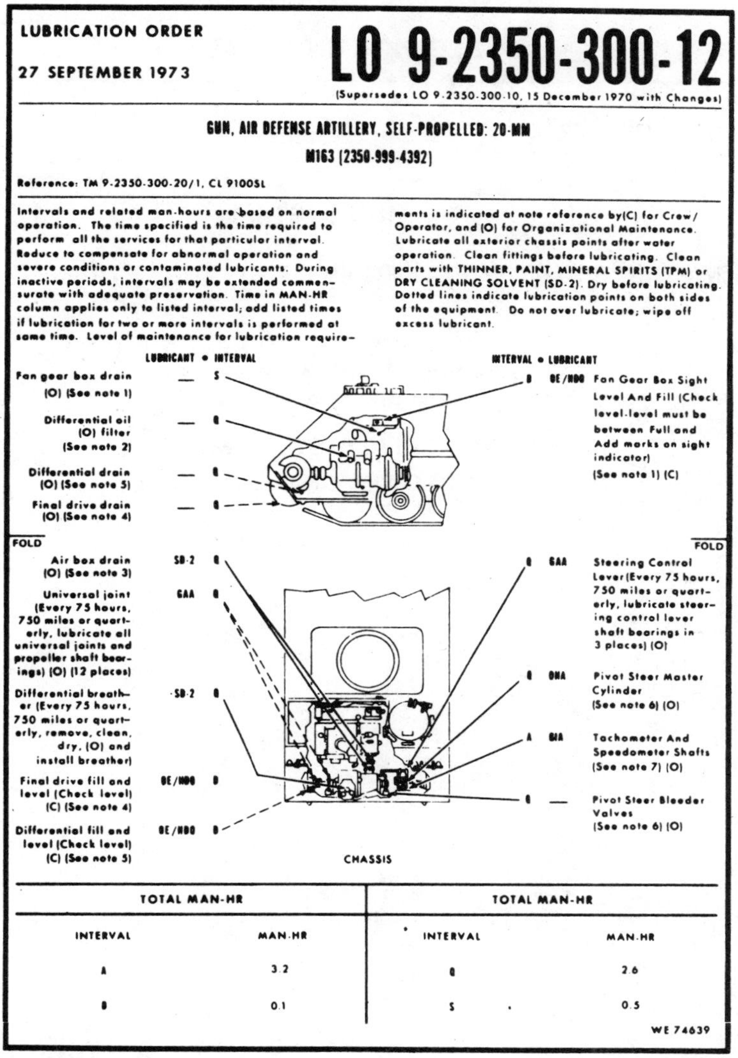

LUBRICATION ORDER

27 SEPTEMBER 1973

LO 9-2350-300-12

(Supersedes LO 9-2350-300-10, 15 December 1970 with Changes)

GUN, AIR DEFENSE ARTILLERY, SELF-PROPELLED: 20-MM

M163 (2350-999-4392)

Reference: TM 9-2350-300-20/1, CL 9100SL

Intervals and related man-hours are based on normal operation. The time specified is the time required to perform all the services for that particular interval. Reduce to compensate for abnormal operation and severe conditions or contaminated lubricants. During inactive periods, intervals may be extended commensurate with adequate preservation. Time in MAN-HR column applies only to listed interval; add listed times if lubrication for two or more intervals is performed at same time. Level of maintenance for lubrication requirements is indicated at note reference by(C) for Crew/Operator, and (O) for Organizational Maintenance. Lubricate all exterior chassis points after water operation. Clean fittings before lubricating. Clean parts with THINNER, PAINT, MINERAL SPIRITS (TPM) or DRY CLEANING SOLVENT (SD-2). Dry before lubricating. Dotted lines indicate lubrication points on both sides of the equipment. Do not over lubricate; wipe off excess lubricant.

TOTAL MAN-HR		TOTAL MAN-HR	
INTERVAL	MAN-HR	INTERVAL	MAN-HR
A	3.2	Q	2.6
B	0.1	S	0.5

WE 74639

Fig. 9-13. Lubrication order.

additional authorized list (AAL) items, and expendable supplies and materials list (ESML). Each of these lists provides the operator complete identification of the required items. Figure 9-18 shows a COEI list that identifies items that are issued with the basic equipment required to perform the equipment missions. The BII list shown in Fig. 9-19 lists the items that are issued with the equipment required to perform or support maintenance. All the items listed on the COEI and BII are issued and must remain with the equipment. The AAL items shown in Fig. 9-20 are required to support operation

TM 9-2350-252-10-2

TROUBLESHOOTING TABLE

MALFUNCTION
    TEST OR INSPECTION
        CORRECTIVE ACTION

25MM GUN SYSTEM

1.   25MM GUN STALLS OR FAILS TO FIRE AND IS NOT A HOT GUN.

NOTE
25mm gun is hot if 100 or more rounds
have been fired within 15 minutes.

Move TURRET DRIVE SYSTEM switch to OFF.

Move TURRET POWER switch to OFF.

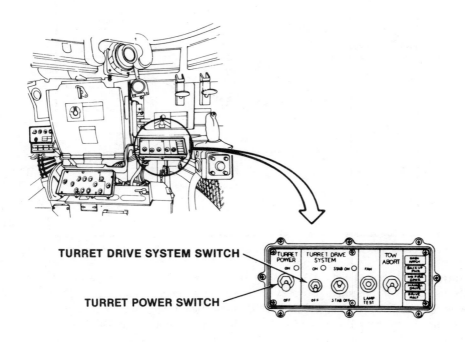

TURRET DRIVE SYSTEM SWITCH

TURRET POWER SWITCH

Fig. 9-14. Troubleshooting table.

TROUBLESHOOTING TABLE

MALFUNCTION
     TEST OR INSPECTION
          CORRECTIVE ACTION

25MM GUN SYSTEM (cont)

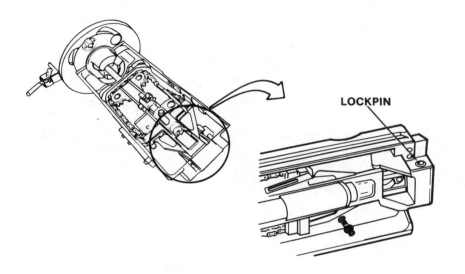

Remove 25mm gun feeder. See task: REMOVE 25MM GUN FEEDER,   page 3-34.

Step 1.   Check that lockpin is not jammed.

Push No. 4 cross-tip screwdriver in lockpin and release.

a. If lockpin does not spring back out, notify organizational maintenance.

b. If lockpin is not jammed, go to step 2.

Fig. 9-14. Troubleshooting table. (Continued from page 126.)

127

REMOVE 25MM GUN BARREL

INITIAL SETUP

Personnel Required:

Gunner
Helper (H)

References:

TM 9-2350-252-10-1

Equipment Conditions:

Engine stopped
(TM 9-2350-252-10-1)
Turret shut down (page 2-423)
25mm gun feeder removed
(page 3-34)

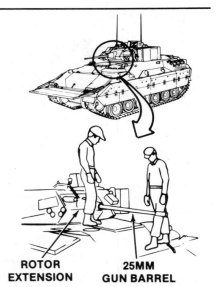

1. LOWER TRIM VANE. See
   TM 9-2350-252-10-1.

2. TAKE POSITIONS TO REMOVE
   25MM GUN BARREL.

   a. Climb on top front of vehicle.

   b. Stand on right side of 25mm
      gun barrel near rotor
      extension.

   c. (H) Stand on left side of
      25mm gun barrel.

ROTOR EXTENSION **25MM GUN BARREL**

WARNING

Gun barrel assembly could burn you. Wear heat protective mittens when you handle 25mm gun barrels and hot gun parts.

3. UNLOCK 25MM GUN BARREL.

   a. Pull out and hold 25mm gun
      barrel release latch.

**25MM GUN BARREL
RELEASE LATCH**

GO TO NEXT PAGE

Fig. 9-15. Maintenance task.

TM 9-2350-252-10-2

CHAPTER 4

MAINTENANCE OF AUXILIARY EQUIPMENT

SCOPE

This chapter tells you where to find maintenance instructions for auxiliary equipment.

COAX MACHINE GUN

See TM 9-1005-313-10.

AN/GRC-160 RADIO

See TM 11-5820-498-12.

AN/PRC-77 RADIO

See TM 11-5820-667-12.

AN/VRC-46 RADIO

See TM 11-5820-401-12.

Fig. 9-16. Maintenance references.

TM 9-2350-252-10-2

25MM AMMO

AUTHORIZED	CLASSIFICATION	IDENTIFICATION (see picture)	FUZE
HEI-T M792	High-explosive, incendiary tracer	Yellow	M758
Dummy M794	Dummy	Gold projectile	None
TP-T M793	Target practice, tracer	Blue projectile	None
APDS-T M791	Armor-piercing, discarding sabot, tracer	Black tip	None
NUMBER AUTHORIZED:	600 rounds stowed, 300 ready, 900 total IFV 1200 rounds stowed, 300 ready, 1500 total CFV		

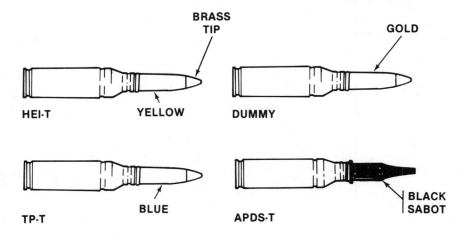

Fig. 9-17. Ammunition information for operator.

Section II.   COMPONENTS OF END ITEM

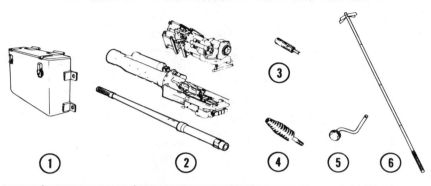

(1)	(2)	(3)		(4)	(5)
		DESCRIPTION			
ILLUS. NO.	NATIONAL STOCK NUMBER	FSCM and Part Number	Usable On Code	U/M	QTY RQR
1	2540-01-096-4559	BOX, VEHICULAR ACCESSORIES STOWAGE (EXTERIOR) (81361) D13-12-40		EA	2
2	1005-01-086-1400	GUN, AUTOMATIC, 25MM, M242 W/BII CONSISTING OF (INTERIOR)		EA	1
3	1005-01-121-2391	BRUSH, BORE, 25MM (19200) 12524014		EA	1
4	1005-01-121-2390	BRUSH, CHAMBER, 25MM (19200) 12524013		EA	1
5	1005-01-119-7269	CRANK, HAND (19200) 12524015		EA	1
6	1005-01-119-7865	ROD ASSEMBLY, CLEANING, 25MM (19200) 12524020		EA	1
NI	5820-01-056-0992	INSTALLATION, HARNESS, ELECTRONIC EQUIPMENT (INTERIOR) (80063) PPL-5233	J38	EA	1
NI	5820-01-054-7175	INSTALLATION, HARNESS ELECTRONIC EQUIPMENT (INTERIOR) (80063) PPL-5234	J63	EA	1

Fig. 9-18. Components of end item list.

Section III. BASIC ISSUE ITEMS

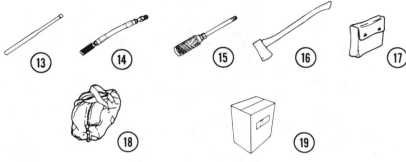

(1)	(2)	(3)		(4)	(5)
ILLUS. NO.	NATIONAL STOCK NUMBER	**DESCRIPTION** FSCM and Part Number	Usable On Code	U/M	QTY RQR
13	1005-01-120-0449	ADAPTER ASSEMBLY, BORESIGHT, 25MM (TOOL BAG) (19200) 12524010		EA	1
14	4930-00-288-1511	ADAPTER, GREASE GUN COUPLING, FLEXIBLE,12 IN. LONG (POWER UNIT COMPARTMENT TOOL BAG) (19207) 6300333		EA	1
15	4930-00-204-2550	ADAPTER, GREASE GUN COUPLING, 6 IN. LONG (TOOL BAG) (19207) 5349744		EA	1
16	5110-00-293-2336	AX, SINGLE BIT, 4 LB HEAD, 31 IN. HANDLE (EXTERIOR, TOP LEFT) (81348) GGG-A-926		EA	1
17	2540-00-670-2459	BAG ASSEMBLY, PAMPHLET (DRIVER'S STATION) (19207) 11676920	J38 J63	EA EA	6 3
18	5140-00-473-6256	BAG, TOOL, SATCHEL, 19 IN. LONG, 6 IN. WIDE, 8 1/2 IN. HIGH (POWER UNIT COMPARTMENT) (19207) 11655979		EA	3
19		CONVERSION KIT, M240C (19207) 12308278		E A	1

Fig. 9-19. Basic issue item list.

Section II.   ADDITIONAL AUTHORIZATION LIST

(1)	(2)		(3)	(4)
NATIONAL STOCK NUMBER	DESCRIPTION FSCM and Part Number	Usable On Code	U/M	QTY AUTH.
6665-00-935-6955	ALARM, CHEMICAL AGENT, M-8 VG W/ ACCESSORIES (81361) D5-15-4444		EA	1
6665-00-859-2215	1 EA ALARM UNIT, CHEMICAL M42E1 (81361) D5-15-4826			
6135-00-935-8738	4 EA BATTERY, DRY CELL, BA3202			
5315-00-450-3528	1 EA BATTERY, DRY CELL, BA3517			
6665-00-859-2201	1 EA DETECTOR KIT, CHEMICAL M43 (81361) D5-15-4400			
6665-00-859-2214	1 EA REFILL KIT, CHEMICAL AGENT, AUTOMATIC ALARM, M229			
5820-00-086-7651	ANTENNA, AT-784/PRC (SQD LDR ONLY) (80058) AT-784/PRC	J38	EA	1
1240-00-930-3833	BINOCULAR, 7X50, M17A1 W/CASE (19200) 10547052		EA	1
1240-00-930-3837	1 EA CASE BINOCULAR			
6145-00-226-8812	CABLE, TELEPHONE, 2 CONDUCTOR, WD-1/TT, 1320 FT (2 EA PLT LDR VEH ONLY) (81349) WD-1/TT		EA	1
	CARTRIDGE, 25MM, 300 READY ROUNDS, 1200 STOWAGE ROUNDS, 30 ROUNDS/CAN CONSISTING OF ANY COMBINATION OF THE FOLLOWING TYPES:	J38	EA	50
1305-01-092-0428	CARTRIDGE, 25MM, M791, APDS-T (19200) 12013720			
1305-01-094-1035	CARTRIDGE, 25MM, M792, HEI-T (19200) 12013722			
1305-01-092-0429	CARTRIDGE, 25MM, M793, TP-T (19200) 12013724			
	CARTRIDGE, 25MM, 300 READY ROUNDS, 600 STOWAGE ROUNDS, 30 ROUNDS/CAN CONSISTING OF ANY COMBINATION OF THE FOLLOWING TYPES:	J63	EA	30

Fig. 9-20. Additional authorized list.

Section II. EXPENDABLE SUPPLIES AND MATERIALS LIST
EXPENDABLE SUPPLIES AND MATERIALS LIST

(1) ITEM NUMBER	(2) LEVEL	(3) NATIONAL STOCK NUMBER	(4) DESCRIPTION FSCM & PART NUMBER	(5) U/M
1	C	8020-00-257-0382	BRUSH, ARTIST'S (81562) 5038-2	EA
2	C	9150-00-190-0907	AUTOMOTIVE GREASE (GAA) (1 LB CAN) (81349) MIL-C-10924	EA
3	C	1005-00-903-1296	BRUSH, CLEANING, BORE, 5.56MM (19204) 11686340	EA
4	C	1010-00-474-5466	BRUSH, CLEANING, BORE, 40MM (19205) 7790665	EA
5	C	1005-00-556-4174	BRUSH, CLEANING, BORE, 7.62MM (19204) 5564174	EA
6	C	1005-00-999-1435	BRUSH, CLEANING, CHAMBER, 5.56MM (19204) 8432358	EA
7	C	1005-00-610-3115	BRUSH, CLEANING, CHAMBER, 7.62MM (19205) 7790452	EA
8	C	1005-00-350-4100	BRUSH, CLEANING, RECEIVER, 7.62MM (19204) 8448466	EA
9	C	1005-00-494-6602	BRUSH, CLEANING, TOOTH, 7 IN. LONG (19204) 8448462	EA
10	C	1005-01-033-3925	BRUSH, GAS CYLINDER (B0897) 3052004839	EA
11	C	6850-00-264-9037	DRY CLEANING SOLVENT (81348) PD-680	QT

Fig. 9-21. Expendable supplies and materials list.

or maintenance, but are not issued with the equipment. The user must requisition AAL items through the military supply system. ESML items illustrated in Fig. 9-21 are consumed through use, therefore expendable, and must also be requisitioned by the user. All BII and ESML items are identified through the maintenance task analysis process discussed in Chapters 5 and 16.

MAINTENANCE MANUALS

Maintenance manuals are prepared for use by each maintenance level that provides support to a system. These manuals are normally used by organizational and intermediate maintenance levels. Figure 9-22 shows an outline for a maintenance manual prepared in accordance with MIL-M-63038B, Technical Manual: Organizational or

---

Front Matter
    Cover
    Warning Page
    Table of Contents
    How to Use This Manual

Chapter 1 - Introduction
    Section 1 - General Information
    Section 2 - Equipment Description
        Equipment Characteristics, Capabilities and Features
        Location and Description of Major Components
        Differences Between Models
        Equipment Data
    Section 3 - Technical Principles of Operation

Chapter 2 - Maintenance Instructions
    Section 1 - Repair Parts, Tools, and Support Equipment
    Section 2 - Service Upon Receipt
    Section 3 - Preventative Maintenance Checks and Services
    Section 4 - Troubleshooting Procedures
    Section 5 - Maintenance Procedures
    Section 6 - Preparation for Storage or Shipment

Appendix A - References

Appendix B - Maintenance Allocation Chart

Index

Fig. 9-22. Maintenance manual (outline per MIL-M-63038B).

MIL-M-63038B (TM)

SECTION II. MAINTENANCE ALLOCATION CHART
FOR
AN/TRN-30 (V)

(1) GROUP NUMBER	(2) COMPONENT/ASSEMBLY	(3) MAINTENANCE FUNCTION	(4) MAINTENANCE CATEGORY					(5) TOOLS AND EQPT.	(6) REMARKS
			C	O	F	H	D		
04	RF CABLE	Inspect		0.1					
		Test		0.1				1	
		Repair			0.4			10 thru 12	
05	DUMMY LOAD DA-639/TRN-30 (V)	Inspect		0.1					
		Test		0.1				1	
		Replace		0.1					
06	AMPLIFIER COUPLER AM-6417/TRN-30 (V)	Inspect		0.1	0.1				
		Test		0.2	0.2			1	
		Replace		0.4				2	
		Repair		0.2	0.5			11, 12 17 thru 22	B
0601	COUPLER ASSEMBLY 3A1	Inspect			0.1				
		Test			0.3			4 thru 6, 13 thru 15	
		Repair			0.4			11, 12	
060101	SWITCHED FILTER ASSEMBLY 3A5	Inspect			0.1				
		Test			0.3			4 thru 6, 13 thru 15	
		Replace			0.5			11, 12	
		Repair			1.0			11, 12	
060102	LOGIC/SERVO CIRCUIT ASSEMBLY 3A1A2	Inspect			0.1				
		Test			0.3			4 thru 6, 13 thru 15	
		Replace			0.4			11, 12	
		Repair					1.0		
060103	DETECTOR CARD ASSEMBLY 3A1A3	Inspect			0.1				
		Test			0.3			4 thru 6, 13 thru 15	
		Replace			0.4			11, 12	
		Repair					1.0		
060104	TUNING COIL ASSEMBLY 3A1A4	Inspect			0.1				
		Test			0.3			4 thru 6, 13 thru 15	
		Replace			0.4			11, 12	
0602	POWER AMPLIFIER- POWER SUPPLY 3A2, 3 & 4	Inspect			0.1				
		Test			0.3			4 thru 6, 13 thru 15	
		Adjust			0.2			4 thru 6, 13 thru 15	
		Replace			0.3			11, 12	
		Repair			0.5			11, 12	
060201	22 VOLT POWER SUPPLY 3A2A1, 3A3A1, & 3A4A1	Inspect			0.1				
		Test			0.3			4 thru 6, 13 thru 15	
		Replace			0.3			11, 12	
		Repair					1.0		
060202	22 VOLT POWER SUPPLY INTERCONNECT BOARD 3A2A2, 3A3A2, & 3A4A2	Inspect			0.1				
		Test			0.3			4 thru 6, 13 thru 15	
		Replace			0.3			11, 12	
		Repair					1.0		

Fig. 9-23. Maintenance allocation chart.

MIL-M-63038B (TM)

SECTION III. TOOL AND TEST EQUIPMENT REQUIREMENTS
FOR
AN/TRN-30(V)

TOOL OR TEST EQUIPMENT REF CODE	MAINTENANCE CATEGORY	NOMENCLATURE	NATIONAL/ NATO STOCK NUMBER	TOOL NUMBER
1	0	MULTIMETER AN/URM-105	6625-00-581-2036	
2	0	TOOL KIT, ELECTRONIC EQUIPMENT TK-101/G	5180-00-064-5178	
3	0	HAMMER HM-1	5120-00-203-4656	XXXXXX
4	F	MAINTENANCE KIT MK-1805/TRN-30		
5	F	MULTIMETER TS-352B/U	6625-00-553-0142	
6	F	OSCILLOSCOPE AN/USM-281A	6625-00-228-2201	
7	F	COUNTER, ELECTRONIC DIGITAL READOUT AN/USM-207	6625-00-911-6368	
8	F	METER AUDIO LEVEL TS-585 (*)/U	6625-00-244-0501	
9	F	CONNECTOR, ADAPTER UG-274/U	5935-00-926-7523	
10	F	MAINTENANCE KIT MK-693/A	5821-00-045-9695	
11	F	TOOL KIT, ELECTRONIC EQUIPMENT TK-105/G	5180-00-610-8177	
12	F	TOOL KIT, ELECTRONIC EQUIPMENT TK-100/G	5180-00-605-0087	
13	F	VOLTMETER, ELECTRONIC ME-30(*)/U	6625-00-643-1670	
14	F	DUMMY LOAD DA-640/TRN-30 (V)	5825-00-124-5094	
15	F	RADIO BEACON TRANSMITTER T-1199/TRN-30 (V)*	5825-00-474-9524	
16	F	DUMMY LOAD DA-75(*)/U	6625-00-177-1639	
17	F	TORQUE WRENCH 5-150 IN.-LB.	5120-00-542-4489	XXXXXX
13	F	TORQUE SCREWDRIVER 4-100 IN. OZ.	5120-00-933-0941	XXXXX
19	F	ADAPTER, 3/8" DRIVE TO 1/4" SQ DRIVE FEMALE SOCKET	5120-00-227-8095	
20	F	SOCKET 3/4" DP, 3/8" DRIVE	5120-00-235-5879	

SECTION IV. REMARKS

Reference Code	Remarks
A	DS will replace phenolic connector and pot for moisture.
B	All repair and replacement of parts performed by organizational maintenance limited to authorized items listed in TM (cite specific TM) -20P.

Fig. 9-23. Maintenance allocation chart. (Continued from page 136.)

Aviation Unit, Direct Support or Aviation Intermediate, and General Support Maintenance. A comparison of this outline with the outline for an operator's manual in Fig. 9-1 shows that the manuals have several sections in common. The significant differences between these manuals is that the maintenance manual does not contain operator-specific information, and the maintenance instructions have been expanded in the maintenance manual. Also, the maintenance manual contains a maintenance allocation chart that is not in the operator's manual.

A maintenance manual is far more technical than an operator's manual and, therefore, is normally much larger. The outline for the maintenance manual shows only one chapter for maintenance instructions where, in actuality, it might have several chapters containing maintenance procedures for different sections or levels of the equipment.

A maintenance manual also contains information on the parts and tools required to support maintenance, and it provides instruction on maintenance tasks that must be performed upon receipt of the equipment or in preparation for storage or shipment.

Maintenance Allocation Chart. The most significant section in a maintenance manual that has not been discussed thus far is the maintenance allocation chart. A maintenance allocation chart (MAC) identifies which level of maintenance is authorized to perform each maintenance task and the tools that are required to support each task. Figure 9-23 shows a MAC. Note that the MAC contains a listing of all maintenance tasks, the standard time to perform the task, identification of the performing maintenance level, and a cross-reference to the appropriate tool or support equipment required to support the task. It is extremely important that the information shown on the MAC matches the SMR code assigned to provisioned spares, otherwise the supply and maintenance systems will not provide adequate support to the equipment. The detailed maintenance task analysis, discussed in Chapter 5 and Chapter 16, is the source for information contained in the MAC.

Depot Maintenance Work Requirements. Manuals for depot-level maintenance are somewhat different from the manuals for organizational and intermediate levels. The depot-level manuals contain extremely detailed information related to the total rebuilding or overhaul of an item of equipment. Depot maintenance work requirements (DMWR) are prepared for specific tasks related to portions or all of an item of equipment, and contain all the information, procedures, processes, tool and tooling requirements, support equipment requirements, inspection criteria, and quality requirements for tasks designed to rebuild, refurbish, or overhaul the equipment. MIL-M-63041C, Technical Manuals: Preparation of Depot Maintenance Work Requirements, contains detailed instructions for preparing DMWRs.

PARTS MANUALS

The final type of manual to be discussed is the parts manual. A parts manual is used as a companion to a maintenance manual to provide a reference for maintenance personnel to identify spare and repair parts. A parts manual might be included as a section of a maintenance manual or as a separate manual. In either case, the parts manual may be referred to as an illustrated parts breakdown (IPB) or as a repair parts and special tools list (RPSTL). Figure 9-24 illustrates how a parts manual contains both an exploded picture of the item and a list of all parts shown. It is significant to note that the SMR code, assigned through the provisioning process, appears as part of the manual information. MIL-M-38807A, Technical Manuals: Preparation of Illustrated Parts Breakdown, and MIL-STD 335, Technical Manuals: Repair Parts and Special Tools List, contain specific instructions on the preparation of these manuals.

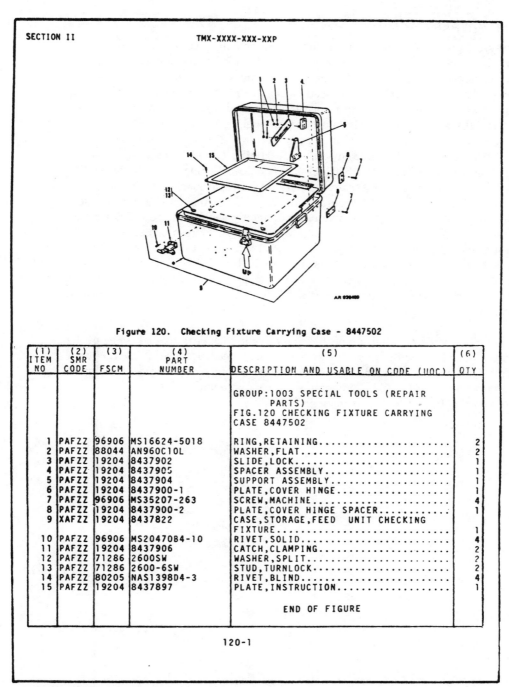

SECTION II   TMX-XXXX-XXX-XXP

Figure 120. Checking Fixture Carrying Case - 8447502

(1) ITEM NO	(2) SMR CODE	(3) FSCM	(4) PART NUMBER	(5) DESCRIPTION AND USABLE ON CODE (UOC)	(6) QTY
				GROUP:1003 SPECIAL TOOLS (REPAIR PARTS) FIG.120 CHECKING FIXTURE CARRYING CASE 8447502	
1	PAFZZ	96906	MS16624-5018	RING,RETAINING......................	2
2	PAFZZ	88044	AN960C10L	WASHER,FLAT........................	2
3	PAFZZ	19204	8437902	SLIDE,LOCK.........................	1
4	PAFZZ	19204	8437905	SPACER ASSEMBLY....................	1
5	PAFZZ	19204	8437904	SUPPORT ASSEMBLY...................	1
6	PAFZZ	19204	8437900-1	PLATE,COVER HINGE..................	1
7	PAFZZ	96906	MS35207-263	SCREW,MACHINE......................	4
8	PAFZZ	19204	8437900-2	PLATE,COVER HINGE SPACER...........	1
9	XAFZZ	19204	8437822	CASE,STORAGE,FEED UNIT CHECKING FIXTURE............................	1
10	PAFZZ	96906	MS20470B4-10	RIVET,SOLID........................	4
11	PAFZZ	19204	8437906	CATCH,CLAMPING.....................	2
12	PAFZZ	71286	2600SW	WASHER,SPLIT.......................	2
13	PAFZZ	71286	2600-6SW	STUD,TURNLOCK......................	2
14	PAFZZ	80205	NAS1398D4-3	RIVET,BLIND........................	4
15	PAFZZ	19204	8437897	PLATE,INSTRUCTION..................	1
				END OF FIGURE	

120-1

Fig. 9-24. Illustrated parts breakdown.

MANUAL PREPARATION

The actual preparation of technical manuals is accomplished through a process of collecting information on all the operation or maintenance tasks required to support an item of equipment. Then, this information is organized to make it easy to use and understand. The LSAR data base discussed in Chapter 17 should be used as the source information for all technical manual preparation. Otherwise, the maintenance tasks, spares and repair parts, and tools and support equipment requirements might be inconsistent, which will impede maintenance. MIL-M-38784B, Technical Manuals: General Style and Format Requirements; MIL-HDBK 63038-1, Technical Manual Writing Handbook; and MIL-HDBK 63038-2, Technical Writing Style Guide provide general guidance and information that is very useful in preparing all technical manuals.

The readability of TMs is extremely important if the manuals are to be used. Government contracts for technical manuals commonly specify the required reading grade level for TMs being prepared. Most TMs are prepared for a target reader with a seventh to ninth grade comprehension level. A standard method for evaluating the reading grade level of a manual is contained in MIL-M-38784B. DoD-STD 1685, Comprehensibility Standards for Technical Manuals, provides detailed information on how to keep manuals simple by using common words and terms in short, understandable sentences.

As stated in the first paragraph of this chapter, the technical manual is one of the most important documents that a contractor prepares, because it is the only documentation that the operator receives. A system's ability to perform its assigned missions depends on how well it is maintained, and maintenance cannot be performed adequately without valid technical manuals.

Chapter 10

Training and
Training Equipment

Equipment operators and maintenance personnel must be properly trained if the equipment is to continually perform its mission. The purpose of a training program is to ensure that the training provided to military personnel is coordinated with the other ILS disciplines, is comprehensive, and contains all the pertinent information required to operate and maintain the equipment when it is in actual operation.

Training can be divided into four categories: (1) operator training, (2) maintenance training, (3) supervisor training, and (4) instructor training. These categories of training can be subdivided into two phases: initial and sustainment. The development of a training program and identification of equipment required to support training is accomplished by training analysts as a part of the total ILS effort. Additionally, contractors might present portions of the training curriculum to military personnel.

This chapter addresses the basic types of training, identification of training requirements, development of a training program, and identification and justification of training equipment required to support the training program.

TRAINING PHASES

The two phases of training, initial and sustainment, can be correlated to the acquisition phases of the equipment being procured by the government. Although the training provided to military personnel in both phases should be identical, the methods of presentation might vary, and the purposes are different. Figure 10-1 shows when each phase of training is normally planned, prepared, and conducted.

Acquisition Phases				
Concept	Demval	FSD	Production	Deployment
Initial Training Planning		Preparation	Conduct	Conduct
Sustainment Planning		Preparation		

Fig. 10-1. Training phases.

Initial Training

The purpose of initial training is to train the operators and maintenance personnel who will work with the new equipment when it is fielded. Initial training is planned and prepared during FSD and early production. It is normally conducted just before and during initial fielding of the equipment. Those chosen to receive initial training are normally selected from units that will be responsible for the new equipment. Because the equipment is new, the initial training program tends to be more flexible and detailed than training that occurs after initial fielding.

Another factor that affects initial training is the lack of historical knowledge and experience to identify key information. This knowledge is accumulated during initial usage and is used to upgrade or change sustainment training. The goal of initial training is to enable full equipment readiness as quickly as possible. The duration of the initial training phase is limited to the time necessary to accomplish this goal.

Sustainment Training

The purpose of sustainment training is to provide qualified replacement personnel to units that operate or maintain the new equipment. Those who receive sustainment training normally receive it prior to being assigned to positions involving the new equipment. Sustainment training tends to be more rigid than initial training; however, it is tailored, based on field experience, to contain the information necessary for the performance of assigned tasks.

Sustainment training starts when initial training ends and continues throughout the life of the equipment. It is normally conducted at permanent training sites maintained by the military service.

TRAINING CATEGORIES

The categories of training-operator, maintenance, supervisor, and instructor—identify both the functions necessary to sustain the operation of the equipment and the types of persons to be trained. The actual curriculum of the training courses developed might overlap extensively. Therefore, it is necessary to understand why the training is categorized in this manner.

Operator Training

Basic operator training is comprised of those functions that the user is required to perform to operate and maintain the equipment. The word "user" is the key to this category. The training is comprehensive to the point that it provides the operator with everything that must be done to place the equipment into operation, perform scheduled and unscheduled operator maintenance, and identify equipment failures.

Maintenance Training

Maintenance training involves those who will staff the organizations responsible for performing maintenance on the new equipment. This might include organizational, intermediate, and depot personnel. The training of maintenance personnel must reflect the maintenance concept and planning developed by the ILS disciplines.

Each level of maintenance training becomes progressively more comprehensive because each level of maintenance is authorized to perform more detailed maintenance. For example, organizational- level maintenance might be required to troubleshoot the equipment and remove and replace faulty modules, so "O" level maintenance training would consist of those tasks required to perform these functions. Training at "I" level would, in many cases, duplicate "O" level training and add the tasks required to repair modules. Depot-level maintenance training would include both "O" and "I" levels and also address how the equipment is overhauled and refurbished.

Training for each level would also include instructions on how to operate test equipment and other support items required to perform maintenance. Equipment operation could also be included at each level if required for troubleshooting, fault verification, or repair checkout.

Supervisor Training

Training for supervisor personnel must be adequate to allow supervision of the activities of subordinate operator and maintenance personnel. This training should include all tasks to be performed and establish criteria for supervisors to determine if the tasks are being accomplished properly. In most cases, supervisors at each level receive training identical to that received by their subordinates, plus additional training on how tasks should be coordinated, scheduled, and supported. Supervisor courses might be required for the operator level and each maintenance level.

Instructor Training

An often overlooked category of training is instructor training. This is very important during initial training and the early stages of sustainment training. Qualified instructors must be well versed in all tasks required to operate and maintain the new equipment. The development of training courses for instructors must be the most comprehensive of the categories of training. If instructors are not trained adequately, the other training will not be effective.

TRAINING CONCEPT

The overall concept of how training will be conducted for personnel who will operate and maintain a new item of equipment is developed in the early phases of the acquisition cycle. The concept might consist of a curriculum developed by a contractor or the military and presented in a combination of methods.

Typically, the contractor who is building the equipment is also required to develop the initial training program. The actual training could be accomplished by the contractor, the military, or a combination of both. There are also alternatives concerning where the training will be accomplished.

Training Development

The most logical approach to development of training requirements is for the contractor to develop them as the equipment is being designed and built. This is normally the most cost-effective method, because the contractor has all the source information that is required to develop the training program. If the government were to procure training services from another source, there would probably be a sizable duplication of effort in developing this material, which would greatly increase the overall equipment life cycle cost.

The government normally provides a requirement for this task in the basic contractual statement of work. As the training material is being generated, the government will normally review, approve, and recommend changes so that the training meets the government's requirements. It is important that the government provide the contractor an initial training concept to use as a guide in developing the training program.

Training Presentation

The decision as to who will actually present initial training is based on several key points, shown in Fig. 10-2. The number of persons to be trained, the duration of training, locations where training must be accomplished, and the priority of the weapon system program are used to determine if a contractor has the resources necessary to support the training effort. Major weapon systems that are fielded globally require extensive resources to conduct the necessary training. In such a case, the military would, necessarily, conduct the majority of the training program. In other situations where deployment schedules or locations allow, the contractor might conduct all initial training. It might be feasible, in some instances, for a combination of contractor and military instructors to present initial training.

The technology being taught and minimum instructor qualifications might dictate that the contractor must present initial training, regardless of the other factors, until military instructors are trained. An example of this situation would be the contractor who trains the military instructors who, in turn, train other military personnel. This type of instruction is called "training the trainers."

The last factor identified is the requirement for training aids. In some cases, such as intermediate- or depot-level maintenance, an actual item of equipment might be required as a training aid in order to adequately present the training course. During

NUMBER OF PERSONS TO TRAIN

DURATION

LOCATIONS

PRIORITY

TECHNOLOGY

INSTRUCTOR QUALIFICATIONS

TRAINING AIDS REQUIRED

Fig. 10-2. Training presentation decision factors.

the initial training phase, when no equipment has been fielded, training equipment is only available at the contractor's facility; therefore, the contractor is the logical choice for presentation of courses when such a constraint is imposed.

An alternate approach to presenting training when multiple locations are involved is the use of a new equipment training team (NETT). The NETT is a group of trainers who provide training at the students' location. This concept of presentation is used when the equipment will be fielded at a number of locations. It is easier and more cost effective to send the trainers to the students than to have all the students come to the trainers.

After initial training has been completed, the government normally assumes responsibility for presentation of sustainment training. Contractors might be required to develop new material or modify initial training material to support sustainment training, or the government might choose to accomplish this task using established training organizations within the military.

TRAINING PROGRAM

Contractors who develop a training program for the government as part of an equipment acquisition should integrate the program with the overall equipment program. The purpose of the training program is to coordinate all training efforts and tasks. MIL-STD 1379C, Military Training Programs, provides detailed guidance as to the processes and analyses that should be accomplished to ensure that the goals of the training program are met. The training program should be designed to provide government personnel with the knowledge and skills required to operate and maintain the new item of equipment being developed.

The training program begins with generation of a training plan and culminates with the presentation of training or delivery of training materials to the government. The training program must include identification of training requirements and appropriate methods for presentation of training material. In all cases, the training program must be integrated with other ILS efforts.

Training Plan

The training plan establishes the foundation for the contractor's training program. It identifies what the contractor intends to accomplish, the resources required, the expected outputs of the program, and a schedule for completion. An example of a training plan outline is shown in Fig. 10-3. The training plan is similar to previously discussed plans. Key points contained in the plan include identification of how training requirements will be developed, personnel and resources required to support training, training equipment requirements, and a schedule for the training program. The government normally reviews and approves the training plan.

Training Requirements

The identification of training requirements is the key ingredient of a successful training program. The systematic method for accomplishing this task is illustrated in Fig. 10-4. Identification begins by obtaining data from sources such as the FMECA, maintenance task analysis, RCM analysis, and operator task analysis. The purpose of this effort is to identify all the tasks required to operate and maintain the equipment.

Using source information generated by other ILS disciplines achieves two objectives: (1) reducing cost by not duplicating effort and (2) integrating the total ILS effort. If training analysts generated the source data from scratch, there would inevitably be differences in tasks, which would cause a disconnection between training and other ILS efforts, and there would be a tremendous increase in cost.

After all operation and maintenance tasks are correlated, a Manpower, Personnel, and Training Analysis (MPTA) is conducted to consolidate all tasks required to support the equipment. MPTA results are then used to develop a Personnel Performance Profile (PPP), which identifies the knowledge and skills that personnel must possess to accomplish the operation and/or maintenance tasks. The PPP is used as the baseline to identify requirements for individual training courses. Training courses can then be developed that provide a vehicle for conveying the knowledge and skills to personnel who will ultimately operate and maintain the equipment.

Although this description of the requirements identification process has been somewhat brief due to space limitations, it must be pointed out that this process is the cornerstone of the entire training program. Figure 10-5 illustrates the flow of information through this process. In the example presented, a single repair task is followed from initial identification on the FMECA to being included in various training courses. If requirements are incorrectly or inadequately identified, the training program will be ineffective. MIL-STD 1379C provides further information on the content and format of the MPTA and the PPP.

Training Methods

The training requirements identified, coupled with the presentation factors discussed previously, provide input to deciding which training methods are most appropriate for the training program. These methods include lectures, demonstrations, performance, on-the-job training, and self-study.

1.0 SCOPE

Briefly describe the training to be developed. Include an explanation of how the training program will be prepared and documented.

2.0 TRAINING COURSES

Identify each training course to be prepared. Provide the following information for each course:

 a. purpose of the course
 b. course prerequisites
 c. course length
 d. class size
 e. milestone schedule
 f. resource requirements
    1. personnel
    2. item of equipment
    3. training equipment
    4. audiovisual aids
    5. training material
    6. technical manuals or other publications
    7. facilities

3.0 TRAINING ORGANIZATION

Describe the training organization that will prepare the training courses. Include any technical or administrative procedures or processes that will be used to streamline the course development.

4.0 SOURCES OF INFORMATION

Describe the sources of information that will be used to obtain the data necessary to develop the proposed training courses. Address the methods that will be used to ensure that the data reflect the delivered equipment configuration.

5.0 INTERFACES

Provide a description of how training personnel will interface with other ILS disciplines during the development of training courses. Key areas that should be addressed are interfaces with maintenance planning, personnel, system safety, human engineering, and LSA efforts.

6.0 COURSE VALIDATION

Describe the method that will be used to validate the accuracy and adequacy of training courses.

Fig. 10-3. Training program and training equipment plan (sample outline).

Lectures. The classical method for instruction is through lectures. In this method, students listen to presentations by instructors and gain knowledge through remembering or taking notes. The lecture method of training should be used judiciously, because it places little demand on students, and its effectiveness can only be measured through extensive testing. Military training programs, especially those dealing with the operation and maintenance of equipment, that rely on extensive use of lectures have proven to be less than desirable.

Performance. The performance method of instruction places the student in direct relationship to the equipment being trained. The instructor acts as a guide in performing the actual tasks that the student will be required to accomplish as a result of the training. This method of instruction, also called "hands-on" training, should be used whenever possible. With this method, students not only have a better retention of training, but they also gain experience and confidence in their abilities to perform the tasks after leaving the classroom.

On-the-Job Training. In some cases, training may be accomplished while students are actually performing the required tasks under the supervision of trained unit personnel. On-the-job training (OJT) is an effective and economical method of training in the appropriate circumstances. This method of training is commonly used when training personnel in new maintenance procedures or on similar equipment to which they are already qualified. By having the student actually accomplish a task as

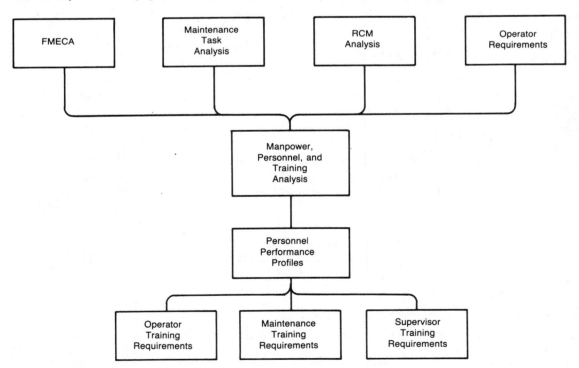

Fig. 10-4. Training requirements identification.

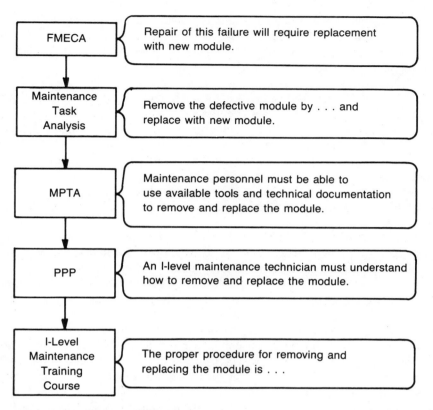

Fig. 10-5. Training requirements information flow.

part of the training program, the service benefits since the work is productive and the student benefits through the learning process. In some instances where the government feels that OJT is an appropriate training method, the contractor may be required to prepare OJT training material and study guides for military personnel to use in administering the training program.

Self Study. In the self-study method, the student is allowed to progress at an individualized pace using a structured set of study materials that provide a step-by-step framework of instruction. This method of instruction for operation and maintenance of equipment has proven to be relatively ineffective because it normally does not provide hands-on access to the equipment. The most logical use of self-study is for personnel who require an overview of the equipment operation and maintenance requirements.

TRAINING MATERIALS

Training cannot be accomplished effectively without adequate supporting material. MIL-STD 1379C explains how training material should be developed as the training program progresses. A key point to remember is that the best training materials are the actual things that the student will ultimately use on the job; specifically, the actual equipment, the applicable technical manual, and the authorized tools and other support equipment required to perform all tasks. Whenever possible, these items should be used in the training program as the primary training material.

Other materials should be used to supplement and augment these items only when necessary. Training materials that a contractor might prepare include instructor guides, student guides, OJT handbooks, course outlines, and course objectives. These would be generated for each training course being prepared, and should stand alone.

Training Objectives. Unlike other types of training, military training is designed to enable every student to pass the training course. At the beginning of any course, the student is provided with a list of the training objectives of the course. They state exactly what the student will be required to learn, and what knowledge and skills should be gained from the course. This list is compiled from the MPTA and the applicable PPP for each course.

Instructor Guide. The instructor guide contains specific instructions that aid the instructor in preparing and presenting training courses. The information in the guide include class and course training objectives, requirements for preparation prior to each class, a time allocation guide, and instructions on how each subject should be presented. The guide usually stresses hazard awareness or safety precaution when the training could pose a threat to students or equipment.

Student Guide. The student guide provides the student with supplementary information and materials about the training course that enhance the student's ability to gain the knowledge and skills presented in the course. The guide might include instruction sheets for each class, supplemental information on the technical documentation provided in the class, study questions, a course outline, training objectives, and recommended additional self-study to reinforce the classroom instruction.

OJT Handbook. When the method of instruction includes an OJT program, the contractor might prepare an OJT handbook. This document is similar to the student guide, but it also contains information to be used by the supervisor responsible for conducting the OJT. The handbook should contain detailed instructions on procedures and actions to be taken by both the student and the instructor to enable the student to become a fully qualified operator and/or maintenance technician.

Training Aids. Another category of training material often prepared by contractors is training aids. These aids are used to supplement or enhance training, or to provide ready references for students. The most common types of training aids include charts and projection transparencies or slides for use by instructors, mock-up or enlarged equipment models, information cards that are retained by students, and class handouts used to highlight a training objective. An innovative training aid that can be used very effectively is a videotape training film. This medium has been used in cases where the training is required at many locations and the subject matter lends itself to such a presentation. The advantages of a videotape are that it can be sent to and retained by the training location. As the need arises, the training can be repeated at no additional cost except the availability of the equipment necessary to show the film.

Testing

The use of testing by military training programs has two purposes: (1) ensure that the student has gained the knowledge and skills presented in the training course and (2) validate the training course. The methods of testing used depend on the subject matter being presented. However, the preferred method of testing involves performance, in which the student demonstrates an understanding of and ability to perform the tasks presented in the course. The basis for preparing test material should always be the Training Objectives List, which identifies the skills or knowledge that the student should learn.

TRAINING SUPPORT

An area that must not be overlooked when preparing a training program is the support required to conduct it. Support includes instructors, training facilities, and training equipment. Each of these items is necessary to conduct a successful training course.

Instructors

The right instructor greatly affects how successful a training course will be. It is important that the qualifications for instructors be established sufficiently in advance of the beginning of training to allow proper selection. Examples of instructor qualifications include experience, knowledge of the subject matter, training, and ability to teach. Too often an instructor is selected solely on experience or knowledge of the subject without regard to the individual's ability to teach. Other times, the determining factor is "who is available" not "who is qualified." These mistakes have been made too often. Selecting the best instructor for a course goes a long way toward having a successful training course.

Facilities

The identification and availability of the appropriate facilities for a training course aid in accomplishing training goals. The facility requirements should include size and arrangement of classroom and work space, environmental needs, lighting and electrical power, toilets, break areas, and reference areas. The training facilities do not have to be luxurious, but they should be adequate to support training and create a positive atmosphere that promotes learning.

TRAINING EQUIPMENT

The selection of training equipment that is compatible with the type of training being conducted is a key factor in the success of the training program. Anything used to support the training effort and aid in student comprehension can be classified as training equipment. This equipment, as defined by MIL-HDBK 220B, Glossary of Training Device Terms, could include training aids (discussed previously in this chapter) and other items such as tools, test equipment, mock-ups, cutaways, simulators, or actual equipment. It is also important to remember that the equipment selected for use in a training course must be operated and maintained.

Equipment Types

Most training equipment can be classified as actual equipment, audiovisual equipment, mock-ups, or simulators. Each type of equipment has its own advantages and disadvantages.

Actual Equipment. The use of actual equipment in training courses is desirable whenever possible. There is no substitute for the real thing. Students who receive training while using or performing maintenance on the actual equipment have a distinct advantage in acquiring and retaining knowledge and skills. There can, however, be disadvantages associated with using actual equipment. First, early in the acquisition of equipment when initial training occurs, the actual equipment might not be available in the quantities required to support training. Second, the use of actual equipment might reduce the feasible instructor-to-student ratio. For example, if a course is being taught on how to perform maintenance inside the turret of a tank, there is not enough room in a single tank turret to accommodate an instructor and 30 or 40 students. Alternative approaches might be required to support this type of training.

Audiovisual Equipment. The lecture method of instruction can be enhanced through the use of audiovisual equipment. Its use is considered a standard method of instruction in military training programs. When used properly, with concise, well-planned transparencies or slides, audiovisual equipment is a definite asset to the training program. However, the constant use of this equipment with monotonous information can prove very detrimental to the training course.

Mock-ups. Mock-ups are items built to represent the real thing. They can be scaled larger or smaller than the actual equipment to illustrate how the equipment operates or must be maintained. The use of mock-ups, in the appropriate circumstances, can be very beneficial to a training program. Such items normally cost much less than the actual item and are just as effective in the classroom. In some cases, mock-ups are the answer to problems caused by using the actual equipment. At the United States Army Armor School, Fort Knox, Kentucky, tank turret maintenance personnel are trained using mock-ups that are really actual tank turrets, modified to allow viewing from the exterior. Using these mock-ups, instructors can teach a greater number of students than would be possible using actual equipment.

Simulators. The use of simulators that provide a realistic representation of the operation or maintenance of an item of equipment has proven to be a cost-effective method of training. Common simulators include trainers for pilots, gunners, and equipment operators. The acquisition of simulators can be expensive; however, the cost savings realized over the life of the equipment can be very substantial. Flight simulators used to train pilots have reached the level of sophistication where there is virtually nothing that an aircraft can do that cannot be duplicated by a simulator. The resulting training is realistic, and that is the purpose of a simulator—to provide realistic training at a substantial cost savings.

Equipment Selection

Training equipment must be selected with caution. The philosophy that should be used is, "What training equipment can be justified for use in supporting the training program, and what items will enhance the ability of students to understand and retain the course objectives?" Maximum use should be made of existing training equipment or, when feasible, modification of existing equipment to support the new training requirements.

The justification for buying new training equipment should be based on the number of individual training tasks that can be supported by an item and the amount of training time that will be required to use the equipment. For example, if an item of equipment is proposed to support training of a single task, then the tasks must merit the expense. This is not normally the case. Most items of training equipment have a broad application to the total training course, and the cost of the item is amortized through the first few uses. For example, in the case of the tank turret mock-up for maintenance training, the student-to-instructor ratio increases, which reduces the overall cost of the training program. The same mock-ups are also used for familiarization training and basic tank crew training. This multiple use of trainers adds to the overall cost of training for the weapon system.

MIL-T-29053A, Training Requirements for Aviation Weapon Systems, contains guidelines for selecting training equipment that can apply not only to aviation, but to all training programs. MIL-T-23991E, General Specification for Military Training Devices, covers the general requirements that must be followed when designing new training devices. These requirements include selection of materials, design considerations, and reliability goals.

Equipment Support

When selecting training equipment, remember that the equipment will require support just like anything else. That means operator training, maintenance planning and procedures, spare parts, and technical documentation. The requirements for preparing and obtaining this support add cost to the overall weapon system program life cycle. While the support might not need to meet the rigid requirements placed on the weapon system, there are basic requirements that must be met.

MIL-G-29011, Preparation of Guides for Operation and Maintenance of Training Aids, is the specification that addresses the requirements for documenting operation and maintenance procedures for several types of classroom training aids. Larger items of equipment, such as simulators, might require the full range of support documentation that normally accompanies a weapon system.

Sufficient support for training equipment must be planned and obtained to ensure proper training results when using the equipment. If a training course is built around the use of training equipment that is inoperative due to lack of spare parts or the inability of the instructor to operate the equipment, then the training will be less than satisfactory.

Chapter 11

Supply Support

Spare parts are required to support both scheduled and unscheduled maintenance. If spare parts are not available, very little maintenance can be accomplished, resulting in an inoperative inventory of military equipment. The objective of supply support is to have the parts available when and where required in the quantities necessary to support maintenance.

Chapter 6 described the process used to identify and obtain the initial quantities of spare parts, and Chapter 12 discusses the physical aspects of getting the spares from the producer to the user. This chapter addresses the overall DoD supply system, determination of requirements for spare parts, and inventory management.

The term "spare parts" is used here to refer to all parts required for maintenance, whether they are actually spares (repairable items), repair parts (items that are nonrepairable and are discarded when they fail), or consumable parts (items that are consumed when used, such as gaskets or adhesives).

DoD SUPPLY SYSTEM

How do maintenance personnel obtain spare parts? One method would be to provide each maintenance organization with a list of the telephone numbers of suppliers of each spare part and let them simply call the supplier and order a part when it is required. As you can imagine, this method would not be realistic. It does not provide critical spares quickly enough, does not ensure that parts are available when needed, and would create a fiscal nightmare when it came time to pay the bills. The DoD supply system, although much maligned, is probably the most organized and well-managed operation of its kind. Just as with other DoD organizations or systems, it has evolved over the years with the single goal of getting the needed parts to the user in the most expedient and cost-effective method possible.

The supply system is multitiered, and has been tailored to fit the roles and missions of each branch of the military. The differences among services are not discussed here because the basic concepts are common to all.

The supply system can be divided into four levels;

- Supplier/DoD interface
- Depot
- Intermediate
- Organization and user

These levels are related much the same as maintenance levels, discussed in previous chapters. Figure 11-1 illustrates the DoD supply system concept.

Supplier/DoD Interface

At the top level of the supply system is the supplier and the DoD organization that deals directly with the supplier to obtain spare parts. During weapon system acquisition, this interface is normally accomplished by the Acquisition Program Office (APO), which is responsible for initial spares provisioning. After an item has been provisioned and fielded, the responsibility for spares might remain with the APO, depending on the size and priority of the weapon system being procured. However, the responsibility for buying replacement spares usually is passed to an item manager located within the normal supply system structure.

The quantities of initial provisioned spares is normally intended to support the weapon system for only two years at the most; many times, less. So, it is up to the item manager to ensure that spares are available for the life of the weapon system. This is a key point. During development, the APO is provided funds by Congress to buy spares specifically for the weapon system being developed. When the responsibility for spares is transferred to an item manager, spares must be procured from a lump sum of funds provided for all the spares managed by the item manager's organization.

It is extremely important that initial provisioning be as accurate as possible and that field usage history be accumulated correctly so that the item manager can buy the appropriate number of spares for the weapon system. This is the point where the weapon system, as a single entity, loses its identity in the supply system, because an item manager is commonly assigned responsibility for spares by NSN, not weapon system. Many different item managers might be responsible for the spares of a single weapon system. Conversely, an item manager might be responsible for one or two spare parts of many different weapon systems.

Item managers are located at various National Inventory Control Points (NICPs) throughout the United States. NICPs are responsible for managing all supplies, including spare parts, at the depot level in the supply system. An NICP may or may not be

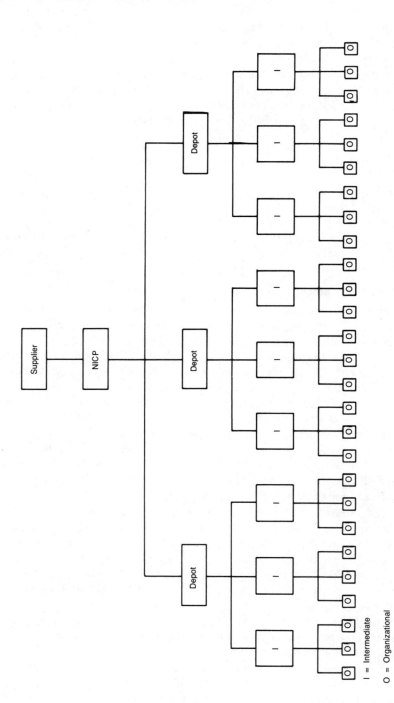

I = Intermediate

O = Organizational

Fig. 11-1. DoD supply system.

physically located at the same installation where the spares are stored after they are received from the supplier.

Depot

Spare parts are stocked at depot level in support of worldwide requirements. It is not uncommon for a single depot to maintain the entire depot-level inventory of a part. The exception to this is in the case of overseas depots that support a geographical area in order to respond more quickly to user needs.

The relevant NICP maintains records of the quantities of spare parts on hand at the depot level. Issues of parts from depot stocks are normally accomplished at the direction of the NICP. Depots, due to their size and the facilities required to sustain operation, are fixed installations. In some cases, depot stocks might consist of parts that are stockpiled to support the services in the event of war. These stocks are referred to as "war reserves."

When parts provisioned on a specific equipment acquisition program are received by the government, they might be located at a single depot or distributed to several, depending on the NSNs assigned to the items and their applications. At the depot level, parts are controlled as unique items with no distinction as to the type of equipment to which they belong. Items stocked at the depot level might be retained for years before being issued to lower-level supply activities, because procurement of replenishment quantities is based on historical usage and economic reorder quantities.

Intermediate

Supply support activities at the intermediate level normally support users within a local geographical area. Stock levels of parts at the intermediate level are determined using historical records to identify the recurring requirements of users. The normal source of replenishment parts for intermediate-level supply activities is the depot level.

While parts may be retained indefinitely at the depot level, inventories at the intermediate level are normally purged of excess material on an annual basis. Stock levels might be limited to items that are continually needed by supported users. War reserve stocks are not normally maintained at the intermediate level except in unique cases in overseas areas.

Organization/User

Spare parts stocked at the organization/user level are those that are most frequently required to support maintenance. Unlike other levels of supply, the organizational/user supply activity must, in most cases, be mobile and capable of moving to support the user in time of war. Therefore, the quantities and types of parts stocked are limited to items that have the highest usage rates.

Requisition

The last topic of general discussion under the DoD supply system is the requisition. This single document makes the whole system work. The Military Standard Requisition and Issue Procedure (MILSTRIP) contains detailed information on the preparation, control, and processing of this document. The basic philosophy of MILSTRIP, shown in Fig. 11-2, begins when the user has a need for a part.

The user requests the part from organizational supply. If the organizational supply has the part, it is issued to the user. This request is not done by formal requisitioning, but informally to conserve time and eliminate paperwork. When organizational supply issues the part or does not have it in stock to issue, a requisition is generated and sent to the intermediate level. If the intermediate level has the part, it is issued to the organizational supply that generated the requisition.

A similar requisition is generated by intermediate when it does not have the part in stock, or when its stock reaches a predetermined level and must be replenished from the depot. Depot accumulates requisitions from intermediate-level supply activities and initiates procurement actions to obtain parts from suppliers.

This process is repeated each time a part is used by maintenance personnel. The requisition is recorded at each level of supply as a single demand for parts on the supply system. This information is used to adjust inventory levels according to the needs

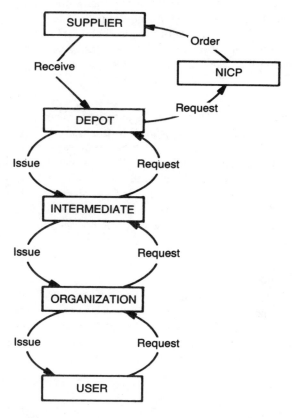

Fig. 11-2. MILSTRIP philosophy.

of the ultimate users. This is discussed in subsequent paragraphs concerning inventory control.

SPARES REQUIREMENTS

Determining the *range* (number of different items) and *depth* (quantity of each item) of spares to be procured and stocked in support of maintenance is a major factor in achieving acceptable equipment availability. There is no magic formula that can be used to identify requirements for spares because there is no method of spare-parts forecasting that can accurately predict the future. The only methods available use past experience to project the anticipated number of spares that will be required for a given period of time in the future.

If it is assumed that all factors that contribute to spares requirements, such as equipment usage, maintenance capabilities, and other resources, remain constant, then these methods have a degree of merit. The problem is that these factors do not remain constant. For example, there is always a fluctuation in equipment usage, environmental conditions, and equipment age. Therefore, no method is without error. The goal is to be as accurate as possible and pad the quantities just enough to allow for some margin of error.

Chapter 6 touched on the Poisson distribution method, which is sometimes used to calculate the number of spares that should be procured to support initial operations. The basis for this method is the equipment or assembly failure rate, which is usually only a prediction based on past experience or best guess.

Other methods can also be used. They all have merit and shortcomings. Therefore, the best way to determine the number of spares required is to use the most recent actual usage information and hope that future requirements do not greatly exceed the resulting estimate. To fully understand the process of obtaining and managing spare parts, there are several topics that must be understood. These include repairable and nonrepairable items, initial spares, demand support spares, investment spares, spare parts pipeline, and spares procurement process.

Repairable Items

Any item that can be fixed when it fails is classified as a repairable item. Normally all items referred to as spare parts are repairable. This becomes significant when predicting the number of spares required to support an item of equipment over a lengthy period of time, because a one-for-one replacement of failed items is not required. As will be shown, the time required to repair a repairable item is used as an input to the spares procurement process.

Nonrepairable Items

Items that cannot be fixed when they fail are termed nonrepairable. These parts are discarded when they fail, so a one-for-one replacement is required each time a failure occurs. There is one catch to this definition: some repairable items might be classified as nonrepairable due to economic considerations. In these cases, an economic trade-off analysis is used to determine whether it is more cost effective to repair a repairable item than discard it when it fails.

Initial Spares

One of the major results of the provisioning process is to identify and procure initial spare parts. The purpose of this initial quantity of spares is to sustain equipment operations until there is sufficient experience to accurately project subsequent spares procurement. Most provisioning is designed to support initial operations for somewhere between one and two years. Hopefully, initial stocks will last two years, because the time required to order and receive major spare parts is normally 18 to 24 months. If initial stock levels are insufficient, the equipment will probably not be adequately supported with spare parts.

Demand Support Spares

Spares that are required frequently in sufficient quantities to generate reliable predictions of future requirements are called "demand supported." That is, there is currently sufficient user demand on the supply system to accurately predict the future demands on the supply system. The demand support concept is a key factor in developing and maintaining inventories of spares. This category of spares includes both repairable and nonrepairable items.

Investment Spares

The government buys some items as spares for which there will probably never be sufficient demand to warrant stocking as normal spare parts. Investment spares are items that are normally extremely expensive, and are procured at the same time as the equipment they are to support. Because they are built concurrently with the prime equipment, they are less expensive. The concept behind this process is to invest money up front to avoid a major expense.

Spare Parts Pipeline

The spare parts pipeline (Fig. 11-3) is the flow of spares from the manufacturer to the ultimate user. It also includes the repair of failed repairable items and their return to inventory for future use.

Spares Procurement Process

The process of obtaining spares after initial provisioning has been completed is called the "normal spares procurement process," although there is nothing really normal about it. Each spare must be managed individually to ensure that the stock of spares will support future user demands. This concept is discussed further in inventory management.

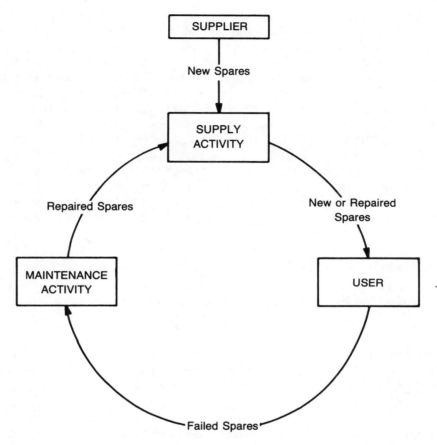

Fig. 11-3. Spares pipeline.

Spares Identification

The detailed identification of the range and depth of spares required to support an item of equipment begins during the FSD phase of the acquisition process. There are two methods that can be used. The first method, *provisioning analysis,* has been used for many years to start spares identification. Using this method, provisioning personnel list every part used in an item of equipment. The failure rate is then applied, along with "good engineering judgment and past experience," to create a recommended list of spare parts and to assign SMR codes. The major deficiency with this method is that it does not take into account the maintenance planning process.

The second method, *maintenance task analysis,* starts with each maintenance task required to fix the equipment when it fails and then identifies each part that will be required to support maintenance. This method identifies only the items that should be required for maintenance, which cuts out unnecessary expense, and justifies the depth of spares by predicting the maintenance frequency that will generate a demand for a spare. An additional benefit of this method is the compatibility of SMR codes to the maintenance planning.

Once initial provisioning is completed and the contractor's provisioning lists have been accepted by the government and put into the DLSC files, it is extremely hard to change SMR codes. If the SMR code does not match the maintenance planning for an item, then the user might not be able to obtain the spare parts required for maintenance. For example, if, during initial provisioning, an item is assigned an SMR code of PAODD (meaning that the item is removed and replaced at the organizational level and repaired at depot), and all of its lower indentured parts were assigned PADDD (meaning that all parts are removed and replaced at depot), the supply system will not honor any requisitions for parts submitted by the intermediate level, even though the maintenance planning was for the item to be repaired at the intermediate level. If at all possible, the second method should be used to identify requirements for spares in order to avoid this problem.

INVENTORY MANAGEMENT

The purpose of inventory management is to maintain the correct number of spares in the appropriate locations to support maintenance. The procedures used to manage spares inventories are the same at each level of supply.

Definitions

Before beginning this discussion, an understanding of some basic inventory terminology is necessary. The following terms are frequently used in inventory management.

- **Inventory Cycle.** The series of events taken to maintain an adequate quantity of spares in inventory is the inventory cycle.
- **Stockage Objective.** The total quantity of a spare that the supply activity should have on hand, on order, or both is the stockage objective.
- **Operating Level.** The operating level is the quantity of a spare that is stocked to satisfy anticipated customer demands.
- **Safety Level.** The safety level is the quantity of a spare that is stocked in reserve to satisfy unanticipated customer demands. (Note: Operating level plus safety level should equal stockage objective.)
- **Procurement Lead Time (PLT).** PLT is the total elapsed time between placement of an order with a vendor and receipt of the new spares.
- **Order Ship Time (OST).** OST is the total elapsed time between a supply activity requisitioning replenishment stock from its next higher supporting supply activity and the receipt of the spares.
- **Reorder Point (ROP).** ROP is the calculated inventory quantity at which an order must be placed in order to receive replenishment spares before the last item in stock is issued.

- **Day of Supply.** The average quantity of a spare that a supply activity issues in one day is known as the day of supply.
- **Economic Order Quantity (EOQ).** EOQ is the quantity of a spare that is the most cost effective to order at one time based on the cost to order compared with the cost of carrying in inventory.
- **Shelf Life.** Shelf life is the number of calendar days that a spare can be maintained in stock without deterioration.
- **Line of Supply.** An item that a supply activity is authorized to stock is commonly referred to as a line of supply, regardless of the quantity of that item that is stocked.
- **Zero Balance.** When a supply activity is authorized to stock a line of supply but has none on hand, the line is said to be at zero balance.

Inventory Cycle

The concept of an inventory cycle is relatively simple. A supply activity maintains a quantity of a spare on hand. As customers submit requisitions and are issued spares, the quantity on hand decreases. The supply activity requests quantities from its source to replenish its inventory. The objective is to maintain an adequate quantity on hand to satisfy customer demands.

In theory it sounds simple, but in reality it never quite works out that way, because of the unpredictable fluctuation in customer demands. That is why the inventory cycle is a series of events designed to react to changes in customer demands so that the necessary spares are on hand when needed. Figure 11-4 illustrates the theoretical inventory cycle.

The key to successful inventory management is to accumulate and maintain a sufficient historical data base to compute the days of supply of a spare that should be stocked to satisfy customer demands. The first point to consider is how to determine what items a supply activity should stock. This is determined by customer demands. A line of supply is not stocked until sufficient customer demands are received to warrant stockage. For example, at the intermediate level, six demands for an item within a 90-day

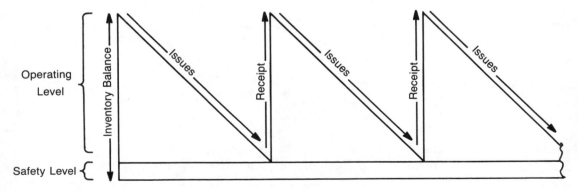

Fig. 11-4. Inventory cycle.

Day of Supply = Number of Demands/Number of Days

Example: 60 Demands in 120 Days
Day of Supply = 60/120 = 0.5

Fig. 11-5. Day-of-supply calculation.

period may constitute a sufficient number of demands to qualify for stockage. Each service has an established method for making this determination. After an item has qualified, a minimum number of demands must be received during each succeeding period to retain the item as an authorized line of supply. This practice limits the number of lines that a supply activity maintains to those really needed to support its customers.

The next question is what quantity should be stocked for each line of supply. Most supply activities maintain stocks to support customers for a predetermined number of days of supply. For example, the authorized days of supply might be 30 days at the organizational level, 90 days at the intermediate level, and 360 days at the depot level. This number of days of supply is the authorized operating level. The safety level would be added as an additional number of days, such as 10 days at the intermediate level and 30 days at the depot level. The organizational level does not normally have a safety level as such, but it might have items that are stocked as "insurance items" that are managed on an exception basis for a critical mission. Figure 11-5 illustrates how one day of supply for a spare is computed. This quantity can then be used to compute the operating level, safety level, and stockage level for the line of supply based on the number of days of supply that an activity is authorized to stock, as shown in Fig. 11-6.

An additional item of information required to complete the inventory cycle is the procurement lead time or order-ship time, depending on how a supply activity receives

Operating Level = Authorized Number of Days × Day of Supply to Stock
Safety Level = Authorized Number of Days × Day of Supply Safety Stock
Stockage Level = Operating Level + Safety Level

Example: Authorized Number of Days to Stock = 60
Authorized Safety Stock = 10% of Operating Level
Day of Supply = 0.5

Operating Level = 60 × 0.5 = 30
Safety Level = 60 × 10% × 0.5 = 3
Stockage Level = 30 + 3 = 33

Result: The Supply Activity Is Authorized to Maintain A Maximum Quantity of 33 Each of This Line of Supply in Inventory.

Fig. 11-6. Computing supply levels.

replenishment spares. Because of the different geographical locations of supply activities and the different modes of transportation used to ship spares, this figure can be different for each activity and each line of supply. Each supply activity uses historical data on supply system performance to determine the applicable PLT or OST. The days of supply and the PLT or OST are used to calculate the reorder point, as shown in Fig. 11-7.

An inventory cycle profile can be prepared for a line of supply, as illustrated in Fig. 11-8, using the authorized days of supply to be stocked, days of supply for a line of supply, and the PLT or OST for that line of supply. This figure shows the theory of demand support inventory control. Actually, the stockage level for a line of supply changes with the fluctuation of customer demands and changes in the PLT or OST. Figure 11-9 shows a typical inventory cycle profile.

Economic Order Quantity

The theory of demand support inventory control does not address the costs that can be incurred when maintaining an inventory and ordering replenishment spares. In many cases, it might be more cost effective to order a larger quantity of items and reduce the number of orders that are placed, or vice versa, in order to reduce the overall cost of the supply system. This is done through determination of the economic order quantity (EOQ). Figure 11-10 shows the concept of EOQ: as the quantity per order increases, the cost per order decreases; and the cost of inventory rises as the quantity on hand increases. The EOQ is the optimum point where the lowest total cost occurs. The formula for computing the EOQ of an item is shown in Fig. 11-11.

The purpose of EOQ is to provide sufficient spares to support anticipated customer demands while reducing the overall cost of maintaining the inventory. EOQ tends to be more applicable in instances where the line of supply has a low dollar-value and high demand, rather than a high dollar-value and low demand.

The primary goal of inventory management is to have the correct types and quantities of supplies at the right place at the right time. This is accomplished through continual review and calculation of the days of supply, order quantities, and PLT or OST. The use of EOQ in applicable cases enables the inventory manager to reduce the overall total cost of doing business while continuing to provide the required support to customers.

Reorder Point   =   (Order Ship Time $\times$ Day of Supply) + Safety Level

---

Example: Order Ship Time = 20 days
Day of Supply = 0.5
Safety Level = 2

Reorder Point   =   (20 $\times$ 0.5) + 2 = 12

Fig. 11-7. Reorder point calculation.

Data: Authorized Operating Level = 60 Days
Authorized Safety Level = 10% of Operating Level
Number of Demands Last 120 Days = 300
Order Ship Time = 20 days

Computations:

$$\text{Day of Supply} = \frac{300}{120} = 2.5$$

Operating Level = 60 × 2.5 = 150
Safety Level = 60 × 10% × 2.5 = 15
Reorder Point = (20 × 2.5) + 15 = 65
Stockage Level = 150 + 15 = 165

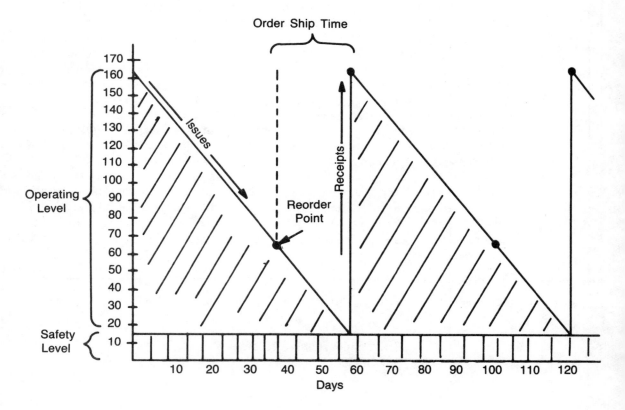

Fig. 11-8. Theoretical inventory cycle profile.

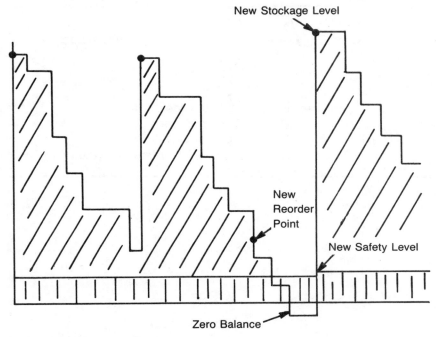

Fig. 11-9. Typical inventory cycle profile.

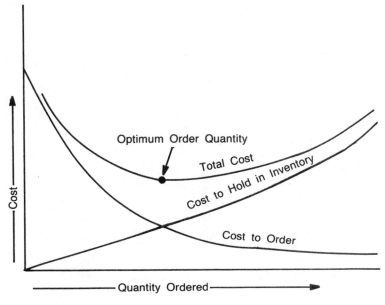

Fig. 11-10. EOQ concept.

167

$$EOQ = \sqrt{\frac{2C_o D}{C_l}} \qquad\qquad N = \frac{D}{EOQ}$$

Where: C_o = Cost to order
D = Demands per year
C_l = Cost to hold in inventory
N = Number of orders per year

Fig. 11-11. EOQ formula.

Chapter 12

Packaging, Handling, Storage, and Transportability

It doesn't matter how good an item of equipment is or how well it performs if the user doesn't receive it in a usable condition. The purpose of packaging, handling, storage, and transportability (PHS&T) is to plan, develop, and manage the activities necessary to ensure that equipment is serviceable when it reaches the ultimate user. PHS&T planning starts in the concept phase and continues throughout the acquisition cycle. This effort is many times overlooked because it does not relate directly to the development of the equipment; however, it plays a key role in achieving supportability goals. The serviceability and timely availability of equipment and spare parts reduces overall life cycle costs.

THE PHS&T PROGRAM

MIL-STD 1367, Packaging, Handling, Storage, and Transportability Program Requirements for Systems and Equipments, establishes standard management requirements for the PHS&T program. It is important that the program interface with other ILS disciplines in determining PHS&T requirements. Unlike previously discussed programs for ILS disciplines, the PHS&T program does not have a set of standard tasks, i.e., reliability, maintainability, etc., that a contractor follows to ensure proper implementation and management of the program. Therefore, it is important that the contractor develop a well-organized program that addresses each aspect of PHS&T adequately. MIL-P-9024G, Packaging, Handling, and Transportability in System /Equipment Acquisition, provides additional requirements and criteria for PHS&T development during the acquisition of equipment.

Definitions

The following definitions from MIL-STD 1367 will help in understanding how the individual areas of PHS&T are interrelated.

- **Packaging.** Packaging includes all the operations and devices required to prepare items for distribution, such as preservation-packaging, packing, marking for shipment, unitizing, and palletizing. It does not, however, include loading of the mode of transportation, e.g., truck, train, aircraft, or ship.
- **Handling.** Handling involves moving items from one place to another within a limited range. It is normally limited to a single area such as between warehouses or storage areas, or movement from storage to the mode of transportation.
- **Storage.** The short- or long-term storing of items. It can be accomplished in either temporary or permanent facilities.
- **Shipment.** Shipment is the transfer of an item for an appreciable distance (several miles or more) using commonly available equipment such as rail cars, trucks, ships, or aircraft.
- **Transportability.** Transportability is the inherent capability of an item to be moved by towing, self-propulsion, or common carrier via highway, railway, waterway, airway, or sea.
- **Modes of Transportation.** The various ways that items are physically shipped, i.e., truck, aircraft, rail, or ship, are called modes of transportation.

Program Requirements

The PHS&T program consists of all activities required to plan, develop, and manage the successful delivery of items to the user. This can only be achieved through a coordinated effort. PHS&T must be involved in all aspects of the acquisition cycle. PHS&T planning starts during the concept phase and continues through production and deployment. It is imperative that a close interface be established and maintained with other ILS disciplines, especially reliability, maintainability, safety, and human engineering. PHS&T must also work closely with configuration management and quality.

Specific areas that the PHS&T program must address include program control, distribution and delivery concepts, special packaging, handling and storage requirements, and development of specific PHS&T design requirements. The preparation of a PHS&T program plan that addresses each of these areas may be required. The plan should be similar in format and basic philosophy to the reliability program plan illustrated in Fig. 3-1.

Program Control

The PHS&T program should provide adequate ways of monitoring and controlling the progress of the program. This includes PHS&T participation in program reviews, involvement of subcontractors, methods for reporting and measuring progress toward achieving program goals, and development of appropriate schedules. Tasks 101,

102, and 103 of MIL-STD 785B can be used as guidelines for developing the PHS&T program.

Distribution and Delivery Concepts

The product specification and/or statement of work normally identifies proposed concepts for distribution and delivery of equipment being developed. Special attention should be given to identification of constraints or limiting factors relative to PHS&T. The concepts are normally determined based on the anticipated equipment use, locations, and maintenance concept.

Figure 12-1 illustrates a distribution and delivery concept for a Navy guided missile. This illustration shows that there are normally several approaches to the concept that must be considered. In this example, the missile could be routed four different ways from the ready issue storage site to the combatant ship. These concepts must be developed early in the program, preferably in the concept phase, in order to allow PHS&T to influence design to achieve maximum transportability of the equipment.

Special Packaging, Handling, and Storage

Requirements for special packaging, handling, and storage should be avoided if at all possible. Such requirements should be identified as early as possible in order to conduct trade-off studies to determine if the need can be fulfilled using standard procedures. Special requirements increase life cycle cost and should be avoided unless absolutely necessary.

Design Criteria

The development of equipment design criteria relative to PHS&T requirements begins in the concept phase. This effort is based on the distribution and delivery concept and requirements for special packaging, handling, and storage. PHS&T engineers evaluate proposed alternative design concepts or recommend changes to designs that will enhance the transportability of the equipment. Trade-off studies might be required to determine the most cost-effective alternative. One of the purposes of such trade-offs is to verify or validate the need for special packaging, handling, or storage, and eliminate them whenever possible.

PHS&T Planning

As stated earlier, the PHS&T program must start as early as possible to provide input to the design process that enhances the transportability of the equipment. The first step in the planning process is to identify the ultimate destination of the equipment. The destination includes the environments where the item will be used, the types of storage and handling capabilities that will be available, the anticipated methods of getting the equipment from the contractor to the user, and the anticipated requirements for movement after the user takes possession. All of these factors are then used to develop

171

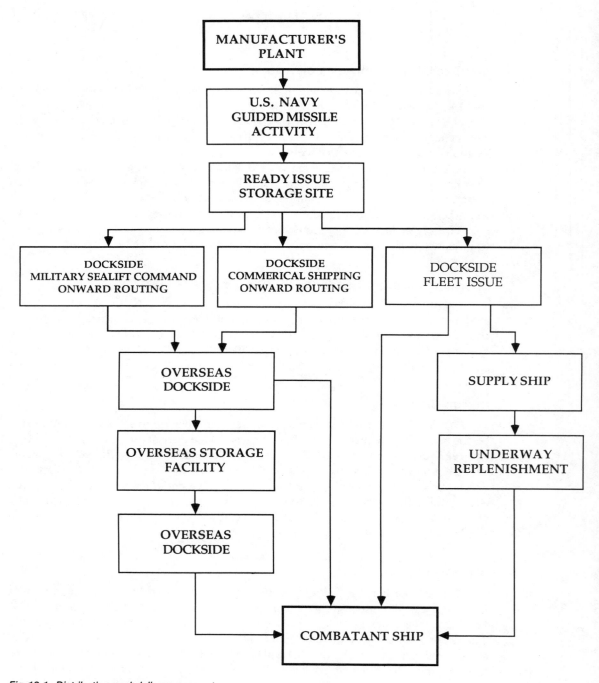

Fig 12-1. Distribution and delivery concept.

a PHS&T profile for the equipment. Figure 12-2 illustrates the types of environmental stress, both natural and induced, that must be considered when planning PHS&T. This is like determining if the hole is round or square before building the peg.

The contractor can use the PHS&T profile to develop design criteria for the equipment, such as limits on dimensions, provisions for handling, modularity of components, and safety requirements. For instance, if the equipment will be used by a unit whose mission requires that all its equipment must be transportable by military aircraft, then the new equipment must not exceed allowable weight and dimension limits for airlift. Or, if the equipment contains sensitive electronics, then it must be adequately protected against damage from handling and shipment.

Criteria are also developed to make maximum use of existing PHS&T technology and standard packaging and handling equipment. When the design criteria have been established, the PHS&T engineers can concentrate on evaluating the design to ensure that it meets the criteria. It is important for the contractor to keep the government apprised of the capabilities of the design to meet the criteria and identify PHS&T problems as they arise.

Many times the product specification contains requirements that negatively affect PHS&T. These problems should be resolved early in the program. The purpose of the planning process is to address and resolve areas of concern before they become problems. PHS&T is important, not only during contractor activities, but also throughout the life cycle of the equipment. Failure to solve PHS&T problems will result in increased life cycle costs during the deployment and use of the equipment, and might impede the ability of the equipment to perform assigned missions.

PHS&T Trade-off Analyses

It is not always possible to design equipment that is easily transportable using standard methods. Some equipment, because of inherent characteristics required to perform its mission, requires special PHS&T. For example, it is impossible to design a main battle tank that can be transported by a standard cargo truck. Likewise, all nuclear weapons require special PHS&T, regardless of the design. In cases such as these, the goal of PHS&T engineers is to reduce the magnitude of the transportability problem through identification of design changes and development of distribution and delivery strategies that minimize the impact of special requirements.

Equipment designs that do not exceed standard size and weight limits and do not inherently require special PHS&T allow the development of alternate PHS&T concepts. These concepts can then be quantified and the results used to conduct trade-off studies to determine which concept provides the best alternative based on use of resources, operational availability, and cost.

Figure 12-3 is a simplified example of two PHS&T concepts for a small item of equipment. Concept A consists of packaging the item in a paper bag and mailing it to the user. Concept B requires making a special container and shipping by air express. Both concepts have advantages and disadvantages, as shown in Fig. 12-4. The final decision on which concept to use would depend on which one best meets the

	Handling & Road Transportation	Handling & Rail Transportation	Handling & Air Transport	Handling & Ship Transport	Handling & Logistics Transport (Worst Route)	Storage, Sheltered (Tent, Shed, Igloo)	Storage, Open
Environmental Stress Generation Mechanisms (Induced)	Road Shock (Large Bumps/Potholes) Road Vibration (Random) Handling Shock (Dropping/ Overturning)	Rail Shock (Humping) Rail Vibration Handling Shock (Dropping/Overturning)	In-Flight Vibration (Engine/ Turbine Induced) Landing Shock Handling Shock (Dropping/Overturning)	Wave-Induced Vibration (Sinusodal) Wave Sine Shock Mine/Blast Shock Handling Shock (Dropping/Overturning)	Road Shock (Large Bumps/Holes) Road Vibration (Random) Handling Shock (Dropping/Overturning) Thermal Shock (Air Drop)	None	None
Environmental Stress Generation Mechanisms (Natural)	High Temperature (Dry-Humid) Low Temperature Rain/Hail Sand/Dust	High Temperature (Dry-Humid) Low Temperature Rain/Hail Sand/Dust	Reduced Pressure Thermal Shock (Air Drop Only)	High Temperature (Humid) Low Temperature Rain Temporary Immersion Salt Fog	High Temperature (Dry/Humid) Low Temperature/Freezing Rain/Hail Sand/Dust Salt Fog Solar Radiation Reduced Pressure	High Temperature (Dry/Humid) Low Temperature/ Freezing Salt Fog Fungus Growth Chemical Attack	High Temperature (Dry/Humid) Low Temperature/ Freezing Rain/Hail Sand/Dust Salt Fog Solar Radiation Fungus Growth Chemical Attack

Fig 12-2. PHS&T environmental stress.

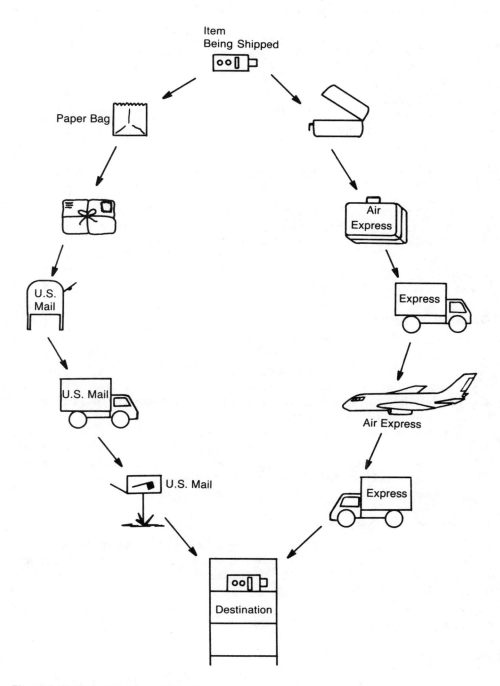

Fig. 12-3. PHS&T concept comparison.

Evaluation Element	Method Providing Best Alternative
Material Cost	A
Shipping Cost	A
Speed	B
Protection	B
Dependability	B
Overall	?

Fig. 12-4. PHS&T trade-off evaluation.

government's requirements. If cost were the only determining factor, then concept A would be selected; however, if speed and protection were the major concerns, then concept B would be the more desirable. Combinations of portions of each concept could also be used to develop other alternatives, such as packaging the item in a special container and shipping by mail. This process of trade-off analysis is repeated until the optimum balance of all PHS&T aspects is achieved.

TRANSPORTABILITY

Transportability, or the inherent capability of an item to be moved, is the foundation for cost-effective PHS&T. It should be a primary consideration in the planning and design of new equipment. The capability of equipment to be moved efficiently by all required modes of transportation will achieve a higher operational availability and lower life cycle cost. Equipment designs that require special or unique methods for PHS&T should be avoided if at all possible.

Incorporating handling, tie-down, and sling points into designs for equipment is necessary to optimize transportability and complement storage, maintenance, and other handling requirements. Equipment must be designed to be handled and transported safely. Sectionalization and disassembly capability for transport, with ease of reassembly for operational use or maintenance, should be a design consideration.

Transportability Characteristics

During the design process, PHS&T evaluates the equipment design to identify and propose solutions for transportability problems. MIL-STD 1319A, Item Characteristics Affecting Transportability and Packaging and Handling Equipment Design, identifies design characteristics that affect transportability. Typical examples of these design characteristics are illustrated in Fig. 12-5. These include designs that result in equipment being oversized, overweight, fragile, dangerous, or hazardous.

Physical Limits

MIL-STD 1366, Definition of Transportation and Delivery Mode Dimensional Constraints, contains data on the standard dimension limits for military equipment.

PHYSICAL PROPERTIES

Width	Net Weight
Height	Gross Weight
Length	Center-of-Gravity

DYNAMIC LIMITATIONS

Acceleration—Allowable acceleration, pulse time, and pulse shape along each of the mutually perpendicular axes.

Vibration—Critical resonant frequencies in plane of shipping attitude.

Deflection—Maximum allowable bending in plane of shipping attitude.

Skin Loading—Maximum allowable skin pressure loading diagrams.

Securing—Maximum allowable dynamic load on tie-down, mounting, and handling fittings along each of the mutually perpendicular axes.

Leakage—Maximum allowable solid, liquid, or gaseous emission rates.

ENVIRONMENTAL LIMITATIONS

Temperature—Critical temperature range requiring controlled conditions.

Pressure—Critical pressure range requiring controlled conditions.

Humidity—Critical humidity range requiring controlled conditions.

Cleanliness and sterilization.

HAZARDOUS EFFECTS

Personnel Safety—Toxicity of fumes or liquids on contact with humans.

Radiation—Control of electromagnetic or radioactive radiation.

Electrostatic—Grounding requirements.

Explosives—Sensitivity to deterioration from impact or penetration.

Etiologic or Biologic—Personal or public health debilitation or lethality potential.

Fig. 12-5. Item characteristics affecting transportability.

Equipment designs that exceed standard limits of eight feet in height, eight feet in width, 32 feet in length, and 11,200 pounds when prepared for transport should be avoided because they will require special handling and transportation. If equipment is to be transported inside cargo trucks or standard cargo shipping containers, the standard limits are reduced to seven feet in height, 6.5 feet in width, 18.5 feet in length, and 10,000 pounds. These dimensions are used as guidelines when planning for transportation of equipment.

Dynamic Limitation

Sensitive or fragile equipment might require special handling or unique packaging due to the type of material used or the design. The dynamic limitations of the design must be identified and compensated for by proper PHS&T planning. Susceptibility to damage from acceleration or vibration during handling or shipment must be considered when selecting packaging and packing material. The addition of fixtures for securing the equipment during transit reduces the possibility of damage from induced stress.

Environmental Limitations

Although military equipment is normally considered rugged and able to withstand extreme environmental conditions, that is not always the case. Many items of equipment can be damaged or become less reliable under extreme temperature, atmospheric pressure, or humidity. Unavoidable design features that cause these limitations require special attention. Special containers that provide environmental control might be required to protect these items.

Hazardous Effects

Equipment that is hazardous or dangerous to transport creates special PHS&T problems, and should be at the top of the list for resolution. The protection of personnel involved with the physical PHS&T aspects of equipment is always a priority issue. PHS&T engineers must coordinate efforts with system safety, maintainability, and human engineering professionals to effectively eliminate hazards or devise methods to reduce them to acceptable levels.

Remember, system safety engineers not only maintain a hazard log that identifies all known equipment hazards, they are ultimately responsible for eliminating hazards. Maintainability and human engineering are also involved in identifying and solving hazards and hazardous actions. This is one area where every ILS discipline is involved in formulating answers to the problems caused by hazards.

PACKAGING

The area of packaging entails all the tasks required to prepare an item for shipment. These tasks include the planning, management, and accomplishment of preservation, packing, and marking. Additionally, packaging engineers determine the level of packaging required, identify the need for and design special containers, and document the packaging requirements for each item.

Levels of Packaging

The degree of protection provided to an item by packaging is dependent on the anticipated destination, mode of transportation to be used for shipment, and type of storage to be used at the destination. Packaging types are divided into three levels: A, B, and C. Figure 12-6 illustrates how the required level of packaging is selected

	LIMITED TENURE OF STORAGE		FAVORABLE STORAGE		SHED OR OPEN STORAGE		UNKNOWN STORAGE		OVERSEA AIR SHIPMENT		OVERSEA WATER SHIPMENT		CONSOL'D or CONTAINERIZATION SHIPMENT		OVERSEA PARCEL POST	
	Pres-Pkg	Packing	Pres-Pkg	Packing	Pres-Pkg	Packing	Pres-Pkg	Packing	Pres-Pkg	Packing	Pres-Pkg	Packing	Pres-Pkg	Packing	Pres-Pkg	Packing
Shipments from contractors or other supply sources in CONUS to:																
a. CONUS requisitioners for immediate use w/no redistribution	C	C														
b. CONUS depots for storage and redistribution to:			See note 3													
(1) CONUS requisitioners			B/C	B/C	A	A	A	A					A	C		
(2) Oversea requisitioners			A/B	B	A	A	A	A								
c. CONUS contractors for assembly (GFP)	C	C	B/C	B/C	B	B	A	A								
d. Oversea requisitioners																
(1) In support of combat operations																
(a) For immediate use									A/B	A/B/C	A/B	A/B	A/B	A/B/C	B/C	B/C
(b) Storage and redistribution anticipated			A/B	A/B	A	A	A	A	A/B	A/B	A	A	A/B	A/B	A/B	A/B
(2) Other than combat operations																
(a) For immediate use	B/C	B/C							B/C	B (See note 3)	A/B	A/B	A/B	B/C	B/C	C
(b) Storage and redistribution anticipated			B	B	A	A			A/B	B	A/B	A/B	A/B	B	A/B	B

NOTES:

1. When a choice of levels is indicated, selection will be based on the following criteria:

a. Level A
  (1) Ultimate destination is unknown; or
  (2) Duration or condition of storage is unknown or cannot be determined; or
  (3) Unfavorable transportation or handling conditions are known or anticipated; or
  (4) Open or shed type storage is known or anticipated; or
  (5) The item/shipment is known or anticipated to require the maximum degree of protection.

b. Level B
  (1) Ocean shipments are intended for immediate use and favorable transportation, storage, and handling conditions are known to exist; or
  (2) Short term favorable storage and movement which do not involve ocean transportation; or
  (3) Storage tenure will not exceed the level B package shelf life when established by specifications; or
  (4) Level B preservation-packaging shall not be used without approval of procuring activity.

c. Level C
  (1) Movement, storage, and handling conditions are known to permit this level; or
  (2) Shipments moving via other than ocean transportation are for immediate use at the first receiving activity.

2. In selecting levels based upon transportation media, consideration will be given to the additional modes of transportation and type of handling to which the item/supplies may be subjected.

3. Shipments to humidity controlled storage facilities with no redistribution required - preserve, package, and pack level C.

4. Weapon System pouch shipments - domestic and overseas, when authorized by contract - preserve, package, and pack level C.

5. Requirements of this table are also applicable to oversea contractors.

6. Repackaging/repacking of items which have been preserved, packaged, or packed to a level which exceeds the requirement specified above is required when delay of shipment is permissible, and savings to be realized in packaging and transportation costs will exceed the total costs of such repackaging/repacking. All changes in preservation, packaging, and packing required in response to (amended or initial shipping instructions) shall be reported to the responsible administrative contracting officer for analysis. Any changes in cost of performance resulting from such instructions shall be processed in accordance with the clause of the applicable contract entitled "changes."

Fig. 12-6. Selection of levels of protection.

based on the factors stated in Fig. 12-5. MIL-STD 2073-1A, Procedures for Development and Application of Packaging Requirements, defines the levels of packaging as follows:

- **Level A.** Maximum protection, designated as Level A, is the level of preservation or packing required for protection of material against the most severe worldwide shipment, handling, and storage conditions. Preservation and packing so designated will be designed to protect material against direct exposure to extremes of climate, terrain, and operational and transportation environments without protection other than that provided by the pack.
- **Level B.** Intermediate protection, designated as Level B, is the level of preservation or packing required for protection of material under anticipated favorable conditions during worldwide shipment, handling, and storage. Preservation and packing so designated will be designed to protect material against physical damage and deterioration during favorable conditions of shipment, handling, and storage.
- **Level C.** Minimum protection, designated as Level C, is used for protection of material under known favorable conditions.

Level A packaging provides the most protection for an item, and is the most expensive due to the extra material and labor required. It is capable of withstanding outdoor storage in all climatic conditions for a minimum of one year with no deterioration of the item contained inside. Level B is less protective, but should be able to withstand storage in a favorable warehouse environment for a minimum of 18 months.

Preservation

The purpose of preservation is to protect the item being prepared for shipping from deterioration due to corrosion, physical damage, or other types of deterioration. MIL-P-116, Methods of Preservation, describes methods and processes for preservation of equipment. Figure 12-7 shows the different methods that can be selected for preservation. Each method consists of cleaning, drying, and wrapping or packing. Cleaning is accomplished by methods such as wiping, immersion, or scrubbing with cleaning solvent; degreasing with vapors; electro cleaning; steam cleaning; abrasive cleaning; or ultrasonic processes.

After cleaning, items must be dried by wiping, draining, compressed air, exposure to infrared lamps, or oven drying. Preservatives are applied by dipping, flow coating, slushing, brushing, filling, flushing, fogging, or spraying. Figure 12-8 shows the types of preservative materials that can be used to protect items. The preservation method should be consistent with the level of protection necessary to protect the item. Guidance for selecting of the appropriate MIL-P-116H preservation method to be used when packaging a specific item is contained in MIL-STD 794, Procedures for Packaging of Parts and Equipment.

Packing

The development of packing requirements for equipment is described in MIL-STD 2073-1A. There are three basic classifications of packing: common, selective, and

I	IA	IB	IC	II	III
Preservative coating (with greaseproof wrap as required)	Watervaporproof enclosure (with preservatives as required)	Strippable compound coating (hot or cold dip)	Waterproof or waterproof greaseproof enclosure (with preservative as required)	Watervaporproof enclosure with desiccant (with preservative as required)	Physical and mechanical protection only
	IA-5 Rigid metal container, sealed	IB-1 Direct application of strippable compound	IC-1 Greaseproof, waterproof bag, sealed	IIa Floating bag, sealed	
	IA-6 Rigid container (items immersed in preservative oil type) sealed	IB-2 Aluminum foil wrap, strippable compound	IC-2 Container, bag, sealed	IIb Container, bag, sealed, container	
	IA-8 Watervaporproof bag sealed		IC-3 Waterproof bag, sealed	IIc Watervaporproof bag, sealed	
	IA-13 Rigid container other than all metal sealed		IC-4 Rigid container other than all metal, sealed	IId Rigid metal container, sealed	
	IA-14 Container, bag, sealed, container		IC-7 Blister pack, single or multiple compartment, individually sealed	IIe Container, bag, sealed	
	IA-15 Container, bag, sealed		IC-9 Skin pack, greaseproof, waterproof, vacuum formed	IIf Rigid container other than all metal, sealed	
	IA-16 Floating bag sealed		IC-10 Skin package, waterproof, vacuum formed		

Fig. 12-7. Methods of preservation.

Type	Description	Specification No.
P-1	Thin film preservative (hard drying, cold application)	MIL-C-16173, Gr. 1
P-2	Thin film preservative (soft film, cold application)	MIL-C-16173, Gr. 2
P-3	Thin film preservative, water displacing (soft film, cold application)	MIL-C-16173, Gr. 3
P-6	Light preservative compound (soft film, hot application)	MIL-C-11796, Class 3
P-7	Medium preservative oil (cold application)	MIL-L-3150
P-9	Very light preservative oil, water displacing (cold application)	VV-L-800
P-10	Engine preservative oil	MIL-L-21260; Type I, Gr. 10, 30 or 50 or Type II, Gr. 10 or 30
P-11	Preservative grease (application as required)	MIL-G-10924
P-14	Corrosion preventive (food handling machinery and equipment), nontoxic	MIL-C-10382
P-15	Hydraulic preservative oil	
P-17	Instrument bearing preservative oil	MIL-L-6085
P-18	Volatile corrosion inhibitor	
P-19	Thin film preservative (transparent, nontacky)	MIL-C-16173, Gr. 4
P-20	Lubricating oil, contact and volatile corrosion inhibitor treated	
P-21	Thin film preservative, water displacing (soft film, cold application low pressure, steam removable)	MIL-C-16173, Gr. 5

Fig. 12-8. Preservatives.

special. The government has developed descriptions for packaging material, contained in MIL-STD 2073-2, Packaging Requirement Codes. Any item that can be packaged using these standard codes is classified as a *common packaging item*. Items that cannot use the predetermined descriptions, but are not so complex that they require special specifications, are classified as *selective packaging items*. Packaging for selective items is normally accomplished by modifying predetermined packaging.

Special items are any items that are not common or selective. They have peculiar characteristics such as weight, complexity, fragility, or other considerations that require the design of special packaging. MIL-STD 2073-1A contains detailed instructions and procedures for identifying and developing packaging requirements for any item.

Reusable Containers

Items identified as candidates for special packaging might require durable containers that can be reused many times. These reusable containers require documentation just like the equipment they carry. The government has data on containers that can be obtained through procedures contained in MIL-STD 1510, Procedures for Use of Container Design Retrieval System. Special reusable containers might require maintenance planning, provisioning, and other documentation, including assignment of an NSN. This must be considered when developing requirements for these containers.

Packaging Data

The government has developed a standard coding method to identify packaging requirements for common items. This coding consists of 13 alphanumeric characters that tell the methods of preservation; cleaning and drying; preservative, wrapping, cushioning, and dunnage material; type of container; and level of protection required for packaging an item. Figure 12-9 illustrates this coding system.

The individual codes used are contained in MIL-STD 2073-2. As contractors develop the packaging requirements for items, the information is recorded on a Preservation and Packaging Data Sheet (PPDS). An example of this data sheet is in Fig. 12-10. The instructions for completing this form are contained in MIL-STD 2073-1A, and the codes used are in MIL-STD 2073-2. When the packaging requirements cannot be adequately documented on the PPDS, a Special Packaging Instruction Data Sheet is used to provide additional information, as shown in Fig. 12-11.

Marking

One of the last steps in packaging an item is marking. MIL-STD 129, Marking for Shipment and Storage, establishes the uniform procedures used for marking military supplies and equipment. Information that must be provided on the exterior of packaged items includes identification data and warnings or cautions. Figure 12-12 shows the standard marking format. MIL-STD 129 also contains information on the selection and placement of warning and caution information.

HANDLING

The movement of items for short distances is normally accomplished either manually or by using material handling equipment (MHE). The rule of thumb for maximum weight of items to be handled manually is a gross weight of 40 pounds. Of course, the dimensions of an item when packaged might preclude manual handling, even if it weighs less than this amount. Packaging engineers must coordinate requirements for manual handling of items with Human Engineering consultants to ensure that allowable limits are not exceeded.

MHE includes any equipment designed for handling packaged or unpackaged equipment. Typical examples of MHE include forklifts, pallet jacks, roller systems, cranes, and dolleys. MIL-STD 137, Material Handling Equipment, provides a detailed discussion and identification of MHE. Every effort should be taken to use standard MHE when planning requirements for handling equipment.

The development of special MHE should be avoided if at all possible because it normally drives up the overall system life cycle cost. When the design of new MHE cannot be avoided, it should be accomplished in accordance with MIL-STD 1365, General Design Criteria for Handling Equipment Associated with Weapons and Weapon Systems. Handling of equipment shipped by military aircraft is accomplished using the 463L materials-handling support system. The 463L system uses a series of equipment sets comprised of roller systems and pallets that allow rapid preparation and loading/unloading of aircraft. It can be used on all military cargo aircraft including the C-130, C-141, and C-5 aircraft in both tactical and nontactical situations.

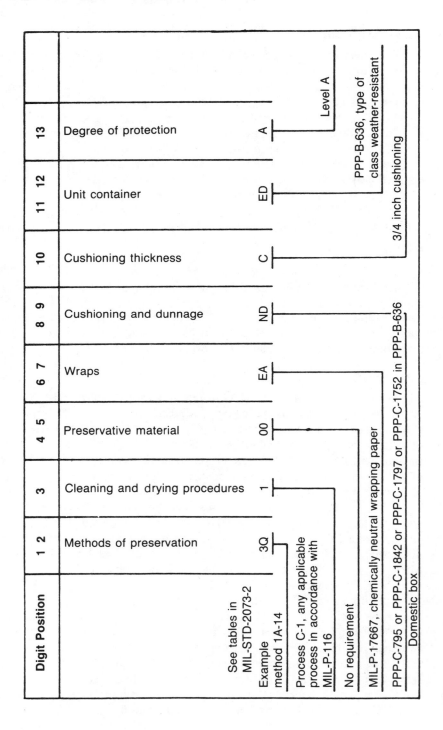

Digit Position	1 2	3	4 5	6 7	8 9	10	11 12	13
	Methods of preservation	Cleaning and drying procedures	Preservative material	Wraps	Cushioning and dunnage	Cushioning thickness	Unit container	Degree of protection
	3Q	1	00	EA	ND	C	ED	A

See tables in MIL-STD-2073-2

Example method 1A-14

Process C-1, any applicable process in accordance with MIL-P-116

No requirement

MIL-P-17667, chemically neutral wrapping paper

PPP-C-795 or PPP-C-1842 or PPP-C-1797 or PPP-C-1752 in PPP-B-636

Domestic box

3/4 inch cushioning

PPP-B-636, type of class weather-resistant

Level A

Fig. 12-9. Packaging code sequence for common items.

PRESERVATION AND PACKING DATA

NOMENCLATURE	APPROVAL STAMP	Form Approved
DESIGN ACTIVITY'S CODE (FSCM)		OMB No. 0704-0188
DESIGN ACTIVITY'S PART NUMBER		Expiration Date: June 30, 1986
CONFIGURATION ITEMS SPECIFICATION NUMBER		

ITEM IDENTIFICATION DATA (TABLE I)

DOC CON						NATIONAL STOCK NUMBER														ITEM WT.					LENGTH				WIDTH				DEPTH				PKG
						FSC			NIIN										ADDL	POUNDS				10	INCHES			10	INCHES			10	INCHES			10	P/C
1	2	3	4	5	6	7	8	9	10	11	12	13	14	15	16	17	18	19	20	21	22	23	24	25	26	27	28	29	30	31	32	33	34	35	36	37	38	39	40
A																																							

CAT WF P		SPEC MKG		QUP				FSCM					DRAWING OR PART NUMBER																									S C I
				QUP		ICQ																																
41	42	43	44	45	46	47	48	49	50	51	52	53	54	55	56	57	58	59	60	61	62	63	64	65	66	67	68	69	70	71	72	73	74	75	76	77	78	79	80

PRESERVATION — PACKING DATA (TABLE II)

DOC CON						NATIONAL STOCK NUMBER														H M	QUP						PRES METH		C D	PRES MTL		WRAP MTL	CUSH DUNN		C T	UNIT CONT
						FSC			NIIN										ADDL		QUP			ICQ													
1	2	3	4	5	6	7	8	9	10	11	12	13	14	15	16	17	18	19	20	21	22	23	24	25	26	27	28	29	30	31	32	33	34	35	36	37	38	39	40
B																																							

D P	INT CRT		U C L	SPEC MKG		PACK-ING			UNIT PACK WEIGHT					UNIT PACK SIZE												UNIT PACK CUBE							IN-THE-CLEAR REFERENCE			O P I	S C I
														LENGTH				WIDTH				DEPTH			WHOLE CUBE				1000TH								
41	42	43	44	45	46	47	48	49	50	51	52	53	54	55	56	57	58	59	60	61	62	63	64	65	66	67	68	69	70	71	72	73	74	75	76	77	78	79	80

SUPPLEMENTAL DATA (TABLE III)

DOC CON						NATIONAL STOCK NUMBER														IN-THE-CLEAR INSTRUCTIONS																		
						FSC			NIIN										ADDL																			
1	2	3	4	5	6	7	8	9	10	11	12	13	14	15	16	17	18	19	20	21	22	23	24	25	26	27	28	29	30	31	32	33	34	35	36	37	38	39	40
C																																							

| 41 | 42 | 43 | 44 | 45 | 46 | 47 | 48 | 49 | 50 | 51 | 52 | 53 | 54 | 55 | 56 | 57 | 58 | 59 | 60 | 61 | 62 | 63 | 64 | 65 | 66 | 67 | 68 | 69 | 70 | 71 | 72 | 73 | 74 | 75 | 76 | 77 | 78 | 79 | 80 |

SPECIAL PACKAGING INSTRUCTION DATA (TABLE IV)

DOC CON						NATIONAL STOCK NUMBER														CODE IDENT					SPI NUMBER								R E V
						FSC			NIIN										ADDL	FSCM														
1	2	3	4	5	6	7	8	9	10	11	12	13	14	15	16	17	18	19	20	21	22	23	24	25	26	27	28	29	30	31	32	33	34	35	36	37	38	39	40
D																																							

SPI DATE					CONT. NATIONAL STOCK NUMBER																																		
					FSC			NIIN																															
41	42	43	44	45	46	47	48	49	50	51	52	53	54	55	56	57	58	59	60	61	62	63	64	65	66	67	68	69	70	71	72	73	74	75	76	77	78	79	80

DD Form 2326, 84 FEB

Fig. 12-10. Preservation and packing data sheet.

185

SPECIAL PACKAGING INSTRUCTION
(MIL—STD—2073)

CODE IDENT	19204
SPI NO.	0841728

PART OR DRAWING NO. 1090094	NATIONAL STOCK NO. 6625 00 084 1728LF	DATE 77001	REVISION A

QUP	ICQ	UNIT PACK WT	UNIT PACK CUBE	UNIT PACK SIZE	
001	000	0266.2	0001.233	014.6 008.8 009.6	SHEET 1 OF 2

PRESERVATION	STEPS	REQD	DESCRIPTION
LEVEL A — MIL—P—116, Method I	*1		Preservative, MIL—P—116, P—11
LEVEL B — MIL—STD—2073—1A	*2	1	Wrap, MIL—B—121, Grade A, Class 3 Type 1, 6"x12"
CLEANING — MIL—P—116, C—1	*3	1	Tape PPP—T—60, Type V, Class 2, 1"x14"
DRYING — MIL—P—116	4	1	Container, PPP—B—636, RSC, WSC, 17"x10"x11" (Inside D) 18"x12"x12" (Outside D)
PACKING	5	1	Pad, PPP—F—320, Grade V11C,16 7/8" x9 7/8"
LEVEL A — MIL—STD—2073—1A	6	2	Blocking, see notes
LEVEL B — MIL—STD—2073—1A	7	2	Closure, PPP—T—76, 2"x20"

MARKING MIL—STD—129

*Preserve and wrap end of pintle with spec-
ified material, securing wrap with specified tape.

NOTES:

1. MATERIAL FIBERBOARD
      SPEC...PPP—F—320
      GRADE........W5C
2. STAPLES, TAPE, OR ADHESIVE MAY BE
USED TO RETAIN FORM. IF ADHESIVE IS
USED, IT SHALL CONFORM TO MMM—A—250
AND COVER NOT LESS THAN 50% OF THE
AREA IN CONTACT. ANY SQUEEZE CUT
SHALL BE REMOVED PRIOR TO SETTING.

3. TOLERANCE ± 1/16

DD FORM 2169
1 JAN 79

Fig. 12-11. Special packaging instruction data sheet.

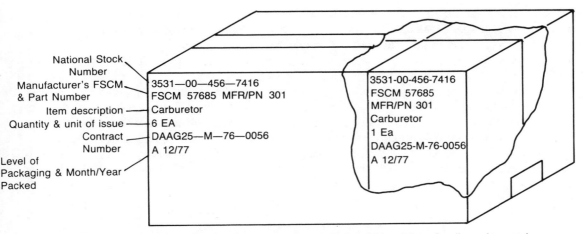

National Stock Number
Manufacturer's FSCM & Part Number
Item description
Quantity & unit of issue
Contract Number
Level of Packaging & Month/Year Packed

3531—00—456—7416
FSCM 57685 MFR/PN 301
Carburetor
6 EA
DAAG25—M—76—0056
A 12/77

3531-00-456-7416
FSCM 57685
MFR/PN 301
Carburetor
1 Ea
DAAG25-M-76-0056
A 12/77

Note: Marking for large items may include additional lines for dimensions and weight.

Fig. 12-12. Standard marking format.

STORAGE

Equipment can be stored in several different ways, i.e., warehouses, covered in open areas, uncovered in open areas, controlled environment, or special facilities. It is important that the storage be compatible with the level of packaging. Requirements for special storage, such as environmentally controlled areas, segregated storage due to the nature of the contents of the package, special security requirements, or other considerations that preclude straightforward handling should be identified as early in a program as possible, so that alternative approaches to storage can be developed. Special storage requirements can have a waterfall effect that generates additional requirements for special facilities, handling, and transportation.

MODES OF TRANSPORTATION

The movement of equipment is accomplished using standard modes of transportation for shipment by land, air, and sea. Modes of transportation include cargo vehicle, rail, ship, and aircraft. The government uses both commercial and military

Mode	Capacity	Speed	Availability
Aircraft	4th	1st	4th
Truck	3rd	2nd	1st
Ship	2nd	4th	3rd
Rail	1st	3rd	2nd

Fig. 12-13. Transportation mode analysis.

assets for transportation of equipment. MIL-STD 1366, Definition of Transportation and Delivery Mode Dimensional Constraints, provides guidelines for planning which mode of transportation is appropriate for a specific item of equipment. Additional considerations for transportation by air are contained in MIL-A-8421, General Specification for Air Transportability.

All transportation of military equipment is requested, coordinated, and documented using the procedures contained in DoD 4500.32R, Military Standard Transportation and Movement Procedures (MILSTAMP). Figure 12-13 illustrates the characteristics of each mode of transportation that should be considered when planning for transportation of equipment. The mode of transportation used depends on the priority of the item being shipped, the availability of transportation assets, and the dimensional constraints.

Chapter 13

System Safety and Human Engineering

Humans operate and maintain military equipment. Very few items procured by the government operate in an environment that does not require human contact or involvement with the equipment. Therefore, humans must be considered an integral part of the total system. System safety engineering and human engineering are responsible for evaluating the equipment design to ensure that it allows human participation to be as safe and efficient as possible.

SYSTEM SAFETY PROGRAM

No matter how good the design of an equipment, if it cannot be operated and maintained safely, it is unacceptable. System safety engineering is charged with developing and implementing a system safety program that continually evaluates the evolving equipment design to identify potential safety hazards. As they are identified, hazards are analyzed to determine ways that they can be reduced or eliminated through design or procedure changes.

Objectives

The system safety objectives of any program are to influence the design using a systematic analysis and evaluation approach that results in equipment that is as safe as possible to operate and maintain. Figure 13-1 shows the general objectives of system safety. Achievement of these objectives is realized through implementation of the design criteria listed in Fig. 13-2. These general design criteria illustrate the areas normally addressed by system safety engineers during equipment design.

MIL-STD 882B

1. Equipment design stresses safety consistent with mission requirements.
2. Hazards associated with the design are identified, evaluated, and eliminated or reduced to an acceptable level.
3. Consideration of historical safety data gathered from other systems.
4. Minimize risk through use of new designs, materials, and production and test techniques.
5. Minimize retrofit by addressing safety concerns in the early phases of acquisition.
6. Recommended design changes are accomplished in a manner that does not increase risk of hazards.

Fig. 13-1. System safety program objectives.

Safety Hazards

A hazard is a situation that, if not corrected, might result in death, injury, or occupational illness to personnel, or damage or loss of equipment. System safety engineers analyze and evaluate the proposed equipment design to identify hazards. These hazards are then classified in terms of severity and probability of occurrence. Figure 13-3 shows the four categories of hazards, and Fig. 13-4 provides an example of how the probability of occurrence can be assigned. The combination of severity and probability can be used to develop a hazard assessment matrix, as illustrated in Fig. 13-5, which can be used to prioritize the effort of system safety by identifying the hazards that are most critical to safe equipment operation.

System Safety Program

MIL-STD 882B, System Safety Program Requirements, provides uniform guidance for the establishment and implementation of a System Safety Program (SSP). It

MIL-STD 882B

1. Eliminate hazards through design.
2. Isolate hazardous substances, components, and operations.
3. Locate equipment to reduce hazards to personnel during operation and maintenance.
4. Minimize risks caused by environmental conditions.
5. Design to eliminate or minimize risk created by human error.
6. Consider alternate approaches to eliminate hazards.
7. Provide adequate protection from power sources.
8. Provide warnings and cautions when risks cannot be eliminated.

Fig. 13-2. System safety design criteria.

MIL-STD 882B

Description	Category	Mishap Definition
Catastrophic	I	Death or system loss.
Critical	II	Severe injury, minor occupational illness, or major system damage.
Marginal	III	Minor injury, minor occupational illness, or minor system damage.
Negligible	IV	Less than minor injury, occupational illness, or system damage.

Fig. 13-3. Hazard category table.

identifies the basic safety requirements that all DoD equipment and facilities must meet. Each SSP is tailored to fit the unique requirements of the equipment being designed and the environment where it will be used.

MIL-STD 882B is formatted similar to MIL-STD 785 in that it is divided into distinct tasks that are an integral part of the overall program. It has two task sections; Task Section 100—Program Management and Control; and Task Section 200—Design and Evaluation. The tasks contained in Task Section 100 describe the procedures and control methods that a contractor is required to implement as part of the system safety program. Those in Task Section 200 require specific actions to carry out the safety requirements of the program. These tasks provide the foundation for the SSP. Figure 13-6 illustrates the tasks contained in each section and the acquisition phase when each task is normally applicable.

System Safety Program (Task 100). This task is unique to the system safety program. Task 100 of MIL-STD 882B requires the contractor to establish a system

MIL-STD 882B

Description	Level	Individual Item	Inventory
Frequent	A	Likely to occur frequently.	Continuously experienced.
Probable	B	Will occur several times in life of an item.	Will occur frequently.
Occasional	C	Likely to occur sometime in life of an item.	Will occur several times.
Remote	D	Unlikely but possible to occur in life of an item.	Unlikely but can reasonably be expected to occur.
Improbable	E	So unlikely, it can be assumed that occurrence may not be experienced.	Unlikely to occur, but possible.

Fig. 13-4. Hazard probability.

MIL-STD 882B

Frequency of Occurrence	Hazard Categories			
	I Catastrophic	II Critical	III Marginal	IV Negligible
A. Frequent	1	3	7	13
B. Probable	2	5	9	16
C. Occasional	4	6	11	18
D. Remote	8	10	14	19
E. Improbable	12	15	17	20

Hazard Risk Index	Criteria
1-5	Unacceptable
6-9	Undesirable
10-17	Acceptable with review
18-20	Acceptable without review

Fig. 13-5. Hazard risk assessment matrix.

safety program. Other programs, such as reliability or maintainability, are invoked by referencing the document that states the program requirements, e.g., "the reliability program will be conducted in accordance with the requirements of MIL-STD 785." The presence of this task illustrates the importance of the safety aspects of an item of equipment.

System Safety Program Plan (SSPP) (Task 101). The purpose of the SSPP is to describe in detail the tasks and related activities that system safety engineers and management will accomplish to identify, evaluate, and eliminate equipment hazards. The plan is prepared and formatted similar to the Reliability Program Plan described in Fig. 3-4. Figure 13-7 lists the safety areas to be addressed in the SSPP. MIL-STD 882B is much more explicit than MIL-STD 785B, and provides detailed guidance on information that the SSPP must contain.

Integration/Management of Subcontractors (Task 102). It is important that the system safety requirements be passed to subcontractors. Task 102 formally requires that this be accomplished. On major weapon system programs that have several layers of subcontractors, the government might require preparation of an Integrated System Safety Program Plan (ISSPP) that describes how the total system safety effort will be integrated to ensure that all requirements are met. Subcontractors are normally required to submit reports to the contractor on safety activity and participate in program reviews. Contractors monitor, review, and approve the compliance of subcontractors with applicable safety requirements.

Task		Concept	DEMVAL	FSD	Prod
100 Series Tasks					
100	System Safety Program	G	G	G	G
101	System Safety Program Plan	G	G	G	G
102	Integration of Subcontractors	S	S	S	S
103	Program Reviews	S	S	S	S
104	System Safety Working Group	G	G	G	G
105	Hazard Tracking	S	S	S	S
106	Test and Evaluation Safety	G	G	G	G
107	Progress Summary	G	G	G	G
108	Key Personnel Qualifications	S	S	S	S
200 Series Tasks					
201	Preliminary Hazard List	G	S	S	N/A
202	Preliminary Hazard Analysis	G	G	G	GC
203	Subsystem Hazard Analysis	N/A	G	G	GC
204	System Hazard Analysis	N/A	G	G	GC
205	Operation Hazard Analysis	S	G	G	GC
206	Health Hazard Assessment	G	G	G	GC
207	Safety Verification	S	G	G	S
208	Training	N/A	S	S	S
209	Safety Assessment	S	S	S	S
210	Safety Compliance Assessment	S	S	S	S
211	Safety Review of ECPs/Waivers	N/A	G	G	G
212	Software Hazard Analysis	S	G	G	GC
213	GFE/GFP System Safety Analysis	S	G	G	G

Applicability Codes:

S - Selectively applicable
G - Generally applicable
GC - Generally applicable to design changes only
N/A - Not applicable

Fig. 13-6. System safety program tasks by acquisition phase.

System Safety Program Reviews (Task 103). As with previously discussed programs, Task 103 requires formal participation of system safety in all program reviews. System safety concerns are normally addresses at all preliminary design reviews (PDR) and critical design reviews (CDR). Additionally, the government might require the contractor to convene special program reviews that deal strictly with safety

1.0 PROGRAM SCOPE AND OBJECTIVES

Describe the overall scope of the program and how safety engineers will participate in the design process.

2.0 PROGRAM TASKS

Describe how each safety task (MIL-STD 882B) will be accomplished. Describe interfaces between safety and other engineering activities.

3.0 SYSTEM SAFETY ORGANIZATION

Describe the system safety organization. Include organizational charts. Identify the responsibilities and authority of safety personnel. Specifically identify who has the decision-making authority to ensure that safety concerns are addressed. Include name, address, and telephone number of the system safety program manager. Describe the personnel and other resources that will be used to accomplish program tasks.

4.0 SYSTEM SAFETY PROGRAM MILESTONES

Include a milestone chart that tells in detail when each safety task will begin and end. Identify critical activities that must be accomplished in order to produce a hazard-free system. The milestone chart should be keyed to overall program milestones.

5.0 SAFETY REQUIREMENTS AND CRITERIA

Describe how the safety requirements and criteria will be integrated into the equipment design. Provide an explanation of the processes that will be used to channel safety concerns to the responsible design engineering activity.

6.0 HAZARD ANALYSES

Describe the analytical processes that will be used to identify system hazards. Explain how the results will be used to eliminate or reduce hazards. Identify the depth to which each analysis will be used.

7.0 SYSTEM SAFETY DATA

Identify what data will be recorded and how they will be managed. Describe the procedures that will be used to process and disseminate information. Address how deliverable data items will be prepared and submitted.

8.0 SAFETY VERIFICATION

Describe the methods of test, analysis, and inspection that will be used to verify that requirements for system safety are met.

Fig. 13-7. System safety program plan (sample outline).

9.0 AUDIT PROGRAM

Describe the internal audit procedures that will be used to oversee the system safety effort to ensure that the program goals are accomplished.

10.0 TRAINING

Describe any safety training that will be provided to engineering, technical, operator, or maintenance personnel.

Fig. 13-7. System safety program plan (sample outline). (Continued from page 194.)

issues. These special reviews might be required to support certification boards in the areas of munitions or aircraft flight readiness.

System Safety Working Group (Task 104). Contracts for acquiring major weapon systems that are expensive, complex, or critical to national defense require the establishment of a System Safety Working Group (SSWG). The SSWG is comprised of representatives from the government, the contractor, and major subcontractors. The purpose of the SSWG is to provide immediate channels for resolving hazardous conditions identified in the design process of the equipment being procured. The level of participation required to support the SSWG must be stated in the contract when Task 104 is imposed.

Hazard Tracking and Risk Resolution (Task 105). Task 105 is imposed by the government to provide a formal method for recording the ongoing efforts of safety analyses. This bookkeeping requirement establishes a log for recording potential hazards as they are identified and provides a method for tracking the resolution and disposition of each hazard. The hazard log is an invaluable tool in managing the system safety effort and providing visibility for critical safety problems. An example of a hazard log format is shown in Fig. 13-8.

Test and Evaluation Safety (Task 106). Safety issues are not limited to the design of equipment. Task 106 requires system safety engineers to monitor and review the safety aspects of all test and evaluation activity associated with the development of equipment. The testing and evaluation of an unproven new equipment design poses unique safety considerations. Safety engineers review test procedures and processes to ensure that tolerances and limits do not exceed safe operating boundaries.

System Safety Progress Summary (Task 107). The government normally requires periodic reports concerning the status of the system safety program. Task 107 formally establishes the requirement for this report. The System Safety Progress Summary provides information on current safety issues, and identifies problem areas and the status of activity relative to program milestones. This report is provided to the government either monthly or quarterly, as stated in the contract.

Personnel Qualifications (Task 108). System safety is a critical consideration during the design process, therefore it is imperative that contractors have qualified safety personnel. Task 108 states the special qualifications and experience that key safety personnel must possess. The qualifications include education, certification, and

Control Number	Part Number	Hazard Description	Responsible Engineer	Date Entered	Date Closed

Fig. 13-8. Hazard log worksheet.

prior work experience. This task is unique in that the government is actually telling a contractor what kind of personnel must be employed to work on a contract. If a contractor cannot comply with the requirements of this task, a waiver must be approved by the government.

Preliminary Hazard List (Task 201). Task 201 requires the contractor to prepare a preliminary hazard list early in the acquisition cycle that identifies the possible safety hazards inherent in the proposed equipment design or use environment. This list forms the basis for further system safety activity, and, as the design evolves, it can be modified as required to guide the SSP and determine the scope of other SSP tasks.

Preliminary Hazard Analysis (Task 202). System safety engineers perform a preliminary analysis of the potential hazards identified in Task 201 to determine the overall level of hazard that exists in a proposed design. This analysis is used as an input to trade-off studies evaluating alternative design and deployment approaches during early acquisition cycle phases.

Subsystem Hazard Analysis (Task 203). Design efforts that integrate major subsystems might require hazard analyses of each segment of the design to determine the potential subsystem hazards. A subsystem hazard analysis is applicable to major weapon system contracts where several pieces of equipment are integrated to complete the system design. Task 203 requires evaluation of hazards related to operation and failures of the subsystem.

System Hazard Analysis (Task 204). The system hazard analysis evaluates the total system operation and failure modes to determine the effect of these activities on the overall system. Special emphasis is placed on the effect of integrating subsystems and the results of failures in one subsystem that could cause a hazard to other subsystems or degrade the performance of the total system. Task 204 is normally performed by the prime contractor, while Task 203 is performed by subcontractors.

Operating and Support Hazard Analysis (Task 205). The purpose of Task 205 is to identify and evaluate operating and support hazards associated with the environment, personnel, procedures, and related equipment involved with the life cycle of the equipment being designed. Activities evaluated include operation, testing, installation, maintenance, storage, and training. It identifies needed design changes to eliminate potential hazards. This task is intimately related with the human engineering effort described later in this chapter.

Occupational Health Hazard Assessment (Task 206). The occupational health hazard assessment identifies design characteristics that pose health hazards to operator and maintenance personnel. The presence of hazardous materials and physical environments (noise, vibration, extreme atmospheric conditions) is identified. This task is also related to the human engineering effort and provides input to design changes to correct or reduce inherent risks to personnel.

Safety Verification (Task 207). As the design matures, the safety status of equipment must be verified to ensure that specification requirements are met. This verification is normally accomplished as a part of the testing process of equipment. System safety participates in the preparation of test requirements and actually conducts tests and demonstrations to verify that the design meets all safety criteria. The results of testing are used to identify further safety considerations or hazards that must be corrected through design changes, changes to procedures, or additional safety controls.

Training (Task 208). Personnel involved with the design, development, testing, and eventual use of the equipment must be properly trained to avoid hazards. System safety is responsible for identifying and preparing the training needed to ensure that all personnel understand the safety requirements of equipment. The output of this task is used by training personnel to develop training courses in the safe operation and maintenance of the equipment being designed.

Safety Assessment (Task 209). A safety assessment is performed near the end of equipment development that tells the user all the unsafe design or operating characteristics of the equipment. The assessment should also contain the controls or procedures required to reduce the risks imposed by these characteristics. It might also identify all the efforts that occurred in order to reduce risks to as low a level as possible. The Safety Assessment Report is essential in conveying to the user the importance of adherence to recommended safety procedures.

Safety Compliance Assessment (Task 210). Task 210 requires the contractor to conduct an assessment of the equipment design to verify that it complies with all military, federal, national, and industry codes imposed either contractually or legally. This is a comprehensive report that includes all safety aspects of the equipment. The results of all the tasks of the SSP are included in the assessment. It also includes identification of special design features, procedures, protective equipment, and support resources required to ensure safe use. In certain cases, where the equipment design is inherently low-risk with regard to safety, this might be the only system safety task that is required contractually.

Engineering Change Proposal (ECP) Review (Task 211). The purpose of Task 211 is to ensure that system safety reviews all proposed engineering changes prior to submittal to the government for approval. This evaluation is necessary to identify possible impacts on equipment safety that might be caused by the proposed change. Correction of a design problem could cause another more serious problem in the area of safety. A representative of system safety must review and sign all proposed engineering changes when this task is imposed.

Software Hazard Analysis (Task 212). System safety conducts an analysis of the proposed system software to ensure that safety specifications are accurately translated into the equipment software. Because many systems rely on software to identify hazardous occurrences during equipment operation and maintenance, it is imperative that the software does not cause hazardous functions to occur or inhibit desired functions. Analysis techniques include software fault tree, software sneak circuit, integrated critical path, and cross-check analysis. This analysis begins when functional allocations are made in early design phases and expands to include examination of the actual software source and object codes of safety critical programs.

GFE/GFP System Safety Analysis (Task 213). Many times a major system will contain both contractor-designed equipment and government-furnished equipment (GFE) or government-furnished property (GFP). When an equipment design contains or must interface directly with GFE or GFP, the contractor may be required to conduct an analysis to identify potential hazards that might occur upon integration. This task allows the contractor to integrate the GFE/GFP items into the system design with full knowledge of the associated hazards and risk controls involved.

The desired result of the system safety program is the design, development, and fielding of equipment that is as free of hazards as possible. Through effective implementation of the system safety program outlined in MIL-STD 882B, the contractor is able to produce equipment that meets the performance and safety specifications established by the government.

HUMAN ENGINEERING PROGRAM

All items of equipment, with few exceptions, require human interaction for operation, maintenance, or both. The goal of human engineering is to optimize this man-to-machine interface. MIL-H-46855B, Human Engineering Requirements for Military Systems, Equipment and Facilities, contains the requirements for including human engineering in the procurement of equipment. It establishes what a contractor will accomplish to ensure that the human aspect is considered in the development of the equipment being designed. MIL-STD 1472, Human Engineering Design Criteria for Military Systems, Equipment and Facilities, and MIL-HDBK 759A, Human Factors Engineering Design for Army Materiel, provide guidelines for the design of military equipment that consider the limitations caused by using humans as part of the system. A formal contractual requirement for human engineering is normally found on only major system design efforts.

The Human Engineering Program

MIL-H-46855 contains detailed guidance for the establishment and conduct of a human engineering program. The program is similar to those required for other ILS disciplines, such as reliability and maintainability, although there are no numbered tasks, i.e., Task 101, 201, etc. The program consists of analyzing human function requirements, participating in the design process to ensure that human factors are adequately addressed, and testing and evaluating the final design to validate its operability. A human engineering program plan, although not discussed in detail in MIL-H-46855, may be required by the government and should be prepared using an outline similar to that of a reliability program plan.

Analysis. The first step in a human engineering program is to identify and analyze the functions that the equipment is required to perform as stated in the procurement specification. Particular interest is paid to the equipment operation and maintenance requirements and to the environment in which the equipment will be used. The functions are assimilated in logical flow and processing sequence to identify exactly how the equipment must perform to accomplish its mission.

Once the functional requirements have been identified, then the man-to-machine interfaces that must be accomplished to meet the functional requirements are quantified. The human tasks necessary to operate and maintain the equipment are the man-to-machine interface. These tasks are not developed in a vacuum by human engineers.

They are the result of analyses performed through a coordinated effort of reliability, maintainability, maintenance planning, human engineering, and other ILS disciplines. Remember, the FMECA starts this analysis process, and it continues through the generation of final technical documentation.

Once the tasks have been initially identified, they are analyzed to determine which ones are critical, in terms of human engineering, to mission accomplishment. Critical tasks are those that, if not accomplished as needed to meet a functional requirement, would result in equipment failure to accomplish the mission, create a safety hazard, or significantly reduce equipment reliability. An analysis of critical tasks is performed to identify key human factors data about each task. Figure 13-9 illustrates typical human factors data that are pertinent to the analysis of critical tasks. The results of this analysis are used as input to the design process to ensure that man-to-machine interfaces are optimized.

Design and Development. Human engineering provides input during the design phase based on the results of human engineering analyses. The design is continually evaluated and changes are recommended to enhance the man-machine interface. Through the use of mock-ups, models, and dynamic simulation, human engineering

Item	Description
Environment	Location and condition of work environment.
Space	Amount of space required to perform task. Amount of space available. Body movements required to perform task.
Information	Amount of information available to operator. Amount of information required to perform task.
Time	Amount of time allocated for task completion. Frequency of task performance. Maximum allowable time for task completion.
Resources	Number of personnel required to perform task. Tools and other equipment required. Instructions and manuals required.
Other	Safety hazards. Interaction required between crew members. Personnel performance limitations. Equipment performance limitations.

Fig. 13-9. Human engineering analysis input data.

- Physical man-to-machine interface (physical, aural, visual)
- Physical man-to-man interface (physical, aural, visual)
- Physical comfort of operator/maintenance personnel
- Equipment-handling requirements (weight, cube)
- Temperature, humidity, etc., to be encountered
- Inclement conditions (rain, snow, mud) anticipated
- Climate (artic, desert)
- Equipment environment (vibration, noise)
- Useable space availability
- Effects of special clothing (gloves, NBC, coat)
- Safety and hazard protection
- Mission-related requirements (tactical environment)

Fig. 13-10. Human engineering design considerations.

develops solutions to interface problems and evaluates design alternatives to achieve the best possible human interface situations. Using the design criteria provided in MIL-STD 1472, human engineering is able to provide quantitative requirements to the design process and the detailed design of work environments. Figure 13-10 illustrates human engineering areas of concern in the design of work environments.

Test and Evaluation. Human engineering test and evaluation of the proposed design are conducted as an integral part of the overall equipment test plan. Normally no separate tests are conducted for human engineering. The tests are conducted concurrently with engineering design and development testing, performance tests and demonstrations, and maintainability demonstrations. The purpose of this testing is to validate that the equipment can be operated and maintained by typical user in the intended operating environment. Critical performance tasks are given special consideration during testing. All failures that occur during equipment design and testing are evaluated to determine if human involvement was the cause. This can be user error, maintenance-induced error, or interface deficiencies. The results of these tests are used to recommend design changes, as required, to meet the human performance requirements of the equipment.

The results of the human engineering program are inputs to Logistic Support Analysis (LSA) and the development of training requirements and technical manuals. All human engineering activities must be coordinated with those of other ILS disciplines to receive the greatest benefit from this program. The human link in the man-machine interface is often overlooked and, therefore, is many times the weak link that significantly reduces the system capability for mission success.

Human Engineering Design Criteria

MIL-STD 1472, Human Engineering Design Criteria for Military Systems, Equipment and Facilities, and MIL-HDBK 759A, Human Factors Engineering Design for Army Materiel, provide the principles and methodology that are used to integrate the human into the design of equipment to achieve mission success. The goal of this integration is to optimize the effectiveness, efficiency, safety, and reliability of the operation and maintenance of equipment. The quantitative nature of the criteria contained in MIL-STD 1472 and MIL-HDBK 759A aids in analyzing and evaluating the human engineering aspects of equipment design. Subsequent paragraphs provide an overview of typical human engineering design considerations.

Controls and Displays. Most operational man-machine interfaces are done through controls and displays. Controls are devices that the operator uses to regulate or give commands to the equipment. Displays provide the operator with information for decision making and monitoring of the results of manipulation of controls. MIL-STD 1472 describes the selection criteria for the most desirable types of controls to fulfill specific functions. It describes how controls should be grouped and arranged to reduce excessive operator movements; methods for coding, labeling, and identifying controls; and design methods that prevent accidental activation. Displays can be either visual or audio, depending on the type of information being provided to the operator. Visual displays range from simple indicator and warning lights to cathode-ray tube (CRT) displays. They also include mechanical scales, counters, meters, printers, and plotters. The type of visual display used is commensurate with the content and format of the information available and the precision of display required.

The use of audio displays is limited to warnings and voice communications. Warnings are short, go/no-go messages that require immediate or time-critical responses. They are used to inform the operator of critical situations or to indicate impending danger to personnel, equipment, or both. The integration and compatibility of controls and displays are critical. Colocation of controls with corresponding displays should be a basic design goal. The controls and displays selected should be complementary. The information provided by the display should be sufficient to allow the operator to effectively use the control, and the control should provide sufficient range to receive maximum operation.

Environment. The environment in which equipment will be operated and maintained must be considered in the design process. The basic physical comfort of personnel is directly related to the ability of the equipment to conduct and sustain operations. Environmental considerations can be grouped into three categories; factors that design can control (interior lighting and temperature), factors that design cannot control (rain, snow, dust, etc.), and inherent design factors (noise, vibration, etc.). Human engineering analyzes the design to identify deficiencies and propose solutions in areas that can be controlled or are inherent design problems. Additional consideration is given to providing alternatives for coping with and minimizing the environmental problems that cannot be directly controlled by the design. Figure 13-11 indicates environmental design considerations.

Anthropometry. Anthropometry is the study of measurements of the human body. Human engineering uses comparative human body measurements to determine the minimum acceptable sizing and dimensioning for man-to-machine interfaces. DoD-HDBK 743, Anthropometry of U.S. Military Personnel, contains data obtained by actual

Fig. 13-11. Environmental design considerations.

- Heating
- Ventilation
- Air conditioning
- Humidity
- Hazardous noise
- Nonhazardous noise
- Vibration
- Lighting

measurement of representative population samples of military personnel. These data are organized into limits of 5% to 95% of personnel measured, and theoretically provide design limits to accommodate 90% of the potential equipment users. As an example, DoD-HDBK 743 shows that only 5% of the persons (male and female) were less than 60 inches tall, and that 95% were 73.1 inches tall, or less. Therefore, for human engineering purposes, designs must be able to accommodate personnel who are between 60 and 73.1 inches tall. Figure 13-12 illustrates the anthropometric data for a standing person.

Work Space. The physical dimensions of work space must be adequate for personnel to operate and maintain equipment. Human engineering uses the anthropometric data discussed above to evaluate the proposed design of equipment, ensuring that the provisions for work space do not impose unacceptable restrictions or hazards to personnel. Specific areas analyzed include space required for standing and seated operations; standard and special console designs; crew compartments; stairs, ladders, and ramps; entrance and exit through doors and hatches; and surface colors. The key to optimizing work space is to design the space around the person, rather than designing the space and then putting the person in it. By designing around the person, costly redesign efforts due to insufficient work space can normally be avoided. Figure 13-13 shows how work-space dimension requirements are determined.

Maintainability. Human engineering can have a tremendous impact on the maintainability of an item of equipment. The mean time to repair (MTTR) predicted by maintainability engineering is based on the capability of personnel to accomplish maintenance tasks within a specified amount of time. The design must consider the human interface requirements for performing maintenance. Human engineering evaluates the design to ensure that each maintenance task is as easy to perform as possible. The two keys to achieving maintainability goals, on which human engineering has a direct effect, are accessibility and standardization. If personnel cannot easily access items requiring maintenance, excessive time might be required to perform maintenance. Standardizing fasteners, connectors, and other items that are repeatedly removed and replaced when maintenance is performed standardizes common maintenance tasks, reducing the possibility of human error and increasing proficiency. Specific areas of maintainability interest are accessibility of items to be removed and replaced; lubrication and test points; standardization of fasteners, connectors, and other hardware; design of covers and cases; design for efficient handling; and ease of using tools and test equipment. Significant improvements in the overall equipment MTTR can be achieved through the application of human engineering criteria during equipment design.

Labeling. An area often overlooked is labeling of equipment. Labels are used to identify levels of equipment, state standard procedures, and identify hazards. Labels must be visible and legible, and they must also be durable. The contents, quality, and location of labels that provide directions for operation or maintenance increase the effectiveness of persons using the equipment.

User-Computer Interface. The use of computers is common in current and planned military equipment. MIL-STD 1472 contains specific human engineering criteria for the user-computer interface. The interface is comprised of data entry, data display, interactive control, feedback, prompts, error management, data protection, and system response time. Each of these areas has significant human factors inherent in the capability of the equipment to accomplish its mission. The thrust of human engineering effort in optimizing the user-computer interface is to limit the range of inputs and outputs that are possible and that must be processed by the user and to increase the capabilities of the computer to aid the user in performing required operation and maintenance tasks.

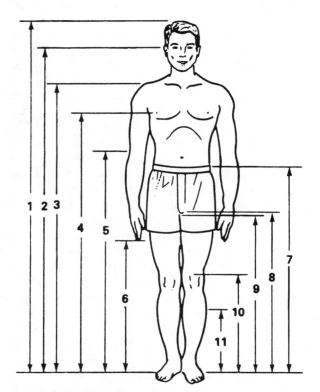

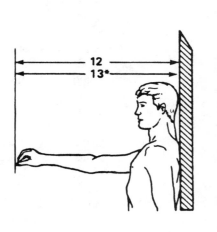

MIL-STD-1472C

*SAME AS 12, HOWEVER, RIGHT SHOULDER IS EXTENDED AS FAR FORWARD AS POSSIBLE WHILE KEEPING THE BACK OF THE LEFT SHOULDER FIRMLY AGAINST THE BACK WALL.

Fig. 13-12. Standing body dimensions.

	PERCENTILE VALUES IN CENTIMETERS					
	5th PERCENTILE			95th PERCENTILE		
	GROUND TROOPS	AVIATORS	WOMEN	GROUND TROOPS	AVIATORS	WOMEN
WEIGHT (Kg)	55.5	60.4	46.4	91.6	96.0	74.5
STANDING BODY DIMENSIONS						
1 STATURE	162.8	164.2	152.4	185.6	187.7	174.1
2 EYE HEIGHT (STANDING)	151.1	152.1	140.9	173.3	175.2	162.2
3 SHOULDER (ACROMIALE) HEIGHT	133.6	133.3	123.0	154.2	154.8	143.7
4 CHEST (NIPPLE) HEIGHT *	117.9	120.8	109.3	136.5	138.5	127.8
5 ELBOW (RADIALE) HEIGHT	101.0	104.8	94.9	117.8	120.0	110.7
6 FINGERTIP (DACTYLION) HEIGHT		61.5			73.2	
7 WAIST HEIGHT	96.6	97.6	93.1	115.2	115.1	110.3
8 CROTCH HEIGHT	76.3	74.7	68.1	91.8	92.0	83.9
9 GLUTEAL FURROW HEIGHT	73.3	74.6	66.4	87.7	88.1	81.0
10 KNEECAP HEIGHT	47.5	46.8	43.8	58.6	57.8	52.5
11 CALF HEIGHT	31.1	30.9	29.0	40.6	39.3	36.6
12 FUNCTIONAL REACH	72.6	73.1	64.0	90.9	87.0	80.4
13 FUNCTIONAL REACH, EXTENDED	84.2	82.3	73.5	101.2	97.3	92.7

	PERCENTILE VALUES IN INCHES					
WEIGHT (lb)	122.4	133.1	102.3	201.9	211.6	164.3
STANDING BODY DIMENSIONS						
1 STATURE	64.1	64.6	60.0	73.1	73.9	68.5
2 EYE HEIGHT (STANDING)	59.5	59.9	55.5	68.2	69.0	63.9
3 SHOULDER (ACROMIALE) HEIGHT	52.6	52.5	48.4	60.7	60.9	56.6
4 CHEST (NIPPLE) HEIGHT *	46.4	47.5	43.0	53.7	54.5	50.3
5 ELBOW (RADIALE) HEIGHT	39.8	41.3	37.4	46.4	47.2	43.6
6 FINGERTIP (DACTYLION) HEIGHT		24.2			28.8	
7 WAIST HEIGHT	38.0	38.4	36.6	45.3	45.3	43.4
8 CROTCH HEIGHT	30.0	29.4	26.8	36.1	36.2	33.0
9 GLUTEAL FURROW HEIGHT	28.8	29.4	26.2	34.5	34.7	31.9
10 KNEECAP HEIGHT	18.7	18.4	17.2	23.1	22.8	20.7
11 CALF HEIGHT	12.2	12.2	11.4	16.0	15.5	14.4
12 FUNCTIONAL REACH	28.6	28.8	25.2	35.8	34.3	31.7
13 FUNCTIONAL REACH, EXTENDED	33.2	32.4	28.9	39.8	38.3	36.5

*BUSTPOINT HEIGHT FOR WOMEN

Fig. 13-12. Standing body dimensions. (Continued from page 204.)

MIL-HDBK-759A

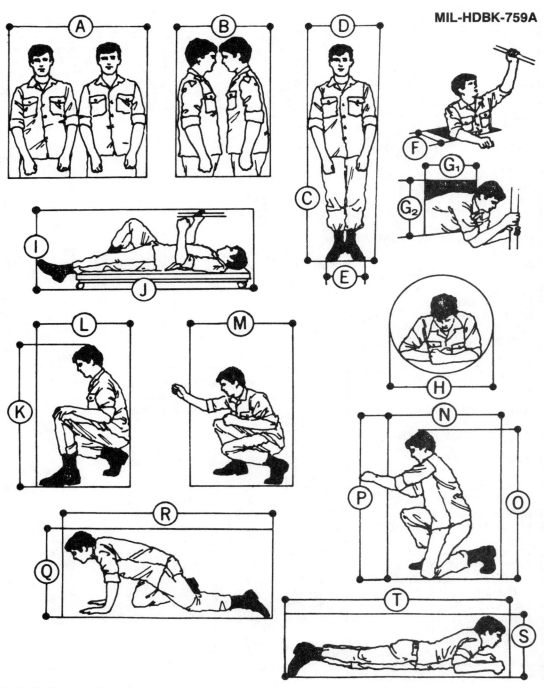

Fig. 13-13. Work space dimensions.

MIL-HDBK-759A

		Dimensions (mm)		
		Minimum	Preferred	Arctic Clothed
A.	Two men passing **abreast**	1.06m	1.37m	1.53m
B.	Two men passing **facing**	760	910	910
Catwalk Dimensions				
C.	Height	1.60m	1.86m	1.91m
D.	Shoulder width	560	610	810
E.	Walking width	305	380	380
F.	<u>Vertical entry hatch</u>			
	Square	459	560	810
	Round	560	610	-
G.	<u>Horizontal Entry Hatch</u>			
	1. Shoulder width	535	610	810
	2. Height	380	510	610
H.	<u>Crawl through pipe</u>			
	Round or square	635	760	810
Supine work space				
I.	Height	510	610	660
J.	Length	1.86m	1.91m	1.98m
Squatting work space				
K.	Height	1.22m	-	1.29m
L.	Width	685	910	-
	Optimum display area	685	1.09m	-
	Optimum control area	485	865	-
Stooping work space				
M.	Width	660	1.02m	1.12m
	Optimum display area	810	1.22m	-
	Optimum control area	610	990	-
Kneeling work space				
N.	Width	1.06m	1.22m	1.27m
O.	Height	1.42m	-	1.50m
P.	Optimum work point		685	-
	Optimum display area	510	890	-
	Optimum control area	510	890	-
Kneeling crawl space				
Q.	Height	785	910	965
R.	Length	1.50m	-	1.76m
Prone work or crawl space				
S.	Height	430	510	610
T.	Length	2.86m	-	-

Fig. 13-13. Work space dimensions. (Continued from page 206.)

The desired result of the system safety and human engineering efforts is to influence the equipment design in order to provide environments that maximize personnel productivity and safety and minimize design-induced human limitation and error. Effective integration of system safety and human engineering with other ILS disciplines increases the ability of the contractor to achieve the overall equipment performance requirements.

Chapter 14

Facilities

The general military definition of a facility is any real property, including parcels of land and buildings, structures, or utilities built on or in the land. When resource requirements for support of operation and maintenance of equipment, training, and storage are being determined, the availability of adequate facilities must be considered as an important portion of overall planning. After all, most operation and maintenance tasks, especially above crew level, require a suitable location for accomplishment. Facilities can be either fixed or mobile. They can also be categorized by their intended use, e.g., maintenance, supply, training, etc.

One point that must be remembered when considering facilities required to support an item of equipment is that normally there is not a one-to-one relationship between an item of equipment and its supporting facilities. Most facilities support numerous types of equipment; therefore, facility planning must also consider factors related to the support of a general population of equipment or similar tasks being accomplished in support of several different programs.

TYPES OF FACILITIES

As stated above, facilities can be described in several different ways. The descriptions refer to either the construction and mobility of the facility or its intended use. The military has established permanent facilities, both in and outside the United States, that provide various types of support to the services. As the missions and types of equipment of the military change, the roles of these facilities change accordingly.

Permanent Facilities

Permanent facilities are required to provide administration, maintenance, supply, and training support for the military. Their design and function are based on the units and missions they support. Because there is a limit to the quantity and types of facilities available, proper use of these resources is critical to ensuring that the military is capable of accomplishing its assigned missions for the most reasonable life cycle cost.

Most facilities are located on military bases and are staffed by DoD personnel, either uniformed military or a civilian work force. Examples of these facilities are maintenance shops, supply warehouses, training centers, maneuver areas, open area storage, office buildings, etc. It is not uncommon to find buildings on a military post that have been modified many times to support different roles, e.g., a building originally built as a maintenance shop might have been transformed through successive changes to support training, administration, supply, and back to maintenance as the needs of the military changed.

Mobile Facilities

Permanent facilities have one limiting factor—mobility. The military is a deployable force, which means it must be capable of moving wherever its assigned missions must be performed. Permanent facilities cannot follow units on the move, so mobile support facilities are required. Common examples of mobile facilities include maintenance shop vans, spare parts vans and containers, special design tents, and enclosures designed to support operations in a field environment.

Maintenance Facilities

Maintenance facilities support the accomplishment of maintenance tasks. Depot maintenance facilities have the ability to rebuild, overhaul, repair, and fabricate equipment. The particular equipment that a specific depot supports is determined by the needs of the military. Each branch of the military has its own network of depot maintenance facilities, which means that an Army depot does not provide support to Navy equipment, and vice versa. In rare cases, where the same type of equipment is used by more than one service, a depot might provide dual service support.

Depots are always permanent facilities. Intermediate-level maintenance facilities can be either permanent or mobile. The determination of whether an "I" level maintenance facility should be permanent or mobile depends on the branch of the military and the type of mission it performs. Intermediate-level maintenance in the Army is provided by direct support units that are, by definition, required to follow the combat units supported; therefore, they must be mobile. The other branches of the military have both permanent and mobile "I" level maintenance facilities, depending on the mission of the units supported. For example, the Navy has shore facilities that are permanent, because harbors do not move very often, and mobile facilities aboard ships that accompany the fleet. Maintenance facilities at the organizational level are either permanent or mobile depending, again, on the organization's mission. The maintenance facilities of deployable units are always mobile.

Supply Facilities

When considering supply facilities, the immediate mental picture is that of huge areas of real estate covered with warehouses and open storage areas crammed with all kinds of supplies. That is a supply depot, and, logically, all supply depots are permanent. In time of war, the military has neither the transportation assets nor the time to move a supply depot. Intermediate-level supply facilities are likewise limited in mobility. In some cases, an "I" level facility might be partially mobile, depending on its location and the units it supports. At the organizational level, supply facilities must be mobile for deployable units. Therefore, inventories must be kept to the essentials. This is where planning is required to make the most of limited storage capacity. For example, if a new weapon system is being designed for installation aboard a ship, the storage of repair parts to support organizational-level maintenance becomes a critical planning factor.

Space allocation for repair parts storage may drive the maintenance concept. Combat units in the Army, by doctrine, must maintain sufficient repair parts on hand to sustain operations for a specified number of days, normally 15 or 30 days, without resupply. This is not an easy task, due to limited mobile repair parts storage at the organizational level.

Training Facilities

Facilities used for training can be categorized as organizational, installation, training center, and contractor. Organizational training facilities are those necessary to maintain personnel proficiency at the unit level. The actual types and configurations used vary among units. Normally training facilities are limited to whatever is available, e.g., space in a motor park, the dining hall, or a spare conference room. Dedicated training facilities may or may not be present.

Most units have some type of training facility contingent on what resources are available. At the installation level (notice there is no intermediate or depot), military bases have dedicated training areas or facilities; each installation is different.

The training facility is usually designed to provide training on numerous subjects, again based on the requirements of units stationed at the installation. Each service has training centers that provide intensive training in a designated area of expertise. For example, the Army has the Armor Training Center, Ft. Knox, Kentucky; the Infantry Training Center, Ft. Benning, Georgia; and the Artillery Training Center, Ft. Sill, Oklahoma. The other branches of the military have training centers for applicable areas of expertise. These training centers provide formal courses of instruction to qualify personnel in the skill specialties needed by the military.

In certain instances, the military might determine that it is desirable to use a contractor's facility for training. This occurs most often during the development of a new item of equipment where the contractor possesses the capability to present the training, and the only place where the equipment is available for training is at the contractor's facility.

Special Facilities

There are some circumstances in which special facilities are required due to unusual requirements. Special maintenance facilities might include a "clean room" for performing maintenance on equipment that must be free of contamination. A clean room is an area having a temperature, humidity, and dust-free controlled atmosphere where maintenance is performed on optics, microelectronic assemblies, or other sensitive equipment.

Another example of a special maintenance facility would be a facility with sophisticated electronic test equipment not normally found at a maintenance facility. Special supply facilities are required when items cannot be stored using standard procedures or facilities. Storage of classified, toxic, oversized, or other sensitive items might require a special supply facility.

REQUIREMENTS IDENTIFICATION

Determining the facilities required to support operations, maintenance, supply, and training for a specific item of equipment is accomplished by analyzing the tasks that will be completed by supporting personnel and comparing the results of the analysis with what facilities are available. The method for this analysis is shown at Fig. 14-1.

In the concept phase, a study must be undertaken to determine the facilities available to support the equipment. The best starting point for this is to identify the facilities that are being used to support the old equipment. The next step is to analyze the tasks required to support the new equipment and determine (1) if the existing facilities can be used without change, (2) if they are inadequate and must be modified, or (3) if a totally new facility is required. This process is not as simple as it might seem.

Maintenance

The source of information on what facilities are required to support maintenance is the detailed maintenance task analysis. As maintenance tasks are identified, the maintenance engineer should have sufficient information available to determine if existing facilities are adequate to support a maintenance task. At this point, the engineer only determines that the existing facility is capable of performing the task. As the maintenance task analysis data base grows, the anticipated annual facility work load for each level of maintenance can be predicted by using the formula shown in Fig. 14-2. This approach addresses both the technical requirements and the capacity requirements for maintenance facilities to support the new equipment.

The final step is a comparison of existing facilities with the new requirements to identify what impact the new equipment will have on existing facilities. Remember, in most instances, a facility will support more than a single item of equipment, so although it might appear that an existing facility can support a new item of equipment, when the work load that is generated by other equipment is added to that of the new equipment, the existing facility might not be sufficient.

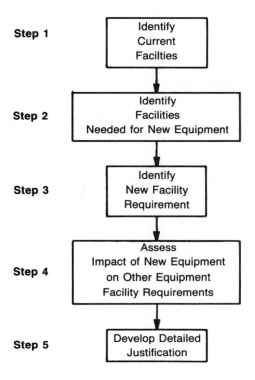

Fig. 14-1. Facility requirements identification.

Supply

Determining the requirements for supply facilities to support a new item of equipment can be a frustrating experience because the equipment is stored in pieces rather than as a complete unit. Therefore, each item that will require spares must be considered a separate thing to be stored. The space requirements of an item depends on the dimensions of the final package, not the size of the item. Additionally, unit packaging must be considered where multiple quantities of an item are packaged in the same container.

$$\text{Annual Facility Workload} = \sum_{1}^{N} (T_{F1} \times T_{T1}) + (T_{F2} \times T_{T2}) + \ldots + (T_{FN} \times T_{TN})$$

$$\textbf{Where: } N = \text{Number of Maintenance Tasks}$$
$$T_F = \text{Task Frequency}$$
$$T_T = \text{Task Time}$$

$$\sum_{1}^{N} = \text{Annual Facility Time Required to Support Maintenance.}$$

Fig. 14-2. Maintenance facility work load prediction.

Figure 14-3 provides a method of computing the space requirements for storage of spares at each level of supply. The formula is designed to accommodate planning for both single packaged and multipackaged spares. Factors for space utilization and fill rate are also included in the formula.

Space utilization makes allowances for unusable space due to dimension limitations and problems with uniformly stacking packages. This factor is always less than 100%, and may be as low as 50%, depending on the type of items being stored. The fill rate factor addresses the anticipated percentage of spares that will be on hand at any time. Realistically, this factor might be a low percentage, but the maximum space that could be required must also be considered. Both of these factors should be agreed on by the contractor and the government.

Training

The space required for training can be determined through a relatively simple process. The factors required are as follows:

- Training courses to be presented
- Frequency of each course
- Students per course
- Type of training area (classroom/work area) and percentage of utilization required for each course
- Average classroom in square feet per student
- Average work area in square feet per student
- Overriding circumstances

$$
\text{Space Requirements} = [(\sum_{1}^{N_S} (S_1 \times C_1) + (S_2 \times C_2) + \ldots + (S_N \times C_N)) + \sum_{1}^{N_M}
$$

$$
(\sum_{1}^{N_M} (M_1/Q_1 \times C_1) + (M_2/Q_2 \times C_2) + \ldots + (M_N/Q_N \times C_N))] \times (F/U)
$$

Where:

$\sum_{1}^{N_S}$ = Sum of space required for single packaged spaces

S_1 = Quantity of spare to be stocked

C_1 = Cubic space required for S_1

$\sum_{1}^{N_M}$ = Sum of space required for multipackaged spares

M_1 = Quantity of spare to be stocked

Q_1 = Quantity of spare per multipack container

C_1 = Cubic space required for multipack container

U = Utilization of space factor

F = Fill rate of spares

Fig. 14-3. Storage space requirements identification.

Training space requirements $= S[(C \times U_C) + (W \times U_N)]$

Where: $S =$ Number of students
$C =$ Classroom space required per student
$U_C =$ Percent of time classroom space is used for training
$W =$ Work area space required per student
$U_N =$ Percent of time work area space is required for training

Fig. 14-4. Training space requirements identification.

The first step is to calculate the space requirements of each course using the formula in Fig. 14-4. The next step is to use the number of courses and course frequency to determine the annual space requirements. The final step is to consider the overriding circumstances such as schedule and dedicated training space requirements that modify the results of the computations from step one.

Scheduling plays a large role in determining training space requirements. If all training must be conducted simultaneously, space requirements are much larger than if the courses are conducted consecutively. Dedicated training areas, such as space for training equipment that is applicable to only one course, will increase the overall requirements.

Special Requirements

In each of the areas just discussed, there can be special circumstances that make the need for a special facility unique for other than work load or utilization reasons. Special facility requirements can be due to many factors; but in most cases, they are the results of technology, mission, and geographical considerations.

The use of state-of-the-art technology in weapon systems can generate requirements for new kinds of repair facilities with capabilities that do not currently exist. For instance, when the Navy started using nuclear power for ships and submarines, the existing maintenance facilities for conventionally powered vessels were not adequate. This change in technology created a requirement for a completely new type of maintenance facility.

The intended mission of a weapon system can require special types of facilities. Mobility requirements are the results of mission-oriented support planning; units must be able to take support facilities as they deploy. There might also be requirements for variations in the types of facilities required to support a weapon system due to the geographical locations in which the equipment will be used. A system that will be used in all types of climates, from the tropics to the arctic, might require several different types of facilities for support of the same tasks. For instance, storage requirements for spares in the tropics and the arctic might be very different due to the physical properties of the items being stored. A different type of storage might be required to provide protection from mildew as opposed to protection from extreme cold. Remembered that there can be, and often are, circumstances that have an impact on the facilities required to support an item of equipment. These special circumstances must be considered during the facility requirements identification process.

REQUIREMENTS JUSTIFICATION

The final process to ensure that adequate facilities are available to support equipment when it is fielded is the preparation of a detailed justification for each facility. The justification must contain sufficient information to validate the requirements and also provide the information necessary to start planning facility identification, renovation, or construction. Detailed justifications should include information on the utilization, design criteria, construction, utilities, and rationale for the required facility.

Utilization

The justification must contain detailed information on the utilization rate of the proposed facility. A justification for a maintenance facility would identify the number of maintenance tasks to be performed, the task frequency, and the number of systems to be supported. Using these data, an annual projected work load can be computed for the facility. The projected work load can be used by the government to determine if an existing facility has the capacity to support the equipment or if a new facility is required.

Design Criteria

If the facility has any unique requirements that must be considered, such as type of construction or utilities, they must be identified. As stated earlier, the government recycles facilities by remodeling or renovation as new uses are identified, rather than constructing new buildings. Unique design requirements must be identified and justified to aid in the decision-making process for determining where the facility assets will be obtained.

Justification Rationale

The final information that must be provided to the government is the rationale used to develop the justification for a facility requirement. The most important questions that must be answered are (1) why a facility is required, and (2) why the recommended facility is the best or most reasonable way to satisfy the requirement.

If the contractor has developed the recommendation through the processes described above, then the government should be able to concur with the recommendation. Failure to sufficiently document the detailed requirements might result in disapproval of the recommendation, which will affect not only the facility planning but other areas such as maintenance planning, technical manuals, and personnel planning.

Chapter 15

Repair-Level Analysis, Reliability-Centered Maintenance, and Life Cycle Cost

Many different types of analyses have been discussed in previous chapters. Each analysis has its own application and merit; however, few analyses have as much impact on ILS as do Repair-Level Analysis (RLA), Reliability-Centered Maintenance (RCM), and Life Cycle Cost (LCC). The determination of where and when maintenance is performed and how much the total system with all its support requirements will cost top the list as drivers in determining what resources are ultimately required to support an item of equipment. RLA is used to determine the level of maintenance that is most cost effective for maintenance actions. RCM determines which maintenance tasks should be conducted on a scheduled basis. LCC evaluates alternatives to identify the most cost-effective utilization of resources over the life of a system.

REPAIR-LEVEL ANALYSIS

Several different names are given to this type of analysis, Repair-Level Analysis (RLA), Optimum Repair-Level Analysis (ORLA), Network Repair-Level Analysis (NRLA), or Level-of-Repair Analysis (LORA); however, the basic concept for which they were developed is the same. An RLA is used to evaluate a maintenance action to determine if it is cost effective and where the task can be accomplished most cost effectively. The easiest way to explain how an RLA works is to refer to the decision tree in Fig. 15-1.

As illustrated by the decision tree, there are actually five decisions made in the RLA process: (1) "O" level repair, (2) discard at point of failure, (3) repair at depot, (4) repair at "I" level; and (5) use RLA cost model. Three of the decisions are based

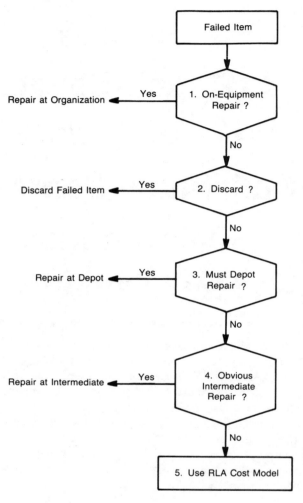

Fig. 15-1. Repair-level analysis decision tree.

NOTES:

1. Corrective maintenance can be performed without removing the item from the system.
2. Discard-at-failure versus repair analysis indicates it is more cost effective to discard than to repair the item.
3. Repair must be accomplished at the Depot level for reasons of technical complexity, high skill-level requirements, or sophisticated facility requirements.
4. Repair at the intermediate level is obvious if skills and support equipment required to accomplish repair are available and also required to perform other intermediate-level tasks.
5. There is no overriding reason for accomplishment of repair at intermediate or depot levels. Use RLA model to determine which level is most cost effective.

on factors not directly related to cost, and two use cost as the determining factor. The source for identification of maintenance tasks to be used as input to the RLA process is the maintenance task analysis (see Chapter 5).

Other information required to complete an RLA is shown in Fig. 15-2. Considerable information is needed to perform an accurate and reliable RLA. Therefore, the use of an RLA in the concept and DEMVAL phases may not be appropriate unless the user can quantify the uncertainties and risks involved. An RLA is most effective when used during FSD when adequate information is available; however, if it is used in the later stages of FSD, the results may be too late to be useful. Therefore, the RLA should be done as soon as possible, depending on the availability of information.

Organization-Level Repair

The most desirable place to perform maintenance is at the organizational level. If all maintenance were done at "O" level, then there would be no requirement for "I" and "D" level maintenance resources, which would significantly reduce the cost for maintenance. On the other hand, it is not realistic to propose that each organization have the capability to completely repair every failure that would occur (refer to the

Maintenance Task Analysis
    Maintenance task identification
    Manhours required per task
    Materials required per repair

Other Sources
    MTBF
    System operating hours
    Life expectancy
    Number of systems in operation
    Number of locations performing maintenance
    Unit cost
    Quantity per end item
    Projected cost of
        Support equipment
        Support equipment maintenance
        Technical documentation
        Training
        Spares
        PHS&T
        Labor
    Number of days for repair turnaround
    Safety-level days of supply

Fig. 15-2. Repair-level analysis input information.

example of maintenance tasks in Chapter 5). The thought of an M1 tank moving out on the battle field towing a trailer full of repair parts and test equipment is pretty scary. So there must be a reasonable middle ground for determining what maintenance tasks should be done where.

The key to identifying the tasks that should be performed by "O" level maintenance is to have a detailed knowledge of the existing capabilities of the proposed user. The tools, test capabilities, training and availability of personnel, and facilities available are used to determine if a task can and should be accomplished at "O" level. If everything needed to accomplish the tasks is at "O" level, then it is obvious that is where the task should be performed.

Discard at Point of Failure

It might be more economical to discard an item when it fails rather than repair it. This decision is based on a comparison of the cost to repair an item with the relative cost to buy a replacement. There are several methods used to determine which items should be discarded rather than repaired. Figure 15-3 illustrates the basic philosophy used in making this decision. This method compares the relative value of a repaired item with the cost to buy a replacement. The factor "N" is a predetermined acceptance level, either supplied by the government or set by the contractor, that is used to establish

---

Formula: IF $(MTBF_2/MTBF_1)\ N < (L + M)/P$ then discard

Where:

$MTBF_1$ = MTBF of new item
$MTBF_2$ = MTBF of repaired item
N = Predetermined acceptance level
L = Labor required to repair item
M = Material required to repair item
P = Unit price of new item

Examples:	A	B	C
$MTBF_1$	1,000	1,000	1,000
$MTBF_2$	1,000	800	600
L	100	20	40
M	100	40	80
P	200	500	200
N	60%	60%	50%
Computation Results	.60 < 1.0	.48 < .12	.30 < .60
Decision	Discard	Repair	Discard

Fig. 15-3. Discard-at-Failure versus repair analysis.

a criterion for derating the value of a repaired item. Basically, it states that if the cost for repair exceeds a given percent of the cost of a new item, the decision should be to discard the failed item.

Depot Repair

Some repair tasks must be accomplished at the depot level due to the technical complexity of the item, high skill requirements to accomplish the maintenance task, or requirements for sophisticated facilities. When such requirements exist, the decision to repair at the depot level overrides all other concerns. Maintenance planning must consider such requirements and establish design criteria early in the equipment design to minimize the need for depot maintenance as much as possible.

Intermediate Repair

Maximum use of existing skills and support equipment at the intermediate level to accomplish maintenance tasks will reduce the cost of repairs. Anytime repairs can be supported at the intermediate level without additional resources, such a repair decision is obvious.

RLA Model

When there is not an overriding reason that dictates accomplishment of a maintenance task at a specific maintenance level, the decision can be made on purely economic considerations. The RLA model is designed to determine which maintenance level will provide the most cost-effective alternative to performing a maintenance task over the life of an item of equipment.

There are several RLA models that can be used for this purpose. In most cases, the models are computerized to efficiently handle the amount of data and calculations that must be made to prepare a complete RLA for an entire weapon system. Figures 15-4 through 15-10 are examples of how an RLA would be used to determine where repairs of a single item would be accomplished. Figure 15-4 provides the basic formulas used to compute the costs for repair at depot and intermediate levels. Remember, the RLA uses cost as the determining factor for where a repair will be accomplished. Notice that the two formulas contain similar components, but the costs may vary between levels as shown in subsequent figures. Figure 15-5 shows the input information that will be used in the RLA computations. The decision tree results show the percentage of maintenance tasks (15%) for the control assembly, not the entire system, that are being modeled. The remaining 85% have already been assigned to a specific level based on the results of the decision tree. Number of repairs per month, calculated as indicated in Fig. 15-5, indicates the volume of repairs that must be accomplished.

Figure 15-6 provides specific input information for the depot-level model, and Fig. 15-7 provides specific input information for the intermediate-level model. Figure 15-8 shows the values, either given or computed, for the depot-level model, and Fig. 15-9 provides similar values for the intermediate-level model. Figure 15-10 is a comparison

$$DC = SE + SEM + TD + TNG + SS + PS + RP + L$$
$$IC = SE + SEM + TD + TNG + S + L$$

Where:

SE = Support equipment
SEM = Support equipment maintenance
TD = Technical documentation
TNG = Training
SS = Safety stock
S = Shipping/stocking spares
PS = PHS&T of failed items
RP = Repair pipeline
L = Labor required to repair item
DC = Depot costs
IC = Intermediate costs

Fig. 15-4. Intermediate versus depot repair analysis.

Item: Control Assembly
Cost: $5,000
QPEI: 2

Aircraft (ACFT): 500
Squadrons (SQDN): 20 (25 ACFT/SQDN)

Life expectancy: 10 yrs
Flight Hrs/Month: 20
Flt Hrs between failure: 10

Decision tree results: 60% Repairs on-equipment
5% Discard
10% Depot must repair
10% Obvious field repair
15% RLA model

RLA Repairs per month: 300

(No. ACFT × Flt Hrs/Mo Flt Hrs between fail) × QPEI × %
Tasks Model = Repairs/Mo

$$(500 \times 20)/(10 \times 2 \times .15) = 300$$

Fig. 15-5. Intermediate versus depot repair analysis (example).

SE = $50,000
SEM = Negligible
TNG = $5,000
TD = None
PHS&T for Failed Item = $150
Safety Stock Level = 15 days
Repair TAT = 60 days
Labor Cost = $12 per hr
Avg M/H per Repair = 2.5 hr

Fig. 15-6. Depot model data.

SE = $100,000 per squadron
SEM = 1% per year
TNG = $30,000 per squadron
TD = $100,000
Repair TAT = 8 days
Cost of Stocking Parts = $120 per repair
Cost of Labor = $5 per hr
Avg M/H per repair = 2.5 hr

Fig. 15-7. Intermediate model data.

SE   = $50,000
SEM = 0
TD   = $5,000
SS   = $750,000

$$\left[\begin{array}{l} \text{Repairables/Mo} \times \text{Safety Level} \times \text{Unit Price} = \text{SS} \\ \qquad 300 \times .5 \ (15 \ \text{days}) \times 5,000 \quad = \$750,000 \end{array} \right]$$

PS = $5,400,000

$$\left[\begin{array}{l} \text{Repairables/Mo} \times \text{Month Expectancy} \times \text{Cost to Pkg \& Ship} = \text{PS} \\ \qquad 300 \times 120 \quad \times 150 \quad = 5,400,000 \end{array} \right]$$

RP = $3,000,000

$$\left[\begin{array}{l} \text{Repairables/Mo} \times \text{TAT} \times \text{Unit price} = \text{RP} \\ \qquad 300 \quad \times 2 \ (\text{month}) \times 5,000 \ = 3,000,000 \end{array} \right]$$

L = $1,080,000

$$\left[\begin{array}{l} \text{Repairables/Mo} \times \text{Months} \times \text{Labor rate} \times \text{Hrs/Repair} = \text{L} \\ \qquad 300 \times 120 \ \times 12 \ \times 2.5 \quad = 1,080,000 \end{array} \right]$$

Fig. 15-8. Depot-level computations.

223

SE = 2,000,000

$$\text{Cost of PSE} \times \text{No. of Squadrons} = \text{SE}$$
$$100,000 \times 20 = 2,000,000$$

SEM = 200,000

$$\text{Cost of PSE} \times \text{Maint Cost \%} \times \text{Yrs Used} \times \text{No. of Squadrons} = \text{SEM}$$
$$100,000 \times 0.01 \times 10 \times 20 = 200,000$$

TD = 100,000
TNG = 600,000

$$\text{Uniting Cost} \times \text{No. of Squadrons} = \text{Tng}$$
$$30,000 \times 20 = 600,000$$

S = 4,320,000

$$\text{Cost of Stocking Parts} \times \text{Failure/Mo} \times \text{No. of Months} = \text{S}$$
$$120 \times 300 \times 120 = 4,320,000$$

L = 450,000

$$\text{Failures/Mo} \times \text{No. Months} \times \text{Labor Cost} \times \text{Hrs/Repair} = \text{L}$$
$$300 \times 120 \times 5 \times 2.5 = 450,000$$

Fig. 15-9. Intermediate-level computations.

Element	Field	Depot
SE	$2,000,000	$50,000
SEM	200,000	0
TD	100,000	0
TNG	600,000	5,000
S	4,320,000	N/A
SS	N/A	750,000
PS	N/A	5,400,000
RP	N/A	3,000,000
L	450,000	1,080,000
Total	$7,670,000	$10,285,000

Fig. 15-10. Intermediate versus depot analysis.

of the values for each model. This analysis shows that, based on purely economic factors, the repairs should be accomplished at the intermediate level.

It is interesting to note that in this example the cost of additional spares and shipping of items to the depot for repair are the driving factors that lead to the decision to repair at the intermediate level. The sensitivity of models for conducting trade-off comparisons between alternatives must also be considered. Figure 15-11 shows that if the reliability of the control assembly could be doubled, from one failure per 10 flight hours to one failure per 20 flight hours, then the results of the RLA model would indicate that the repairs should be accomplished at the depot level, and the total cost for either level would be significantly less than the costs determined in Fig. 15-10.

The RLA is a useful tool when trying to determine where repairs should be accomplished based on cost. It can also be used to make trade-off decisions of proposed alternative support approaches. As can be seen in the above illustration, the RLA can be a complex task when modeling an entire weapon system and, therefore, should be done by computer. MIL-STD 1390, Level of Repair, provides a detailed process for accomplishing an RLA. The formulas contained in this document are comprehensive and give specific guidance on how an RLA should be conducted.

RELIABILITY-CENTERED MAINTENANCE

The purpose of reliability-centered maintenance (RCM) is to develop a scheduled maintenance program that increases the availability of an item of equipment by identifying failures or potential failures before they degrade equipment effectiveness. An RCM analysis is conducted to determine what maintenance tasks would provide increased equipment reliability over the life of the equipment based on a logical selection criteria. This analysis technique is applicable to any equipment development program. MIL-STD 2173, Reliability-Centered Maintenance Requirements for Naval Aircraft, Weapon Systems and Support Equipment, contains detailed instructions on how to conduct an RCM analysis.

Element	Intermediate	Depot
SE	$2,000,000	$50,000
SEM	200,000	0
TD	100,000	0
TNG	600,000	5,000
S	2,160,000	N/A
SS	N/A	375,000
PS	N/A	2,700,000
RP	N/A	1,500,000
L	225,000	540,000
Total	$5,285,000	$5,170,000

Fig. 15-11. Intermediate versus depot analysis (adjusted).

Preventative Maintenance

Maintenance tasks performed on a scheduled, periodic basis to prevent failures while an item of equipment is in operation are termed preventative maintenance tasks. These tasks should not be confused with other scheduled maintenance tasks that are required to sustain operation such as lubrications or adjustments.

Preventative maintenance tasks can be divided into two categories: scheduled inspections and scheduled removals. A scheduled inspection can be accomplished at any level of maintenance to identify failures that have occurred or impending failures. Scheduled removals are conducted to recondition items that have reached a predetermined usage or that have reached an anticipated statistical useful life. Maintenance tasks resulting from scheduled inspections are either on-condition or failure-finding tasks. Scheduled removals generate either scheduled rework tasks or discard of the removed item. The purposes of on-condition, rework, and discard tasks are to prevent single-point failures where failure-finding tasks are to prevent multiple failures. Figure 15-12 describes each of these tasks in detail.

Analysis Process

The RCM analysis process considers the significant items that comprise an equipment item. It uses information generated by the FMECA to identify the items most critical to the reliability of the equipment and where a failure would have the greatest effect on availability. Figure 15-13 illustrates the analysis process and shows how both design and field data are used during the analysis process to identify preventative maintenance tasks. The analysis has proven to be effective in planning

---

MIL-STD 2173

Scheduled inspections:
1. **On-condition task**—A scheduled inspection, test, or measurement to determine whether an item is in, or will remain in, a satisfactory condition until the next scheduled inspection.
2. **Failure-finding task**—A scheduled inspection of a hidden function item to find functional failures that have already occurred but were not evident to the operating crew.

Scheduled removals:
1. **Rework task**—Scheduled removal of units of an item to perform whatever maintenance tasks are necessary to ensure that the item meets its defined condition and performance standards.
2. **Discard task**—Scheduled removal of an item to the card the item or one of its parts at a specified life limit.

Fig. 15-12. Preventative maintenance tasks.

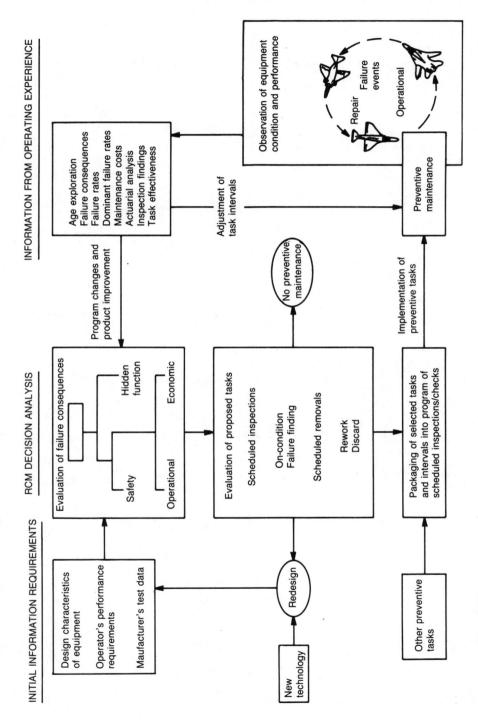

Fig. 15-13. Reliability-centered maintenance analysis process.

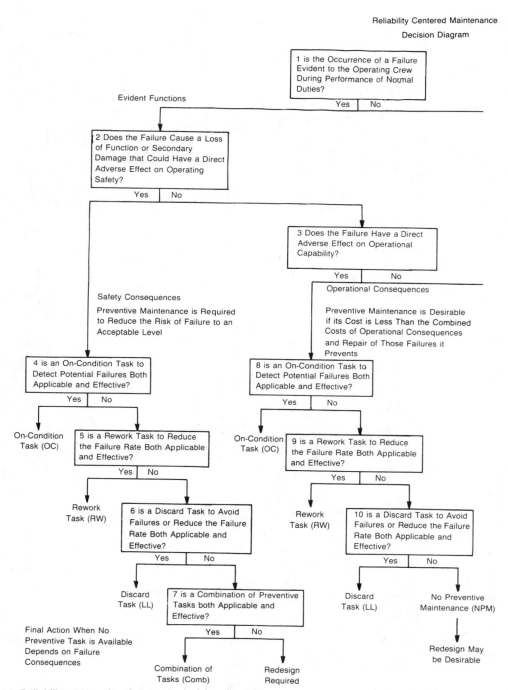

Fig. 15-14. Reliability-centered maintenance decision diagram.

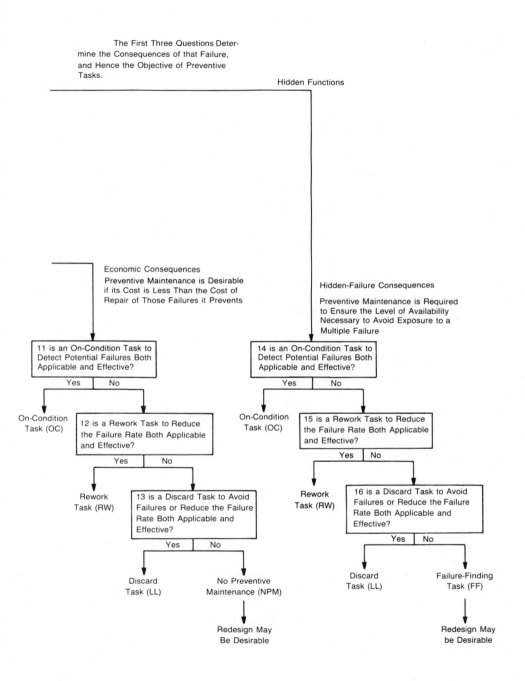

Fig. 15-14. Reliability-centered maintenance decision diagram. (Continued from page 228.)

preventative maintenance programs for new systems during the development process and when upgrading an existing maintenance program for a system that has experienced significant field use.

Decision Logic

The key to the RCM analysis is the RCM decision logic shown in Fig. 15-14. Using this decision tree as a guide, a complete analysis of each significant item can be conducted. The results of the analysis provide a clear decision as to what preventative maintenance tasks should be developed to support a system. As shown in the decision diagram, there is a step-by-step process consisting of 16 yes or no questions that lead the analyst to decide which type of task, if any, is required. In cases where no clear cut information is available for making the decision, MIL-STD 2173 provides default answers that can be used to complete the analysis. The 16 questions that lead to an RCM decision are the summary output for recording the results of the analysis process.

This iterative process is used to evaluate each maintenance-significant item to determine if a preventative maintenance task is warranted. The consolidated results of the RCM analysis process forms the preventative maintenance program for the system. Note that the decisions are divided into four areas: safety, operational, economic, and hidden failure detection. Each area is related to the activities of several ILS disciplines that should be involved in the RCM analysis process.

Age Exploration

The RCM process does not cease when the system is fielded. A continuing analysis of the preventative maintenance program is conducted to identify areas for improvement. This extension of the process is called age exploration. Because many preventative maintenance programs are initially planned and implemented using insufficient data, this technique is necessary to achieve the maximum benefits of RCM.

Age exploration is a systematic gathering of actual field operation data used to refine the preventative maintenance program. Each service has established data-gathering systems to accumulate sufficient field data to thoroughly review and refine the preventative maintenance program.

Documentation

MIL-STD 2173 contains a set of data sheets used to record the results of the RCM analysis process. RCM Worksheet Number 1, which provides a summary of the results for a specific item, is shown in Fig. 15-15. Note that block number 4 contains the answers to the applicable questions of the decision tree shown in Fig. 15-14. The other worksheets provide detailed backup for the summary contained on this worksheet. This information can also be documented in the LSAR data base, discussed in Chapter 17.

LIFE CYCLE COST

The prediction of the total costs that will be incurred throughout the life of a weapon system, or any other equipment, procured by the government serves an important role

RCM Worksheet No. 1

Summary Data

1) Page ____ Of ____

a) System/Subsystem Nomenclature

b) LSACN/WUC

c) Part Number/FSCM/Model Number

d) Reference Drawing

e) Indenture Level

f) End Item Nomenclature/Type, Model, Series

g) Revision No./Date

h) Prepared by/Date

j) Reviewed by/Date

k) Approved by/Date

1) Item LSCAN/WUC

2) Item Nomenclature

3) List the Functional Mode Code (FFMC)

4) RCM Logic Question Answers (Y or N)
1 2 3 4 5 6 7 8 9 10 11 12 13 14 15 16

5) Description of Maintenance Requirements From RCM Task Worksheets

6) Task Number

7) Inspection Interval (Packaged)

8) LSA Task Code

Fig. 15-15. RCM analysis summary worksheet.

in the acquisition process. The total life cycle cost of a weapon system encompasses every conceivable direct and indirect cost that will be related to the acquisition, operation, support, and disposal of the system. This is an extremely useful tool during the acquisition process when determining the best alternatives for design configurations, operation concepts, maintenance concept, production schemes, and logistics support policies. MIL-HDBK 259, Life Cycle Cost in Navy Acquisitions, provides a detailed description of the applications for life cycle cost analysis.

Cost Elements

The predicted total life cycle cost is determined by combining all relevant cost elements associated with the costs incurred for acquisition, operation and support, and disposal of a weapon system. These costs, both direct and indirect, reflect the government's cost of ownership. Figure 15-16 shows the relative portion that each of these elements contributes to total life cycle cost. The figure illustrates the often overlooked fact that the majority of costs for a weapon system are due to operation and support costs rather than acquisition costs. This is why it is of utmost importance that every effort be made during the acquisition process to design a system that optimizes total life cycle cost rather than acquisition cost. Remember, ILS is responsible for

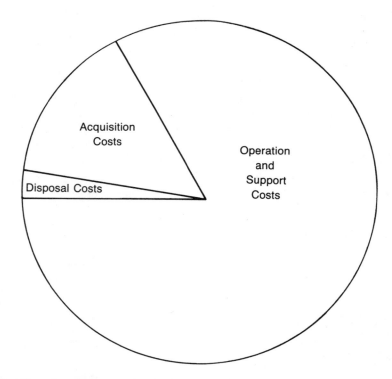

Fig. 15-16. Life cycle cost element contributions to total cost.

influencing the design to produce a system that is as supportable as possible for the lowest cost.

Acquisition Costs. All costs incurred by the government from the start of the concept phase until the end of the production phase are normally considered acquisition costs. This includes funds expended both internally and externally by the government, not just the funds paid to contractors to design and develop the weapon system. Acquisition costs are further divided into research and development (R&D) costs and investment costs as shown in Fig. 15-17.

Research and Development Costs. The costs incurred by the government during concept, demonstration and validation, and full-scale development phases of the acquisition process are categorized as R&D costs. The products of R&D are completion of the detailed documentation, e.g., engineering drawings, specifications, and plans, necessary to enter the production phase. All costs associated with the R&D effort, including planning, management, engineering, test, evaluation, special equipment, and facilities, whether incurred by the government or contractors, make up this cost element.

Investment Costs. The costs of actually producing the weapon system, procuring and developing the necessary initial support, and establishing an initial operating capability are considered investment costs. They include the cost of building the system, initial spare and repair parts, training operation, maintenance and supervisor personnel, support and test equipment, documentation, maintenance facilities, and initial PHS&T.

Investment costs often overshadow other life cycle costs and distort the perspective of attempting to reduce costs because they are considered the price paid by the government to contractors to build a weapon system. Weapon system procurement contracts can exceed a billion dollars, which seems like an enormous amount of money, but it represents only a small percentage of the total cost that the government will pay for ownership of the weapon system over its complete life cycle.

Research and development costs	Investment costs
Planning	Production
Management	Planning
Engineering	Management
Test	Initial spares
Evaluation	Training
Equipment	Support equipment
Facilities	Technical manuals
	Engineering
	Test
	Facilities
	Initial PHS&T

Fig. 15-17. Acquisition cost elements.

Operation and Support Costs. The largest percentage of costs incurred over the life cycle of a weapon system are due to operation and support (O&S) costs. Included in O&S costs are direct and indirect costs necessary to sustain the weapon system. Figure 15-18 illustrates typical O&S costs elements.

Direct O&S Costs. Any cost that has a direct relationship to the operation or support of a weapon system is considered a direct cost. Costs for personnel include operators, maintenance personnel, and supervisors, both military and civilian, who are responsible for the operation or maintenance of the weapon system or components. This can include support personnel who are directly related to the mission scenario of the weapon system. A subelement of personnel cost is the cost of specific training related to operation and maintenance of the weapon system, which can be initial SSC-related training, OJT training, or sustainment training.

Consumables are any items required to sustain operation or maintenance, e.g., fuel, lubricants, maintenance materials, expendable supplies, or repair parts. The purchase of spare parts needed to replace initial provisioned items or increase the range and depth of spares is a direct cost. The cost of maintenance of support and test equipment and procurement of replacement items is also a direct cost.

Any facility costs that are incurred for support of operation or maintenance activities are direct, but this does not include the construction or modification costs, which are considered an investment cost. Facility costs also include the cost of water, power, and other utilities directly related to maintenance or operations. Maintenance of supplies and equipment is a direct cost, but the labor requirements for maintenance are included in personnel costs and should not be duplicated in this cost element. PHS&T costs include all movements of the weapon system due to operation or maintenance needs after initial delivery and the movement of spares and repair parts between maintenance facilities, supply facilities, and the user. Technical data are initially procured as an investment cost; however, the maintenance and updating of the data are direct O&S costs.

Direct costs	Indirect costs
Personnel	Personnel
Consumables	Facilities
Replacement spares	Training
Support equipment	
Facilities	
Maintenance	
PHS&T	
Technical data	
Supply management	
Modifications	

Fig. 15-18. Operation and support cost elements.

Supply management costs are attributed to the unique spare and repair parts of the weapon system that must be stocked to support operations or maintenance. These costs are incurred at all levels of supply, from organizational to depot, and also include the administrative costs of maintaining DLSC and NICP records for the items. All engineering changes and other modifications to the weapon system that occur after deployment are direct O&S costs. Modification costs are considered sustaining investment costs necessary to enhance the reliability, maintainability, supportability, or operational capabilities of the weapon system.

Indirect O&S Costs. Those costs incurred for relevant services, support personnel, and noninvestment items that are necessary to sustain operations or maintenance, but cannot be directly related to a specific weapon system, are categorized as indirect O&S costs. These costs can include a broad range of cost elements such as military installation facilities, medical facilities, maintenance of real estate, and initial training costs. Personnel costs classified as indirect may include medical personnel, initial training instructors, and personnel administrators and managers. Indirect facility costs consist of real property maintenance and upkeep, installation maintenance, base exchanges and commissaries, and other facilities required to indirectly support either the personnel or operation and maintenance of the weapon system. Initial training cannot be attributed to a specific weapon system, but it is required to produce trained operation and maintenance personnel.

Disposal. An often ignored cost element is the cost of disposing of a weapon system as it becomes obsolete or is replaced. In some instances, the equipment might have salvage or resale value to offset the cost of disposal; however, costs can be incurred.

Figure 15-19 shows typical costs incurred during disposal. Spare and repair parts that are unique to the weapon system being disposed must be purged from the active supply inventory, which may constitute a significant expense. If the weapon system, such as an aircraft, has many lines of supply that must be disposed of, then the cost for such an operation should be identified. PHS&T costs are incurred to physically move the weapon system from its operational sites to a disposal site.

As a part of the disposal effort, the data collected during the life cycle of the system must be closed and dispositioned. Significant data related to operations, reliability, maintainability, performance, or other information that has other uses are reviewed and forwarded to the appropriate destination. If the weapon system is to be sold as foreign military sales (FMS) or redistributed to other users, then refurbishment or

Fig. 15-19. Disposal cost elements.

Inventory closeout
PHS&T
Data management
Refurbishment
Demilitarization
Waste management

overhaul might be required. A portion of this cost might be recouped after transfer; however, portions might be charged as a part of disposal.

Demilitarization is the act of rendering an item useless for military purposes. Government regulations require that certain classes of items be demilitarized before disposal. If the weapon system being disposed of requires such actions, then the costs are accrued as disposal costs. Weapon systems, or their components, that contain dangerous or hazardous materials require special handling for disposal. If the item contains nuclear materials or dangerous chemicals, then the disposal process might be lengthy and very costly. Such costs can have a significant impact on the predicted life cycle cost.

MODELS

Life cycle cost models are extremely complex if they are to be of any value. As described above, there are a myriad of cost factors that must be considered when attempting to predict the life cycle cost of a weapon system. Therefore, all modeling should be done by computer due to the number of possible elements and variable inputs. There are several LCC models available for use by contractors; each service has a preferred method that should be used.

In each case, the models have been developed to address specific situations related to life cycle cost, and proper use depends on an understanding of the intent of the model. Some models are actually a series of submodels that address certain aspects of the life cycle.

Modeling Concept

The basic concept of life cycle cost modeling is illustrated in Fig. 15-20. Each of the major cost elements can be expanded to include several hundred subelements and variables. Figure 15-21 shows how the concept of Fig. 15-20 can be expanded using only the subelements identified in the previous section of this chapter. This refinement process can be repeated until the resulting model contains elements that address every cost that can be associated with a weapon system. That is why it is important to use computer models for this task.

Model Characteristics. If a life cycle cost model is to be a useful tool in analyzing the total cost of a weapon system, it should contain certain characteristics. MIL-HDBK 259 contains a list of desired LCC model characteristics, shown in Fig. 15-22. Regardless

---

$$C_T = C_A + C_I + C_{OS} + C_D$$

$$\text{Where: } C_T = \text{ Total life cycle cost}$$
$$C_A = \text{ Total acquisition cost}$$
$$C_I = \text{ Total investment cost}$$
$$C_{OS} = \text{ Total operation and support cost}$$
$$C_D = \text{ Total disposal cost}$$

Fig. 15-20. Life cycle cost concept.

$$C_T = C_{RP} + C_{RM} + C_{REN} + C_{REV} + C_{REQ} + C_{RF} + C_{IPR} + C_{IPN} + C_{IM} + C_{IS} +$$
$$C_{ISE} + C_{TM} + C_{IE} + C_{IF} + C_{IP} + C_{ODP} + C_{OC} + C_{ORS} + C_{OSE} + C_{ODF} +$$
$$C_{ODM} + C_{OP} + C_{OTD} + C_{OSM} + C_{OM} + C_{OIP} + C_{OIF} + C_{OIT} + C_{DI} + C_{DP} +$$
$$C_{DDM} + C_{DR} + C_{DD} + C_{DW}$$

Where:

C_T = Total life cycle cost
C_{RP} = R&D planning costs
C_{RM} = R&D management costs
C_{REN} = R&D engineering costs
C_{REV} = R&D evaluation costs
C_{REQ} = R&D equipment costs
C_{RF} = R&D facilities costs
C_{IPR} = Investment production costs
C_{IPN} = Investment planning costs
C_{IM} = Investment management costs
C_{IS} = Initial spares costs
C_{ISE} = Initial support equipment costs
C_{TM} = Technical manual costs
C_{IE} = Investment engineering costs
C_{IF} = Investment facilities costs
C_{IP} = Initial PHS&T costs
C_{ODP} = O&S direct personnel costs
C_{OC} = O&S consumables costs
C_{ORS} = O&S replacement spares costs
C_{OSE} = O&S support equipment costs
C_{ODF} = O&S direct facilities costs
C_{ODM} = O&S direct maintenance costs
C_{OP} = O&S PHS&T costs
C_{OTD} = O&S technical data costs
C_{OSM} = O&S supply management costs
C_{OM} = O&S modification costs
C_{OIP} = O&S indirect personnel costs
C_{OIF} = O&S indirect facilities costs
C_{OIT} = O&S indirect training costs
C_{DI} = Disposal inventory closeout costs
C_{DP} = Disposal PHS&T
C_{DDM} = Disposal data management costs
C_{DR} = Disposal refurbishment costs
C_{DD} = Disposal demilitarization costs
C_{DW} = Disposal waste management costs

Fig. 15-21. Basic life cycle cost model.

1. The model should be useful to the acquisition management process as well as to the review process.
2. The model should be sensitive to management control factors, design changes, and varied operational and logistics support scenarios.
3. All significant cost drivers that are relevant to the issue under consideration should be incorporated into the model as clearly as possible.
4. The development, alteration, updating, and operation of the model should be as inexpensive as possible.
5. The model should be sensitive to design parameters or acquisition characteristics that affect the cost of investment alternatives.
6. Valid, relevant input data should be readily available.
7. The model should be flexible and capable of accommodating the growing complexity of an acquisition; and it should allow for adjustment of inflation, discounting, and learning curve factors.
8. The model should be separated into interactive modules for easier modification.
9. Inputs and outputs should be expressed in terms that are familiar to users and that can be verified to ensure credibility.
10. Outputs should be reliable, i.e., results should be repeatable.

Fig. 15-22. Life cycle cost model characteristics.

of the type or origin of the model chosen, it should be capable of providing comparisons and evaluations for trade-off analyses of alternative options, identifying risks, and establishing a baseline for sensitivity analyses throughout the acquisition process.

Modeling Problems. When LCC models are used as a tool for making critical design and support decisions, problems can occur that distort the model's utility. These problems should be considered when choosing the model and interpreting the resulting predictions. Common problems include (1) use of invalid assumptions when insufficient data exist, (2) changes in production schedule or order quantities, (3) lack of uniformity in categorizing cost elements, (4) inadequate description of the life cycle, (5) use of obsolete data, (6) inappropriate cost element structure, and (7) use of inaccurate inflation or discount rates. Any combination of these problems can invalidate the results of an LCC model.

Another common problem associated with using the results of LCC models is for analysts to focus too much attention on the cost aspects rather than on the limited availability of some critical resources. Sometimes cost may not be the driving factor for making critical support decisions; it may be the optimum use of limited critical resources.

Data Sources. LCC modeling requires an enormous amount of input data from many different sources in order to produce a reasonable prediction. Figure 15-23 shows typical data sources for LCC modeling. Much of the data comes from the government through direct or indirect sources. The Office of Management and Budget (OMB) provides current and projected costs for government cost elements. The Defense Logistics Agency (DLA) can be a source for data related to projected supply and maintenance activities. Other significant information from the government includes the force structure of the military, when and where the weapon system will be deployed, and operational scenarios.

Contractor- supplied information is the results of analyses and data collection efforts by ILS disciplines over the course of supporting a specific contract that consolidates data from engineering design, manufacturing, and other development costs. In cases where valid data do not exist, the contractor must develop ground rules for assumptions to be used until actual data are available.

It should be pointed out that the majority of the costs predicted for the weapon system will occur after the contract for development and production is complete, so

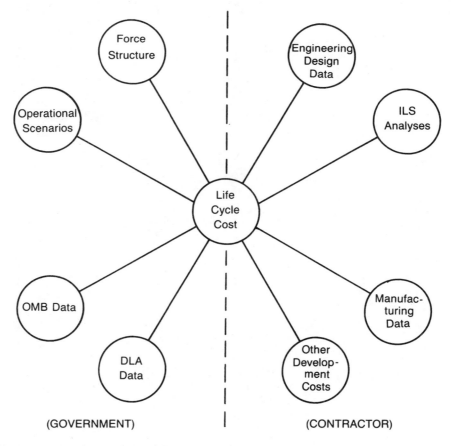

Fig. 15-23. Life cycle cost data sources.

the contractor must rely heavily on the government for valid data to predict the operation and support and disposal costs for the weapon system. During the concept phase, the quantity and quality of detailed design data are limited so the bulk of information used for modeling must come from assumptions or estimates. As the design matures near the end of full-scale development, the data should be much more accurate, making the LCC model results as precise as possible.

Estimating Techniques

There are three accepted techniques for estimating life cycle costs when data for modeling are not available: analogies, parameters, and engineering estimating. MIL-HDBK 259 contains a detailed explanation of these techniques and their application. Each method has varying degrees of application to the phases of a weapon system life cycle. They can be used independently or interactively to produce estimates of the predicted life cycle cost.

Chapter 16

Logistic Support Analysis

Each of the ILS disciplines discussed in previous chapters has a specific role in the overall logistic support planning process. One of the biggest problems that has faced ILS is how to coordinate the activities of these disciplines to achieve the best logistic support package possible. There is an endless supply of stories about how the ILS disciplines failed to coordinate information during the design of a weapon system, which resulted in technical manuals that did not match the equipment, spare parts that were not interchangeable with the original equipment, training courses that did not address the actual equipment design, and useless or unnecessary support equipment. In fact, there was no established method for the disciplines to formally communicate, so it is easy to understand how errors such as these occurred. Another problem for ILS disciplines was that it was next to impossible to have any input into the design process because of the disjointed methods of collecting and analyzing support information. Because of this and other reasons, the process known as logistic support analysis (LSA) was developed.

Generally speaking, any analysis method or technique that addresses logistics support or that is used to identify logistics support resources is a logistic support analysis. However, the term LSA now has a more specific meaning. The LSA process was developed with four goals in mind, as shown in Fig. 16-1.

The first goal is to use the results of the LSA process to influence the equipment design process to consider supportability requirements. That is, use LSA to identify ways of making the weapon system easier to support. The second goal is to identify the support problems and items that drive the cost of support early enough in the design process to change the design to fix or eliminate support problems. The third goal is

- Cause logistic support considerations to influence design.
- Identify support problems and cost drivers early.
- Develop logistic support resource requirements for system life.
- Develop a single logistic support data base.

Fig. 16-1. Logistic support analysis goals.

to develop a complete set of projections of the total support resources that will be required to support the weapon system or equipment over its complete life cycle. The final goal of LSA is to develop and use a single data base for all analyses.

Prior to LSA, each ILS discipline collected, analyzed, and stored data for its own use. This resulted in a mismatch of information or lack of continuity that caused some of the problems noted above. By using a single data base, each discipline can be assured that the information being used is the same that others are also using, and the results that one discipline generates are readily available to others.

The LSA process can only be successful if applied to a program where these goals can be achieved. There are two methods of planning logistic support: sequential and integrated. As shown in Fig. 16-2, the difference between these two methods is when the support system is designed in relationship to the design of the equipment.

LSA cannot be effective if applied using the sequential method because the first two goals, the influence of support in equipment design and the early identification of support problems and cost drivers, are not possible since the system design is complete before support planning begins. Therefore, the integrated method must be used to realize the full benefit of the LSA process.

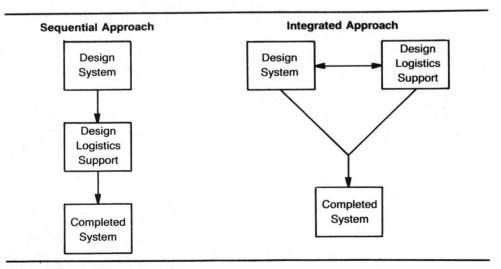

Fig. 16-2. System design approaches.

There are two distinct areas in LSA: doing the analyses and recording the results. Too many times, logisticians get caught up in the documentation part of LSA and forget that the real purpose of LSA is to perform the analyses. The LSA program is structured in a manner that allows detailed identification of specific requirements for each program. This enables tailoring of requirements to match the specific complexity of weapon system or equipment being designed, and it encourages emphasis on analyses rather than on merely filling in the boxes on data sheets. LSA has proven to be a significant step forward in ILS planning and the development support resource requirements.

THE LOGISTIC SUPPORT ANALYSIS PROGRAM

The establishment and implementation of a logistic support analysis program is a government contract requirement for design and development of weapon systems and other equipment. MIL-STD 1388-1A, Logistic Support Analysis, contains detailed descriptions of the requirements of an LSA program and the tasks that must be performed. The LSA program consists of a series of 15 interrelated tasks that are divided into five sections: program planning and control, mission and support systems definition, preparation and evaluation of alternatives, determination of logistic support resource requirements, and supportability assessment, as shown in Fig. 16-3. This organization

---

MIL-STD 1388-1A

Task section 100—Program planning and control
    Task 101—Early LSA strategy
    Task 102—LSA plan
    Task 103—Program and design reviews

Task section 200—Mission and support systems definition
    Task 201—Use study
    Task 202—Mission hardware, software, and support system standardization
    Task 203—Comparative analysis
    Task 204—Technological opportunities
    Task 205—Supportability and supportability-related design factors

Task section 300—Preparation and evaluation of alternatives
    Task 301—Functional requirements identification
    Task 302—Support system alternatives
    Task 303—Evaluation of alternatives and trade-off analysis

Task section 400—Determination of logistic support resource requirements
    Task 401—Task analysis
    Task 402—Early fielding analysis
    Task 403—Post-production support analysis

Task section 500—Supportability assessment
    Task 501—Supportability test, evaluation, and verification

---

Fig. 16-3. Logistic support analysis tasks.

is similar to that of MIL-STD 785B and MIL-STD 470, where tasks are provided for specific actions; however, because of the broad application of MIL-STD 1388-1A and the many aspects of ILS that the standard addresses, most tasks are divided into subtasks that can be tailored to fit a specific application of LSA. MIL-STD 1388-1A is also unique in that it provides not only a description of each task, but it also identifies the inputs required to perform a task and the outputs that performing a task generate.

As each task is discussed in this chapter, the task inputs and outputs are illustrated to show how each ILS discipline must participate in the LSA process in order to achieve the goals of the program. (Note that in many cases the information required to complete a task must be supplied by the government. This is an important point since prior to LSA there was little definition as to what information a contractor could expect to be supplied by the government.)

For years, ILS has accomplished many tasks as a matter of course with little or no detailed guidance. Many of these tasks have significant impact on the overall logistics resources required to support a weapon system, but detailed directions did not exist that forced the analyst to consider them each time. This led to a haphazard approach to logistic analyses. If a contractor follows the guidance of MIL-STD 1388-1A to the letter, the resulting equipment design and logistic support package will be the best possible balance among system performance, supportability, and life cycle cost.

Program Planning and Control (Task Section 100)

The purpose of the program planning and control section is to provide a standard method for LSA program initiation, control, and management. Task Section 100 is made up of three tasks: strategy, planning, and reviews. These tasks are applicable to any phase of an LSA program and should be required for initiation of the program.

Early LSA Strategy (Task 101). The first task of the LSA program is unique in that there is no other task contained in any document mentioned heretofore that required someone to stop and think about how a program should best be accomplished. That is the purpose of Task 101, to require the government, the contractor, or both to make a conscious decision, based on facts, as to what should be done with regard to LSA before any other tasks are started. This task should be done by the government before the RFP is written. It surely should be done before any contract is awarded; otherwise, the LSA program may not provide the best return on the investment possible. Figure 16-4 shows the inputs and outputs of this task.

Subtask 1—Develop a Strategy. The strategy developed for the LSA program must be consistent with the proposed design, maintenance concept, and operational scenario for the new weapon system. The key to developing the strategy is to identify the tasks required to achieve the specific goals of the acquisition program and which of them will provide the best return on investment. Another significant decision that must be made in developing the strategy is to identify who will accomplish the LSA tasks and when the tasks should be done. The completed strategy should address each point of the design and operation of the equipment and how the LSA program should be conducted in order to receive maximum benefit.

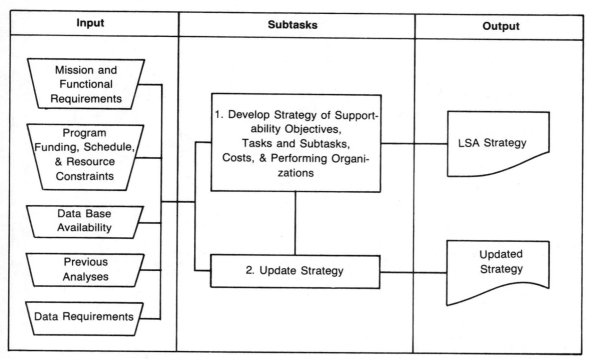

Fig. 16-4. Task 101. Early LSA strategy.

Subtask 2—Update. The second part of this task is to update the strategy as the program proceeds. All too often, a strategy is never reviewed to determine if changes are required based on better information. The LSA strategy should be updated, at a minimum, each time the program changes due to schedule modifications, major design changes, changes in funding, or as indicated by the results of analyses.

LSA Plan (Task 102). The purpose of the LSA plan is to document the processes and procedures that will be used to manage and control the LSA program. Figure 16-5 shows the inputs and outputs of Task 102. The plan must address each facet of the program and identify detailed responsibilities for accomplishment. It is not unusual for a contractor to be required to submit a draft LSA plan to the government as a part of the overall response to an RFP. Therefore, the contractor must understand the LSA process and be able to apply this understanding to the specific requirements of an RFP before any contract is awarded.

Subtask 1—Prepare Plan. An LSA plan (LSAP) provides a detailed description of the complete LSA program. A key requirement for an acceptable plan is that it must address each task, explaining how the task will be accomplished, the schedule for accomplishment, information required to complete the task, and how the output will be used to meet the LSA program goals. Additionally, the plan must integrate the activities of all ILS disciplines in a manner that streamlines the ILS process while

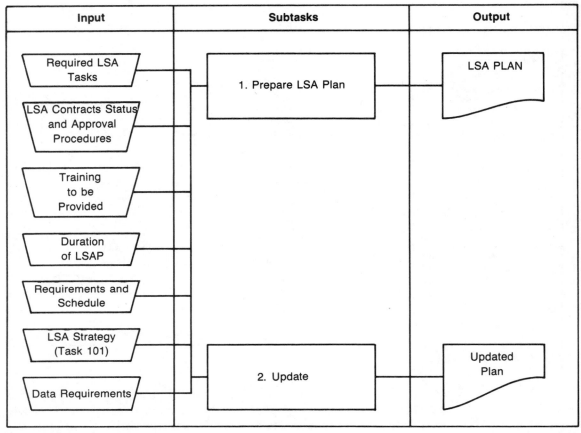

Input	Subtasks	Output
Required LSA Tasks		
LSA Contracts Status and Approval Procedures	1. Prepare LSA Plan	LSA PLAN
Training to be Provided		
Duration of LSAP		
Requirements and Schedule		
LSA Strategy (Task 101)		
Data Requirements	2. Update	Updated Plan

Fig. 16-5. Task 102. LSA plan.

receiving the maximum benefit from each discipline. Figure 16-6 provides a basic outline for an LSAP. Note that the outline covers the complete spectrum of activities related to ILS. The management portion of the plan is essential to integrating the efforts of each ILS discipline. Figure 16-7 lists the points that should be addressed in an LSAP.

Subtask 2—Update. The LSAP should be updated as changes occur to the program. The dynamic nature of the LSA program causes changes in the procedures and processes identified in the LSA plan; therefore, the LSAP must be continually updated to account for them. Most contracts require that the LSAP be updated quarterly throughout the LSA program. Remember that the LSA plan is the single controlling document for the detailed mechanics of how the LSA program is conducted.

Program and Design Reviews (Task 103). The LSA program and its progress toward meeting the program goals should be a topic of discussion at every meeting between the government and the contractor. The impact of the LSA program should be felt in every other activity concerned with the overall design effort. The purpose of Task 103 is to formally require that LSA be addressed at each program and design

1.0 Introduction
    1.1 Purpose of plan
    1.2 Scope of plan
    1.3 Reference documents
    1.4 Summary of plan
    1.5 Plan updating process
2.0 System Description
    2.1 Hardware and software description
    2.2 Support equipment description
        2.2.1 Common support equipment (CSE)
        2.2.2 Peculiar support equipment
    2.3 Maintenance concept
        2.3.1 Crew-level maintenance
        2.3.2 Organization-level maintenance
        2.3.3 Direct support-level maintenance
        2.3.4 General support-level maintenance
        2.3.5 Depot-level maintenance
        2.3.6 Military occupational specialty
        2.3.7 Support equipment maintenance
3.0 LSA/LSAR Process
    3.1 Purpose and scope
    3.2 LSA application
        3.2.1 System-level LSA
        3.2.2 ILS element-level LSA
        3.2.3 Supportability assessment and verification
    3.3 LSA process
        3.3.1 LSA input
        3.3.2 LSA tasks
        3.3.3 LSA modeling techniques
    3.4 Logistic support analysis record (LSAR)
        3.4.1 Input data sheets
        3.4.2 Sample input sheets
    3.5 LSA control numbers
        3.5.1 LCN structure
    3.6 Selection of LSA candidates
    3.7 Automated data processing (ADP)
    3.8 LSAR summaries
    3.9 LSAR updating
    3.10 LSAR data delivery
4.0 LSA Program
    4.1 Responsibilities for LSA

Fig. 16-6. LSA plan (contents).

4.1.1 Data sheet responsibility
4.1.2 Input data responsibility
4.2 Government interfaces
    4.2.1 Organization
    4.2.2 ILS organization
    4.2.3 LSA communication
    4.2.4 Utilization of data
4.3 Subcontractor/vendor interface
    4.3.1 Subcontractor control
    4.3.2 Government review of subcontractor data
4.4 Government interface
    4.4.1 Government reviews
4.5 LSA program schedule

Fig. 16-6. LSA plan (contents). (Continued from page 247.)

review held by the government or the contractor. This gives LSA visibility equal to that given design throughout the program. Figure 16-8 shows the inputs and outputs of Task 103.

Subtask 1—Establish Review Procedures. Prior to conducting any formal reviews, it is necessary to establish accepted and agreed-upon procedures for how the meetings will be conducted, how the agendas will be determined, and who is responsible for recording the events that occur during the meetings. The significant thing that must occur relative to the LSA program is identifying exactly how the results of the analysis process will be introduced and addressed at reviews. It is important that sufficient emphasis be placed on reviewing and using the results of the LSA process to improve overall supportability rather than trying to micromanage the LSA program. The topics that should be addressed, as appropriate, at reviews are listed in Fig. 16-9.

Subtask 2—Design Reviews. The LSA program should be a topic of discussion at all design reviews, especially during the preliminary design review (PDR) and the critical design review (CDR). These reviews are where the government approves the contractor's proposed equipment design. This is where the LSA program input to the design process should be discussed with supportability and supportability-related topics having equal merit as design performance capabilities.

- Address task 102 requirements in detail.
- Be specific whenever possible.
- Identify general approach when specifics not clear.
- Identify what *will* be done and how.
- Identify what will *not* be done and why.
- Identify what is expected from DoD.

Fig. 16-7. Keys to preparing a good LSA plan.

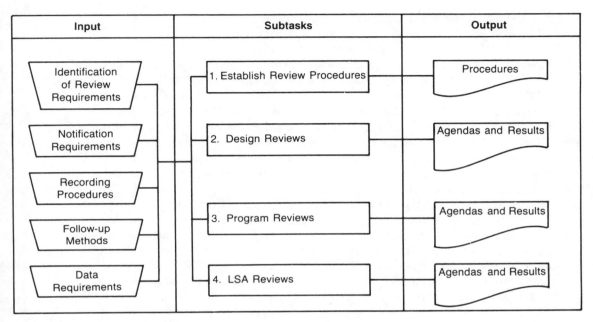

Fig. 16-8. Task 103. Program and design reviews.

Subtask 3—Program Reviews. Program reviews are normally held monthly during the early stages of a program and later may be held quarterly when the design becomes more firm. LSA should always be a subject for discussion at these reviews. Additionally, the contractor should hold program reviews with subcontractors to integrate the total program effort. Each subcontractor having responsibility for conducting an LSA

1. Status of LSA program by task and WBS element
2. Supportability assessment of proposed design
3. Support alternatives under consideration
4. System alternatives under consideration
5. Evaluation and trade-off analysis results
6. Comparative analysis results
7. Design or redesign actions proposed or taken
8. Review of supportability-related design requirements
9. Progress toward achieving supportability goals
10. Design problems affecting supportability
11. Scheduling or analysis problems affecting supportability

Fig. 16-9. LSA topics for program and design reviews.

program should be required to include the program results in any program review. The areas of interest should be the same as those for design reviews.

Subtask 4—LSA Reviews. Specific reviews are held to assess the progress of the LSA program. These LSA reviews are normally held quarterly at the contractor's facility. The topics of discussion are the same as mentioned above, but the level of detail of the discussions should be greater than occur at a program or design review. It is important to remember that the purpose of these reviews is to determine how the LSA process is being used to verify or improve the supportability of the equipment being designed. Many times the LSA review becomes sessions of arguing about how a specific LSA document should be completed, rather than addressing and pursuing the goals of the LSA program. Such occurrences are counterproductive and produce less than desirable results from the LSA reviews.

Mission and Support Systems Definition (Task Section 200). The tasks contained in Task Section 200 are designed to identify the mission or missions that the new weapon system or equipment will be required to perform, quantify the supportability goals of the design program, and provide inputs to the tradeoff analyses used to determine the optimum support system for the weapon system. As stated previously, two of the goals of the LSA program are to influence the design and to identify the cost drivers and problem areas early so that they can be fixed or eliminated whenever possible. These tasks are intended to accomplish those goals.

Use Study (Task 201). The use study provides the basis for all ILS planning and readiness analyses of the new weapon system. In general terms, the use study identifies how, when, and where the new system will be used. It identifies all the qualitative supportability factors required for other analyses. This task should be accomplished for every LSA program. Remember that in each of the previous chapters, considerable information was required to perform analyses or computations to determine the logistic resources required to support a system.

One of the problems that continually plagues the ILS disciplines is the lack of creditable information early in the program for use in making these analyses. When the information is not available, the only solution is to make assumptions based on whatever information is available. This is not an easy task, and the results will vary considerably among disciplines when assumptions must be made about the same points. This causes even more confusion because the changes in assumptions or available data will invalidate any consolidated results of analyses.

The purpose of the use study is to conduct a single analysis to identify all the information that will be needed by any ILS discipline to complete required analyses. With only one document serving as the source for all quantitative information, the task of creating a support system for an equipment is much simpler and effective. Figure 16-10 shows the inputs and outputs of Task 201.

Subtask 1—Supportability Factors Identification. The first step in conducting a use study is to identify and document the pertinent supportability factors related to the intended use of the new weapon system. The factors that should be considered include mission and deployment scenarios, mission or sortie frequency and duration, mobility requirements, basing concepts, projected service life, operation and storage

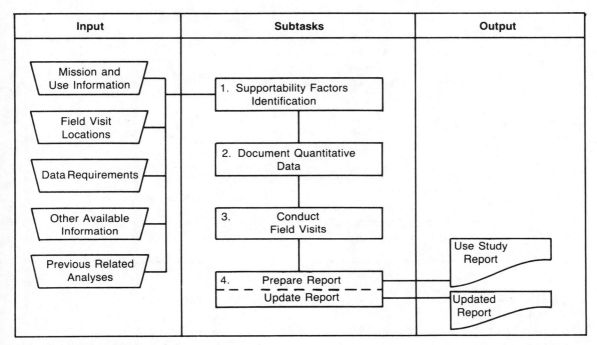

Fig. 16-10. Task 201. Use study.

environments, human interaction and limitations, and operations in conjunction with other systems or equipments. This portion of the use study might consist of a collection of statements that address the areas mentioned above. The point is that each item of consideration must be adequately defined so that supportability requirements can be determined.

Subtask 2—Document Quantitative Data. The next step in preparing a use study is to develop and document discrete quantitative supportability data on the new weapon system. These are the data that will be used for analyses and calculations by the ILS disciplines. Figure 16-11 shows typical quantitative information prepared in the use study process. As can be seen, the quantitative data are used by virtually every analysis that determines requirements for logistic support resources.

Subtask 3—Conduct Field Visits. In some cases, it is advantageous to conduct field visits to operational units or sites to gather information for the use study. When a contractor is required to conduct such fact-finding visits, the government is responsible for access requirements to the necessary information sources. In actuality, this subtask is normally performed by the government, rather than placing the requirement on a contractor. If the government conducts site visits or similar events to gather information, the contractor who is performing the LSA program must be given the results for preparation of the use study.

Subtask 4—Prepare Report. The final step in conducting the use study is to prepare a report of the results of the task. The use study report should document the findings

1. New system mission
2. Mobility requirements
3. Deployment scenario
4. Expected usage (hours, miles, sorties, etc.)
5. Expected useful life
6. Basing concepts
7. Interfaces and supporting systems
8. Operational and storage environments
9. Number of systems
10. Special mission or environmental requirements
11. Maintenance concept
12. Existing system being replaced
13. Existing maintenance and support system
14. Resources required to support existing system

Fig. 16-11. Use study quantitative information.

of the previous subtasks for use by all ILS disciplines. The publication and dissemination of the report is extremely important to ensure that everyone uses the same source information for analyses and calculations. As the information in the use study changes, the impact of the changes will be readily identifiable across all disciplines if the use study report was the source for data used in the original effort. This is how trade-off analyses can be effective; when the baseline information has been documented, all changes can be studied and evaluated in context of the overall system rather than on a chance basis. Figure 16-12 is the basic outline for a typical use study report.

Mission Hardware, Software, and Support System Standardization (Task 202). The purpose of Task 202 is to develop criteria for the new equipment design that will make maximum use of the existing or planned logistic support resources. The thrust of this task is to make the most of resources already available, rather than having to develop a whole new set of resources specifically for the new equipment. Standardization has proven extremely cost effective since it allows the new system to be supported with currently available resources that are already used by other systems. A contractor might be required to develop a detailed standardization program in accordance with MIL-STD 680A, Contractor Standardization Program Requirements, which should be integrated into the performance of this task. Figure 16-13 shows the inputs and outputs of Task 202.

Subtask 1—Identification of Existing and Planned Resources. The first part of this subtask is to identify the existing and planned resources that will be available to support the new system. The use study report, an output of Task 201, should be the starting point for this process. A set of candidate design criteria should be prepared for each alternative system design being considered. The purpose of the criteria is to identify design features that should be tailored to fit the resources that are already available. For example, a design criterion could be that the new design should be capable of being tested using an existing item of test equipment. In this example, designers would have to include whatever design considerations are necessary to use the existing test equipment.

Subtask 2—Support, Cost, and Readiness Information. Identification of resources for standardization criteria development leads to the next step of using other related information to develop the final design criteria. Information on support, cost, and readiness must be used as inputs to the final criteria preparation in order to develop alternative standardization approaches that meet the requirements of the product specification. All this information is necessary to complete the standardization planning package. In some cases, existing technology may be inadequate to support new weapon systems and any attempt to standardize that aspect would be counterproductive.

1.0 GENERAL
    1.1 Scope and purpose
    1.2 System description
    1.3 System mission profile

2.0 QUANTITATIVE SUPPORTABILITY FACTORS
    2.1 Operating requirements
    2.2 Number of systems supported and fielding plan
    2.3 Transportation factors
    2.4 Maintenance factors
    2.5 Environmental factors

3.0 SUMMARY OF SYSTEM BEING REPLACED
    3.1 Operating requirements
    3.2 Number of systems supported and locations
    3.3 Transportation factors
    3.4 Maintenance factors
    3.5 Environmental factors

4.0 EXISTING SUPPORT AVAILABLE FOR NEW SYSTEM
    4.1 Maintenance capabilities
    4.2 Supply support
    4.3 Personnel
    4.4 Facilities
    4.5 Support equipment
    4.6 Test equipment
    4.7 Technical data

5.0 OTHER AVAILABLE SUPPORTABILITY INFORMATION

Fig. 16-12. Use study report (outline).

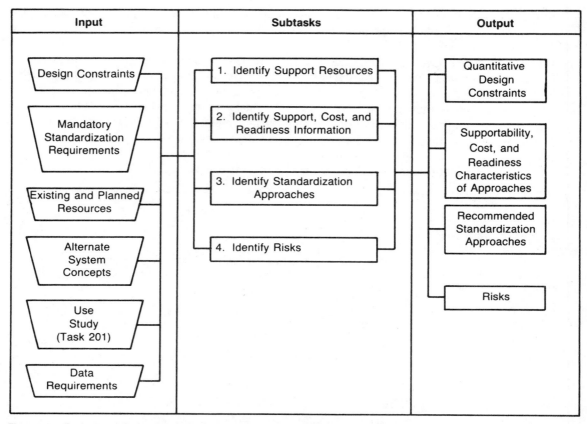

Input	Subtasks	Output

Fig. 16-13. Task 202. Mission hardware, software, and support system standardization.

The key to this subtask is to quantify each standardization approach under consideration using all available information about each design alternative.

Subtask 3—Identify Recommended Standardization Approach. The purpose of this subtask is to choose the best standardization approach based on the information generated in preceding subtasks. Life cycle cost estimating should be used whenever possible in completing this requirement since the long-term worth of each alternative must be identified in order to make a creditable decision. There might be other requirements that must be met that cannot be evaluated based on lowest cost. For example, there might be minimum performance or support requirements that must be met regardless of cost. In these cases, additional information should be used when making a final decision. The output of this subtask should be the standardization approach that will be used during the system design process.

Subtask 4—Identify Risks. The last step in completing Task 202 is to identify the risks associated with the standardization approach chosen for the design. These risks can include limited resource availability, technology-related problems, overloading existing facilities, etc. Virtually every resource has some limiting factor that must be considered when planning for its use or availability. The purpose of this subtask is to force the analyst to make a conscious decision as to what risks are being taken when choosing a standardization approach. This allows the decision to be reconsidered if the situation changes in the future or better information becomes available that affects the availability of resources.

Comparative Analysis (Task 203). The purpose of Task 203 is to use experience and information gained from previous or existing systems to identify areas that should be targeted for improvement in the new system. Additionally, analyses of previous systems should be used to identify drivers in the areas of supportability, cost, and readiness that could be improved in the design of the new system. Basically, this task stresses learning from past history rather than making the same mistakes again. Figure 16-14 shows the inputs and outputs of Task 203.

Subtask 1—Identify Existing Systems For Comparison. The first step in making a

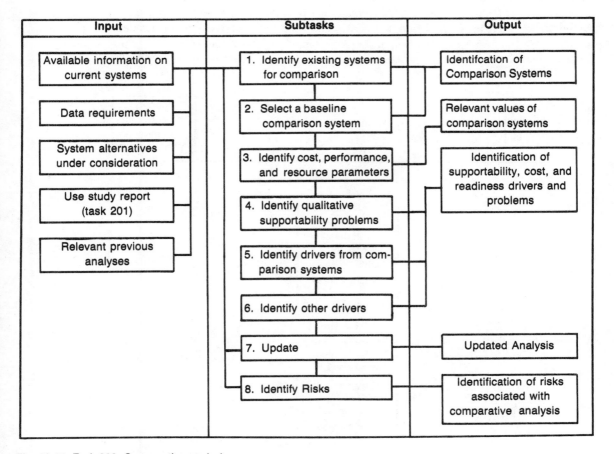

Fig. 16-14. Task 203. Comparative analysis.

comparative analysis is to identify existing or similar systems that have merit for comparative analysis purposes. Several different comparative systems may be considered, each having differing applicability based on design, mission, support systems, etc. If the proposed design of the new equipment has several alternatives, then a comparative system will probably be required for each alternative. The more similar the comparison system to the new system, the more creditable the results of the analysis.

Subtask 2—Select a Baseline Comparison System. In some cases, because of technological advances or other circumstances, there may not be an existing system that can be used for a comparative analysis. When this occurs, it may be possible to develop a composite model using subsystems or sections of several systems that results in a baseline comparison system (BCS) having the overall traits and characteristics of the new system. The BCS can then be used to make a comparative analysis for the new system. Figure 16-15 illustrates how a BCS might be developed. This process is invaluable in the development of analyses when the new system does not have a comparable predecessor.

Subtask 3—Identify Cost, Performance, and Resource Parameters. After the comparison system or BCS has been identified, the next step in the comparative analysis process is to identify the cost, performance, and resource parameters of these systems. This process identifies every possible piece of information available relative to the operation

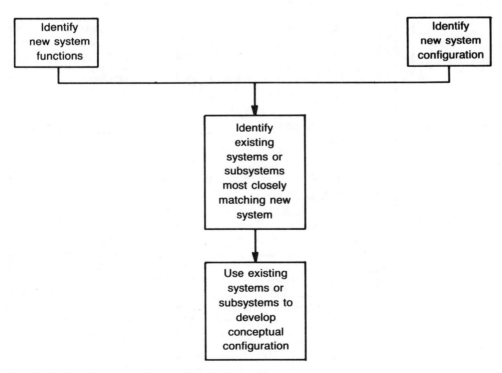

Fig. 16-15. Baseline comparison system development.

and support costs, requirements for logistics support resources, reliability and maintainability values, and performance parameters of the comparative systems. Information compiled might include MTBF, MTTR, TAT, number of spares, cost for spares, number of repair actions per assembly per year, critical failure items, cost for support per year, manpower requirements, support equipment requirements, training requirements, performance of the system in speed, accuracy, availability, and projected life cycle cost parameters. The objectivity and accuracy of this information is critical to performing a comparative analysis. The government might prove to be the best source for this information if the contractor was not involved with the design or production of the comparative system.

Subtask 4—Identify Qualitative Supportability Problems. In addition to the quantitative information collected in subtask 3, pertinent qualitative information about the supportability of the comparative system should be documented. This information might come from field visits during preparation of the use study or from other analyses. In either case, all other information about the comparison system should be taken into account in identifying supportability problems that have occurred in the past on similar systems that should be prevented on the new system.

Subtask 5—Identify Drivers from Comparison Systems. Using the results of each previous subtask, an analysis should be performed to identify cost, support, and readiness drivers of the comparative systems. Each aspect of the system should be reviewed to determine exactly what design feature contributed to the limitations of the system. Drivers should be identified in every conceivable area where possible. At a minimum, the design features or segments that experienced the highest failure rates, elements in the life cycle cost that were higher than projected, requirements for support resources, and continual or unplanned maintenance problems should be identified. These are then analyzed to determine which are applicable to the new system. Those applicable to the new system must be categorized as potential drivers and should receive top priority for resolution. Addressing these areas should give the highest return when applying limited resources to improving the supportability of the new system. In other words, if these potential drivers are improved, all other things remaining equal, the new system should be an overall better system than the comparison system.

Subtask 6—Identify Other Drivers. In some cases, there will not be a comparable system nor anything that can be used to model a BCS. When this occurs, an attempt should still be made to quantitatively and qualitatively identify the potential drivers of the new system. Because of a lack of information, the analysis will probably be more qualitative than quantitative; however, a conscious decision must be made about what the perceived drivers might be in order to focus the analysis process in a direction that has the potential to increase the overall system effectiveness.

Subtask 7—Update. As better information becomes available concerning either the new system or the comparative systems, the comparative analysis should be updated to determine if the results of previous analyses are still valid. The updating of this task is critical to keep the efforts of ILS activities focused on those design areas that provide the greatest potential for improvement.

Subtask 8—Identify Risks. Quantifying the risks involved in each comparative analysis is necessary to understand the limitations that should be placed on using the results of the comparative analysis process. Although a comparative system might closely resemble the new system, it is not the new system, and the differences, even subtle, might be sufficient to make any analysis result questionable. Therefore, the risks inherent in the process and the specific analysis must be identified.

Technological Opportunities (Task 204). The purpose of Task 204 is to identify technologies that can be applied to the new design to improve supportability. By using state-of-the-art technology in the design of the new system, overall supportability and life cycle cost should be improved. Figure 16-16 shows the inputs and outputs of this task.

Subtask 1—Identify Design Approach. The use of new technology to improve supportability starts with identification, first, of the areas that need improvements and, second, of applicable technologies. Basically, this task looks for ways of increasing the supportability of a new system by using new technology in the design. The best place to start is to use the drivers identified in Task 203 as the first targets for improvement. Once the targets for improvement are identified, then the available technologies can

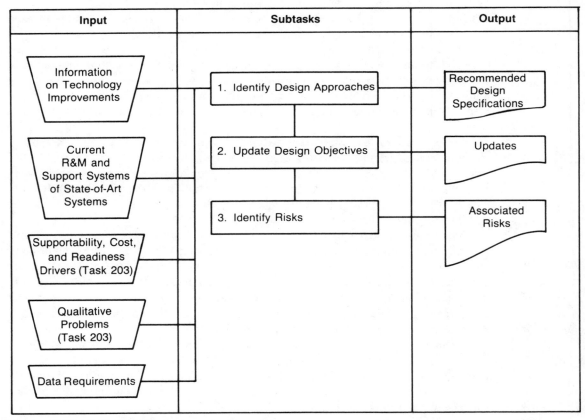

Fig. 16-16. Task 204. Technological opportunities.

be analyzed to determine which, if any, are applicable to the new system. In some cases, the use of a new technology is not cost effective or efficient, so each possible alternative should be studied to see if eliminating the driver warrants use of the new technology. The results of this subtask should be identification of the technologies that do provide a better, more cost-effective method of achieving the supportability goals of the new system.

Subtask 2—Update. The results of this task should be continually updated. The technological advances that occur over the course of a system development provide a constant source of potential applications to new systems. As the detailed design evolves, new requirements for applying new ideas occur, which require this task to be repeated.

Subtask 3—Identify Risks. It is important to identify and document the risks involved with applying any new technologies to the new system design. The use of new or unproven features in a new system can create risks either with the production of the new system or with the supportability when fielded. Identification of these risks allows development of alternatives or work-arounds should the risk prove unacceptable.

Supportability and Supportability-Related Design Factors (Task 205). The purpose of this task is to use the results of the previous tasks to develop a complete set of support and supportability characteristics for the new system. These characteristics will be used in the design process as guidelines for producing a system that has the desired supportability characteristics. Outputs from this task can also be used in future contracts or other documents related to equipment design or supportability. Figure 16-17 shows the inputs and outputs of Task 205.

Subtask 1—Identify Supportability Characteristics. The first step in this task is to quantitatively identify the supportability characteristics in terms of support concepts, reliability and maintainability parameters, operation and support costs, and logistic support resources required for the new system. This is accomplished using the results of previous LSA tasks and forms the initial data to be used by all ILS disciplines for analyzing the new system.

Subtask 2—Establish Supportability Objectives. After the supportability characteristics have been identified, the quantitative supportability objectives for the new system can be established. These objectives are expressed in terms of support, cost, and readiness levels the new system design should achieve. Objectives can include repair turnaround time, mean time to repair, support equipment utilization rates, manpower requirements per maintenance action, etc. In other words, the results of this subtask define supportability goals and objectives for the new system. These should be a combination of the goals and objectives of each ILS discipline, as discussed individually in previous chapters.

Subtask 3—Establish Design Constraints. In addition to the design goals established above, the design constraints for the new system, within which the goals must be met, must be established. These constraints again should be the summation of all the constraints developed by each ILS discipline, as discussed in previous chapters. The government can also impose constraints on the design through contractual requirements.

Subtask 4—Identify NATO Constraints. As a member of the North Atlantic Treaty

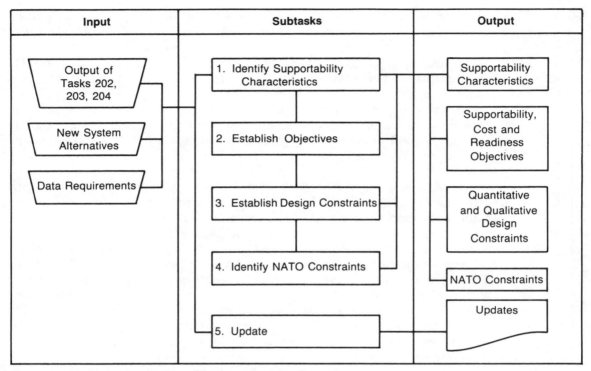

Input	Subtasks	Output
Output of Tasks 202, 203, 204	1. Identify Supportability Characteristics	Supportability Characteristics
New System Alternatives	2. Establish Objectives	Supportability, Cost and Readiness Objectives
Data Requirements	3. Establish Design Constraints	Quantitative and Qualitative Design Constraints
	4. Identify NATO Constraints	NATO Constraints
	5. Update	Updates

Fig. 16-17. Task 205. Supportability and supportability-related design factors.

Organization (NATO), the United States is obliged to plan for interoperability of new equipment with that of other NATO members. The LSA process should consider the applicability of NATO standard items for fulfilling requirements as applicable. Included in the identification of design constraints should be a decision as to the applicability of existing NATO items for use in the design of the new system.

Subtask 5—Update. The results of Task 205 should be continuously updated as the design process and logistic support analysis process evolve. This task uses the results of other tasks and therefore should be kept current with the results of other interrelated tasks since the design objectives and constraints established by this task form the base for all subsequent analyses and tradeoff decisions. The purpose of the tasks contained in Task Section 200 is to define the missions and support systems of the new system. This process is iterative; tasks are continually updated throughout the design process to ensure that the logistics support package developed for the new system meets the actual needs that will be encountered when the system becomes operational.

Preparation and Evaluation of Alternatives (Task Section 300)

The tasks contained in Task Section 300 are designed to develop and choose the alternative best suited for supporting the new equipment. Normally, all three tasks are accomplished sequentially in order to fully develop each proposed support alternative and provide the best opportunity for proper selection. These tasks are iterative, and start the process of generating detailed information that is recorded in the logistic support analysis record (LSAR).

Functional Requirements Identification (Task 301). The purpose of Task 301 is to first identify the functions that the new equipment must perform and, second, identify all the operation and maintenance tasks that must be performed to support the equipment in its intended environment. This is a significant task in the development of the overall logistic support package for the new equipment. Figure 16-18 illustrates the inputs and outputs for this task.

Subtask 1—Identify System Functions. The first step in the functional identification process is to identify exactly what the new equipment must do in order to accomplish intended missions or tasks. This should result in a rather lengthy list of functions. For example, an aircraft would have to take off, fly, land, carry ordnance, provide life support to the crew, communicate with other aircraft or the ground, navigate, take pictures, use radar, etc. The list of possibilities, as can be observed, is rather large. However, this process is important so that it is completely understood what functions must be supported. This might seem like a task that has little bearing on ILS, but without knowing what an item of equipment is supposed to do and the intended environment,

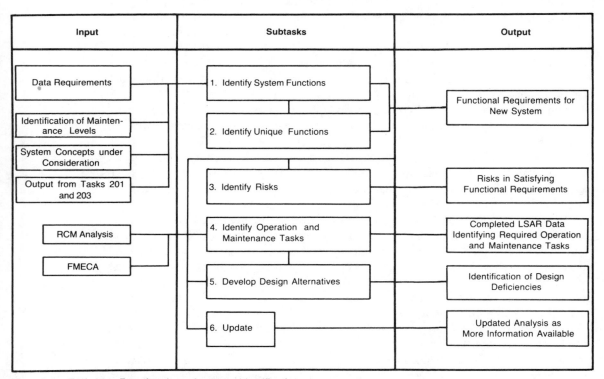

Fig. 16-18. Task 301. Functional requirements identification.

it is much more difficult to plan for its support or recommend design changes that do not degrade its ability to perform critical functions.

Subtask 2—Identify Unique Functions. The next step in accomplishing this task is to identify those functions from subtask 1 that the new equipment must perform that are unique. This is because new functions often create new support problems or require a new method of support. A function may be unique due to new technology in the design or new operational concepts. These unique functions should receive special attention when planning support for the new equipment. By recognizing early in the equipment design that there are unique functions the equipment must perform, planning for alternatives to support these functions can be initiated early to resolve supportability problems.

Subtask 3—Identify Risks. The identification of functions, especially unique functions, focuses attention on the ability of the current, or planned, support system to be able to provide the necessary support to the new equipment. This allows the early identification of risks involved with the supportability of the equipment due to functional requirements. Early identification of supportability risks allows time for design changes or modifications to the functional requirements to eliminate or reduce the risks. The key is to identify the risks early so there is adequate time to address the potential consequences before the design is completed.

Subtask 4—Identify Operation and Maintenance Tasks. The second part of Task 301 is to identify all the operation and maintenance tasks that must be performed in order for the new equipment to be able to accomplish the functions identified in Subtask 1. Subtask 4 is one of the keys to developing a detailed logistic support package for the new equipment. This subtask requires detailed analysis of every facet of the new equipment. In essence, the results of this subtask should be a list of everything that has to be done to keep the equipment operating or fix it when it breaks. For a large weapon system, this list could consist of several thousand maintenance tasks. The place to start this analysis is the failure modes, effects, and criticality analysis (FMECA). The FMECA (see Chapter 3) is an in-depth analysis of the total equipment that identifies all the ways it can fail. Common sense says that a maintenance task should be required for every failure mode. This is not a one-for-one relationship, because one maintenance task might be able to correct more than one failure mode; however, every failure mode should be addressed by a maintenance task.

In addition to the FMECA, the reliability-centered maintenance (RCM) analysis (see Chapter 15) is used to identify the preventative maintenance tasks required to support the new equipment. Another method used to identify tasks is to analyze the functions identified in subtask 1 to identify operation and other support tasks that must be performed, which were not covered in either the FMECA or the RCM analysis. Remember, this subtask should identify every operation and maintenance task required to support the new equipment. The results of this subtask are recorded in the LSAR.

Subtask 5—Develop Design Alternatives. The identification process accomplished in subtasks 1 through 4 normally identifies design deficiencies that cause supportability problems. The purpose of subtask 5 is to formally require ILS disciplines to actively participate in the design process to develop alternative design approaches for solving these problems. Remember, one of the goals of the LSA program is to affect the design and make changes that improve supportability. This is one of the critical places where it is done.

Subtask 6—Update. Task 301 must be continually updated throughout the design process. Each design change must be analyzed to identify new functions and operation and maintenance tasks, which will, in turn, drive other analyses discussed later in this chapter. This is the iterative part of LSA, because as more information is available on the detailed design, more functions and tasks can be identified; and each new level of definition provides more information for analysis until the design is complete.

Support System Alternatives (Task 302). Task 302 is the next step in developing the support system for the new equipment. In Task 301, the functional requirements of the new equipment were identified, along with all the operation and maintenance tasks required for support. The purpose of this task is to develop alternative methods for providing the necessary support, at the system level, to accomplish the functions and tasks identified in Task 301. Figure 16-19 shows the inputs and outputs of Task 302.

Subtask 1—Develop System-Level Support Concepts. The concepts for system-level support of the new equipment must address how the overall support will be provided. This is where every possible alternative must be considered. Not all alternatives considered progress further than the initial discussion stage, but concepts that have merit and appear to provide the basic support required should be given further consideration. The concepts may include number of levels of maintenance to be used, possibility of having contractor support, a combination of military and contractor maintenance support, different sparing techniques, different testing or support equipment, or

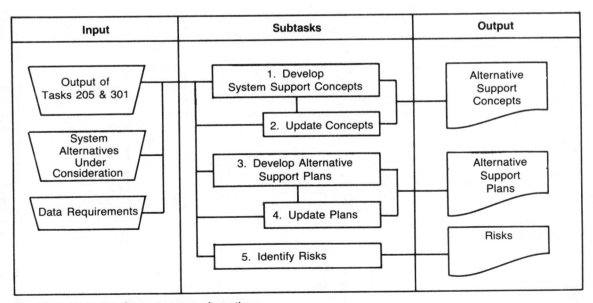

Fig. 16-19. Task 302. Support system alternatives.

combinations or all of these. The purpose is to investigate any feasible alternative for providing the necessary support to keep the new equipment operational.

Subtask 2—Update Concepts. The evolution of the design and the availability of more detailed information as the program progresses allows the concepts to be updated and refined throughout the program. The update process should better define the advantages and disadvantages of each alternative concept and provide more usable information to the decision of which alternative is most desirable. This update is essential since the original concepts are normally based on previous history rather than detailed information about the new equipment. Failure to update the concepts could result in the wrong concept being selected due to deficient or insufficient information.

Subtask 3—Develop Alternative Support Plans. The support concepts developed in subtask 1 should be refined into support plans as detailed information becomes available. Subtask 3 requires that a support plan be developed for each support alternative. The plan should tell how the concept will be implemented and the ways that each ILS element will be fulfilled. Initially, the plan will be little more than an elaboration on the concept; but as the design matures, the plan becomes a detailed information document that addresses each aspect of support for the new equipment covering all levels of maintenance and all operation and maintenance tasks for both hardware and software.

Subtask 4—Update Plans. As the alternative support concepts are updated as required by subtask 2, the corresponding support plan must also be updated. However, only those plans that are still under consideration are updated. The results of trade-off analyses can be used as inputs to update these plans. The final support plan for the equipment is generated through this update process.

Subtask 5—Identify Risks. An additional requirement of this task is to identify any risks involved with the alternative support concepts. These risks can be either qualitative or quantitative, but should be usable in the decision-making process when evaluating the alternatives. Such things as shortages of certain resources critical to the success of a support concept or new and unproven methods can develop into concerns that will end in the rejection of a concept. The identification of risks, or "what if's," should be a significant contribution to the decision of what concept is chosen.

Evaluation of Alternatives and Tradeoff Analysis (Task 303). Task 303 is one of the most complex tasks of the LSA process. This task covers the complete spectrum of ILS disciplines and their impact on the total system being designed. The purpose of Task 303 is to evaluate each alternative support concept developed in Task 302 to determine the preferred method to be used to support the new system. The evaluation process requires the use of trade-off analyses to determine the best alternative that meets the support, design, and operation requirements while also having the best balance among cost, schedule, performance, readiness, and supportability. Figure 16-20 shows the inputs and outputs of this task.

Subtask 1—Evaluation and Tradeoff Process. The purpose of subtask 1 is to establish the criteria for each evaluation or trade-off conducted in Task 303. It is extremely important that the process for conducting evaluations and trade-offs be strictly controlled in order to provide reliable results. The baseline information should be standard throughout the process. This allows a complete analysis of proposed changes as

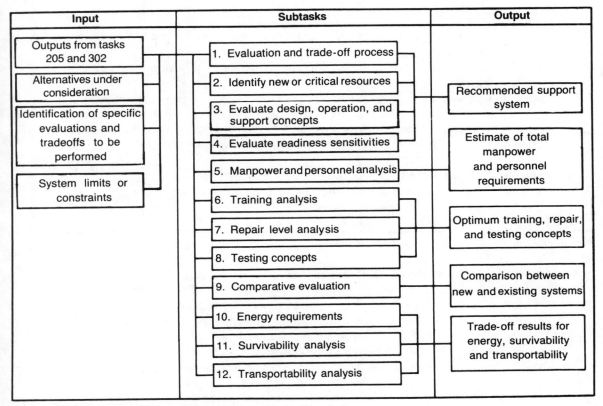

Fig. 16-20. Task 303. Evaluation of alternatives and trade-off analysis.

they occur. Figure 16-21 indicates requirements for conducting evaluations and trade-off analyses. These requirements apply specifically to all evaluations conducted in this task; however, the requirements are applicable to any analysis that a logistician performs and should be used as a standard guideline when conducting evaluations and trade-off analyses.

Subtask 2—Identify New or Critical Resources. A key consideration in the selection of a support alternative is the requirement for new or critical resources generated by the selection of that alternative. As has been discussed in previous chapters, the availability of scarce or limited resources required to support a new system is a driving factor in the selection of the most viable support alternative. One of the areas that should always be considered when looking at critical resources in the area of personnel. There is not an unlimited supply of personnel; and within the personnel structure, there are limited job classifications that can be used to support new systems. Other new or critical resources might include test equipment, support equipment, precious metals, additional facilities requirements, etc.

Subtask 3—Evaluate Design, Operation, and Support Concepts. The purpose of this

Step 1. Identify the quantitative and qualitative criteria to be used to select best alternative.

Step 2. Choose or construct the appropriate model or relationship for conducting the evaluation or trade-off analysis.

Step 3. Conduct the initial evaluation or trade-off analysis of each alternative under consideration (using the model or relationship developed in Step 2) and select the most appropriate alternative (based on the criteria identified in Step 1).

Step 4. Conduct the sensitivity analysis to determine the risks or cost drivers of each alternative.

Step 5. Document the results of the evaluation or trade-off analysis.

Step 6. Update the evaluations and trade-offs as more detailed and accurate information becomes available.

Fig. 16-21. Conducting evaluations and trade-off analyses.

subtask is to identify the need to evaluate the new system support alternatives with regard to the proposed design, operation, and support concepts. The contractor accomplishes this by conducting trade-off analyses of all alternatives and identifying which one provides the best balance for cost, schedule, performance, readiness, and supportability. Subtasks 4 through 12 support the accomplishment of subtask 3.

Subtask 4—Evaluate Readiness Sensitivities. System readiness is the ultimate goal of the LSA process. It is important to know and understand the readiness parameters that are sensitive to variations in design and support parameters, and it is also important to know those parameters that are not sensitive to changes in design or support. The purpose of this subtask is to identify those readiness parameters that can be influenced by variations in either design or support parameters. Basically, this identifies those areas where changes in design or support parameters can increase readiness. Knowing the things that can be improved allows emphasis to be placed where effort will have the most effect. Reliability factors, such as MTBF or MTBCF, and maintainability factors, such as MTTR and MMH/OH, are examples of design and support parameters to which readiness has a high sensitivity level.

Subtask 5—Manpower and Personnel Analysis. An analysis of the manpower and personnel requirements of each alternative is conducted to determine the total numbers of personnel, skill specialties, skill levels, and experience that will be required. A trade-off analysis can then be conducted using the results of this analysis to determine the alternative that optimizes use of personnel (see Chapter 8).

Subtask 6—Training Analysis. This subtask analyzes each alternative to identify the optimum training methods required to implement each alternative. The training can consist of a combination of formal, informal, and on-the-job training (see Chapter 10). Additionally, trade-off analyses can be performed to determine the feasibility and desirability of creating new skill specialties or shifting tasks between existing skill specialties to support the new system.

Subtask 7—Repair-Level Analysis. A repair-level analysis (RLA) is conducted for each alternative to determine the optimum utilization of support resources. Chapter 15 contains detailed information on conducting an RLA.

Subtask 8—Testing Concepts. Each alternative is evaluated to determine the optimum testing concept to be used. The concept may be composed of combinations of BIT, off-line testing, manual testing, and automatic testing. The key to this evaluation is to determine which alternative makes the best use of resources to accomplish testing. There is always a trade-off between the cost and accuracy of manual testing versus the cost of sophisticated test equipment. Chapter 7 provides further information on this process.

Subtask 9—Comparative Evaluation. A comparative evaluation should be made between the exhibited capability of an existing system to achieve its supportability objectives and the projected ability of the new system to do likewise. The purpose of this subtask is to ascertain the feasibility of each alternative's achieving the goals established for the new system. The BCS developed for Task 203 can be used in this subtask when no comparable system exists.

Subtask 10—Energy Requirements. An analysis of the projected energy costs for each alternative is conducted to identify any problem areas that might arise due to fluctuations in the cost or availability of energy sources. Specific attention should be paid to petroleum and petroleum products that are required to support the new system. Sensitivity analyses might be required when the costs of petroleum are considered significant drivers for supportability.

Subtask 11—Survivability Analysis. A survivability analysis is performed to determine which alternative provides the best characteristics for battle damage repair and other considerations of survivability in a combat environment. While this might seem like a requirement that is out of the normal realm of ILS concerns, remember that all systems are developed for wartime use where survivability is the primary consideration. Battle damage repair should always be a prime consideration in any system developed for use in a combat environment.

Subtask 12—Transportability Analysis. Ease of transportability is always a consideration for design of a new system. A transportability analysis is conducted to determine which alternative optimizes the use of transportation resources (see Chapter 12 for further discussion).

Determination of Logistic Support Resource Requirements (Task Section 400)

Task Section 400 addresses three areas: identification of detailed logistic support resource requirements for support alternatives, assessment of the impact the new system will have on other existing systems, and planning for support after the end of the production phase of the acquisition process for the system. These tasks, especially Task 401, can be very costly to accomplish due to the manpower required to perform the tasks and the tremendous amount of documentation generated.

Task Analysis (Task 401). Performance of a complete maintenance task analysis will result in the identification of all the logistic resources required to support the new system. The task analysis process is the most labor intensive and, therefore, most expensive requirement contained in the LSA program. However, if done correctly, it is the single most accurate method of identifying logistic support resource requirements, and it eliminates guess work and rule-of-thumb estimates of resource requirements. The extra expense up front of performing a complete task analysis pays for itself many times over throughout the life of the system by eliminating waste and misuse of resources. Figure 16-22 shows the inputs and outputs of Task 401.

Subtask 1—Conduct Detailed Task Analysis. A detailed maintenance task analysis is conducted to identify all the resources required to support the system. The task analysis process actually starts with the identification of required operation and maintenance tasks by LSA in Task 301. These tasks are then analyzed to identify required support resources. Figure 16-23 illustrates the task analysis process. By repeating the analysis for every operation and maintenance task required to support the new system, the types and quantities of resources can be identified. This process provides the most reliable identification of required support resources, because it is based solely on the performance of operation and maintenance tasks.

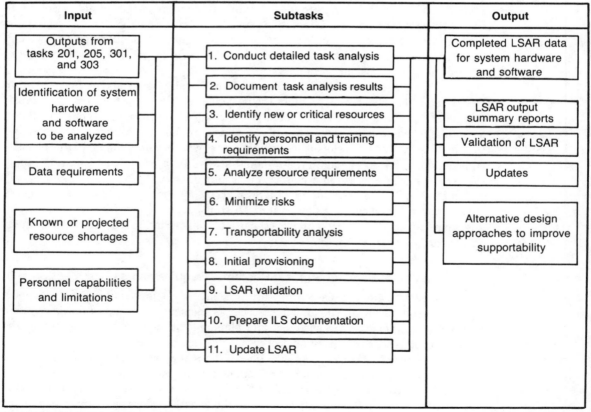

Input	Subtasks	Output
Outputs from tasks 201, 205, 301, and 303	1. Conduct detailed task analysis	Completed LSAR data for system hardware and software
Identification of system hardware and software to be analyzed	2. Document task analysis results	
	3. Identify new or critical resources	LSAR output summary reports
	4. Identify personnel and training requirements	Validation of LSAR
Data requirements	5. Analyze resource requirements	Updates
	6. Minimize risks	
Known or projected resource shortages	7. Transportability analysis	Alternative design approaches to improve supportability
	8. Initial provisioning	
	9. LSAR validation	
Personnel capabilities and limitations	10. Prepare ILS documentation	
	11. Update LSAR	

Fig. 16-22. Task 401. Task analysis.

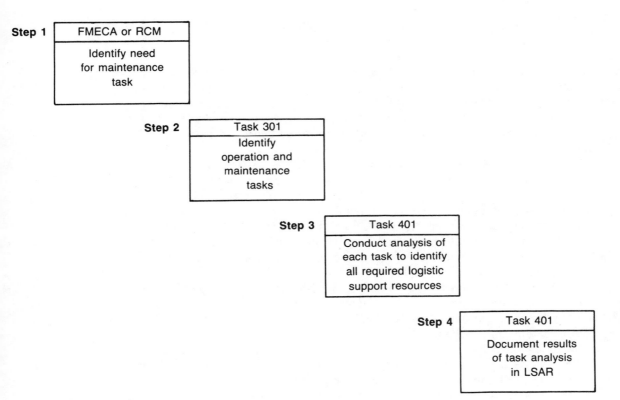

Fig. 16-23. Task analysis process.

Subtask 2—Document Task Analysis Results. The purpose of subtask 2 is to formally require that the results of subtask 1 be recorded in the logistic support analysis record (LSAR), discussed in Chapter 17.

Subtask 3—Identify New or Critical Resources. Identification of new or critical resources is a continual process in LSA. Previously, Task 303, subtask 2, required a similar identification; however, that task deals with the overall system requirements. This subtask can now address the detailed requirements for new or critical resources since the maintenance task analysis identifies all the resources necessary to support the system. The results of this subtask should be a quantitative analysis where the results of Task 303, subtask 2, tend to be qualitative.

Subtask 4—Identify Personnel and Training Requirements. The maintenance task analysis results in identification of the total personnel requirements to support the new system. Additionally, the training requirements for operator and maintenance personnel are also identified. Personnel requirements are based on the number and types of personnel required to accomplish all tasks, and training requirements are based on knowing the exact tasks that personnel must be trained to accomplish. This process of identifying personnel and training requirements results in extremely accurate and usable information for use in acquiring necessary personnel and development of viable training programs.

Subtask 5—Analyze Resource Requirements. The maintenance task analysis process normally identifies many areas where the supportability of a new system can be optimized or where a design change is necessary to meet minimum supportability goals. Subtask 5 establishes the formal requirement for the logisticians performing LSA to provide design engineers with information that will improve supportability. This is one of the major goals of the overall LSA program, to influence the design to achieve optimum supportability.

Subtask 6—Minimize Risks. Several previous tasks contained subtasks that required the identification of risks associated with alternatives or courses of action. The purpose of subtask 6 is to determine if the results of the maintenance task analysis supports the risks or if the detailed information resulting from the task analysis reduces or eliminates the risk. If the risk still exists, then the results of the task analysis should allow the risk to be quantified. This information can then be used by management to specifically address the critical risks of the program that are related to support resources and supportability of the new system.

Subtask 7—Transportability Analysis. An analysis is conducted to determine the transportability requirements of the new system. The purpose of the analysis is to identify transportability problems caused when the new system exceeds established transportation limits. Chapter 12 contains a detailed discussion of transportability planning and analysis.

Subtask 8—Initial Provisioning. This subtask requires that items that require initial provisioning be documented appropriately in the LSAR. No other method of document-ing provisioning requirements should be used. The identification of items for initial provisioning is accomplished through maintenance task analysis. When a maintenance task requires that an item be removed and replaced, then a spare is required. Repair of an item may require a repair part. The end result of this process is a fully justified provisioning list of only the items necessary to support maintenance. Prior to LSA, provisioning was accomplished by listing 100 percent of the items contained in a system on the provisioning documentation. In many cases, this resulted in waste since unnecessary items were also provisioned. Through the LSA process, only items needed to support maintenance or operation of the new system are provisioned. Chapter 6 provides a detailed discussion of the provisioning process.

Subtask 9—LSAR Validation. The LSAR should be validated whenever possible to ensure that the data base is consistent with the system design. The design process is actually a series of changes that occur to the system over a rather lengthy period of time. When changes occur to the system that are not reflected in the LSAR, the logistic support resources requirements are inaccurate. In many cases, uncontrolled documentation is the biggest single problem of an LSA program. Validation can be accomplished during the performance of any review, audit, or demonstration. Chapter 17 provides a detailed discussion of LSAR validation.

Subtask 10—Prepare ILS Documentation. The LSAR serves as the single source of logistics data for documentation purposes. Therefore, ILS documentation must be prepared using the LSAR data base whenever possible. This process consists of generating reports from the data base that are used to fulfill the contractual requirements of ILS reporting. The method of preparing these reports is discussed in detail in Chapter 17.

Subtask 11—Update LSAR. LSA is an iterative process in which the analyses, especially maintenance task analysis, are conducted over and over until the design is finalized and all the resources required to support the system have been identified. Subtask 11 requires that the LSAR be updated throughout the life of the new system in order to maintain an adequate and accurate record of the required support resources. Updates continue as new and more detailed information becomes available.

Early Fielding Analysis (Task 402). The purpose of an early fielding analysis is to analyze the new system in relationship to other systems that already exist in the field. Until this point, planning for resources was limited to the specific requirements of the new system without regard for the needs of existing systems. Additionally, the early fielding analysis addresses the resource needs for the new system when placed in the projected combat environment. Figure 16-24 illustrates the inputs and outputs for Task 402.

Subtask 1—Assess Impact on Existing Systems. Once fielded, the new system will compete for limited assets with other systems. The purpose of subtask 1 is to assess how the introduction of the new system in the field will affect the existing support

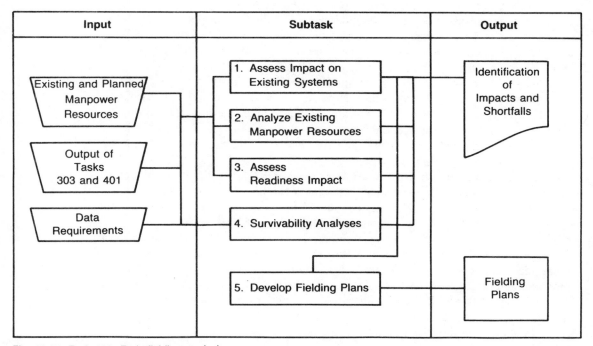

Fig. 16-24. Task 402. Early fielding analysis.

resources, such as personnel, maintenance facilities, support equipment, depot workloads, supply support, training facilities, etc. The results of this subtask should identify any changes to existing resource levels necessary to support the new system when fielded.

Subtask 2—Analyze Existing Manpower Resources. A special consideration is the availability of adequate labor to support the new system. Subtask 2 specifically addresses the impact that the new system will have on the pool of available labor to support both existing systems and the new system. Trade-off analyses should be used to provide feedback to the maintenance task analysis process, where necessary, to adjust manpower requirements based on limited availability. If the new system significantly degrades the ability of the existing system to perform assigned missions, then labor requirements for the new system should be identified as one of the readiness drivers and treated accordingly throughout the LSA process.

Subtask 3—Assess Readiness Impact. The purpose of this subtask is to assess the consequences if the resources identified as necessary to support the new system are not available in the quantities required. This "what if" scenario should lead to the identification of alternative approaches to resource requirements or areas for optimization or streamlining that will minimize the impact. Take care that this subtask does not duplicate any analyses done in Task 303.

Subtask 4—Survivability Analysis. Support resource requirements for a combat environment are much more severe than for peacetime. The purpose of this survivability analysis is to determine unique resource requirements that might be generated when the new system is placed in a combat environment. Such items as predetermined combat spares and battle damage repair procedures and facilities should be considered when appropriate. This analysis should not be a duplication of the survivability analysis of Task 303.

Subtask 5—Develop Fielding Plans. An early fielding plan is generated to develop and implement solutions to the problems identified in subtasks 1 through 4. It is essential to the readiness of the new system that any problems or concerns raised during the early fielding analysis be addressed prior to deployment. Failure to adequately address these problems will degrade not only the new system but also existing systems that will have to compete for limited support resources.

Post-Production Support Analysis (Task 403). Production facilities for a new system do not remain operational throughout the life of the system. Therefore, a post-production support analysis is conducted to determine the source of logistic support resources after production ceases. The period of time between the end of production and the disposal of the system may be as much as 20 years, so the post-production support problems are not insignificant. Items to be considered include sources for replacement spares and repair parts, maintenance and overhaul facilities, reprocurement data, preplanned product improvement, and configuration management. The final step of this task is the preparation of a post-production support plan that assures adequate availability of support resources after production is complete. Figure 16-25 shows the inputs and outputs of Task 403.

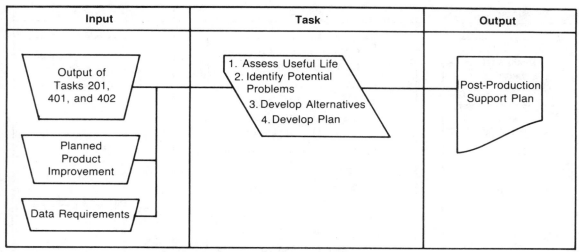

Input	Task	Output
Output of Tasks 201, 401, and 402 Planned Product Improvement Data Requirements	1. Assess Useful Life 2. Identify Potential Problems 3. Develop Alternatives 4. Develop Plan	Post-Production Support Plan

Fig. 16-25. Task 403. Post-production support analysis.

Supportability Assessment (Task Section 500)

The effectiveness of the overall ILS effort and the adequacy of the LSA program are assessed by actions contained in Task Section 500. There are two types of assessment covered in this task section: assessment of supportability during the development of the new system and assessment after the system is deployed. Assessment during development is necessary to identify and correct deficiencies prior to fielding the new system. Postdeployment development is essential in developing product improvements for the system and in documenting long-term supportability problems and solutions for use on future programs. During the early phases of an acquisition program, testing and evaluation may be accomplished using models or simulation to verify the progress toward supportability goals. Toward the end of FSD, testing and evaluation should be done in an environment that matches as closely as possible the actual environment where the system will be deployed.

Supportability Test, Evaluation, and Verification (Task 501). The testing, evaluation, and verification of the supportability of a system are continuous efforts. Historically, testing and evaluation, if any, have been limited to some testing of the support system just prior to full production. This is unacceptable. The support package for the new system must be repeatedly tested and evaluated throughout the acquisition cycle to verify that system supportability requirements are being met. Testing, evaluation, and verification of support must be accomplished during every phase of the acquisition cycle. Figure 16-26 illustrates the inputs and outputs of Task 501.

Subtask 1—Test and Evaluation Plan and Strategy. Prior to the start of each acquisition phase, a plan and strategy for test and evaluation of achievement or progress toward achievement of supportability goals are developed. The purpose of the

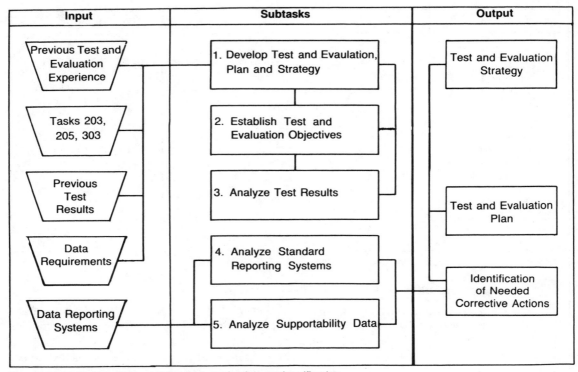

Input	Subtasks	Output

Fig. 16-26. Task 501. Supportability test, evaluation, and verification.

plan is to organize testing and evaluation efforts to focus on critical supportability areas that should be addressed in the next acquisition phase. The plan should include identification of resources required to accomplish testing, a schedule for accomplishment, and the desired results of the tests. The strategy for the testing and evaluation should address how each test fits into the overall verification of system supportability.

Subtask 2—Test and Evaluation Objectives. Detailed test plans and procedures are developed that implement the results of subtask 1. Quantitative objectives for the testing must be established to facilitate evaluation of system performance. Included in the detailed plans should be the availability of resources, data sources, and resource utilization for testing.

Subtask 3—Analyze Test Results. The results of supportability tests are analyzed to identify deficiencies discovered during tests and to verify that corrective actions implemented to correct previously identified deficiencies were adequate to resolve the deficiencies. Figure 16-27 identifies additional uses for supportability test results.

Subtask 4—Analyze Standard Reporting Systems. The supportability of the new system is also evaluated and verified after the system is deployed. Information for evaluation is collected through standard reporting systems in place within each service. The standard reporting systems must be assessed to determine if adequate

1. Correct deficiencies
2. Update system readiness, cost, and resource projections
3. Identify improvement needed to meet readiness goals
4. Identify achievement of contractual goals
5. Update LSAR
6. Data base for future acquisitions

Fig. 16-27. Supportability test data uses.

information will be available through this medium after deployment to conduct an adequate evaluation and verification of supportability. If the standard reporting system is determined to be inadequate, then alternative or supplemental reporting schemes should be considered for implementation.

Subtask 5—Analyze Supportability Data. The analysis of supportability data obtained from actual field use of the system can provide significant information for product improvement or support system enhancement. Additionally, a comparative analysis between supportability projections generated during system development and actual field data can provide significant information for use on future acquisition programs to better project cost, supportability, and readiness.

PROGRAM IMPLEMENTATION

The scope of the LSA program must be commensurate with the needs of each specific acquisition program. Therefore, it is important to understand how a program is implemented to achieve supportability requirements. First, no task should be accomplished on a stand-alone basis. The LSA process is designed to enhance the interrelationship between design and ILS and internally among ILS disciplines. Second, the phases of the acquisition process lend themselves to certain tasks based on the ability of the tasks to accomplish the goals of an LSA program. Last, MIL-STD 1388-1A is designed to be tailored to match the unique requirements of each program. Very rarely will an acquisition program require or justify the expense of accomplishment of all tasks and subtasks.

LSA Task Relationships

The LSA process is based on the relationship among individual tasks and the flow and availability of information. The inputs and outputs of each task are designed to support the iterative LSA process. Figure 16-28 illustrates how each task is supported by significant external information or the results from other tasks and how subsequent tasks build on previous information. The iterative nature of LSA requires that each change to source information be passed through the same process as the original information. If the iterative process is not followed, the resulting data generated in the LSAR will not be compatible with the final equipment design, and the support system will not make optimum use of support resources.

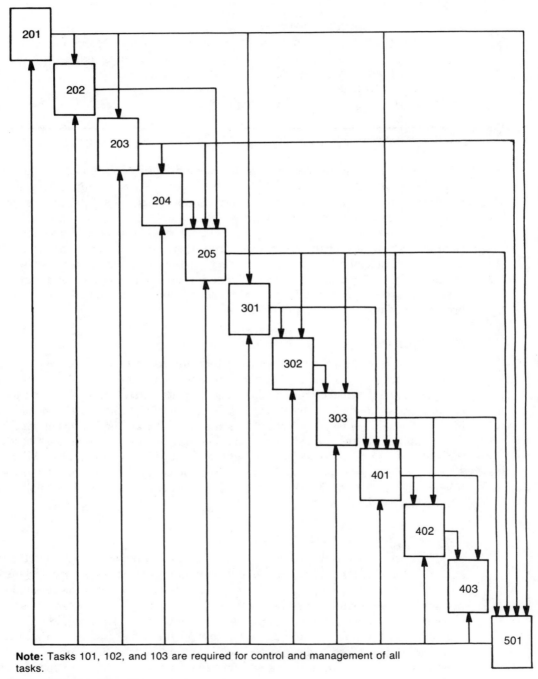

Note: Tasks 101, 102, and 103 are required for control and management of all tasks.

Fig. 16-28. LSA task relationship.

LSA Task Performance by Acquisition Phase

All LSA tasks are not applicable to every acquisition phase. It is important to understand that only those tasks that provide significant supportability input to the overall program should be accomplished. For example, Task Analysis, Task 401, should never be required during the concept phase where there is little or no detailed design in existence. Likewise, LSA Strategy, Task 101, should never be required during the production phase when it would be too late to have an effect on the program. Figure 16-29 shows the relationship between task accomplishment and acquisition phases.

Task	Pre-concept	Concept	Demval	FSD	Prod
100 Series tasks					
101 Early LSA strategy	N/A	G	G	S	N/A
102 LSA plan	N/A	G	G	G	G
103 Program and design reviews	N/A	G	G	G	G
200 Series tasks					
201 Use study	G	G	G	G	N/A
202 Mission hardware, software, and support system standardization	N/A	G	G	G	C
203 Comparative analysis	G	G	G	G	N/A
204 Technological opportunities	N/A	G	G	S	N/A
205 Supportability and support-related design factors	N/A	G	G	G	C
300 Series tasks					
301 Functional requirements identification	N/A	G	G	G	C
302 Support system alternatives	N/A	G	G	G	C
303 Evaluation of alternatives and trade-off analysis	N/A	G	G	G	C
400 Series tasks					
401 Task analysis	N/A	N/A	S	G	C
402 Early fielding analysis	N/A	N/A	N/A	G	C
403 Post-production support analysis	N/A	N/A	N/A	N/A	G
500 Series tasks					
501 Supportability test, evaluation, and verification	N/A	G	G	G	G

Applicability codes:
S— Selectively applicable
G— Generally applicable
C— Generally applicable to design changes only
N/A— Not applicable

Fig. 16-29. LSA program tasks by acquisition phase.

- Amount of design freedom
- Time and resources available
- Work already done
- Past experience
- Historical data availability

Fig. 16-30. LSA tailoring considerations.

Tailoring LSA Requirements

Each LSA program must be tailored to meet the specific requirements of the acquisition program. Tailoring is accomplished by selecting the tasks and subtasks that have the potential for providing the most return on investment in achieving the supportability goals of the program. The government normally tailors the LSA program prior to issuing a request for proposal; however, a contractor can recommend changes to this tailored program in response to the government.

The significant areas that should be considered when tailoring a program are shown at Fig. 16-30. If the design cannot be changed on a specific program, then the LSA program should not contain any tasks whose sole purpose is to provide input to the design process. Timing is always considered when selecting tasks. As stated, not all tasks are applicable to every phase of the acquisition cycle. Work already done should not be repeated just for the sake of accomplishing an LSA task. This information should be used as input directly into the next LSA task. Past experience and historical data should indicate the most appropriate tasks for a specific program.

Chapter 17

Logistic Support Analysis Record

The results of the logistic support analysis (LSA) process are recorded in a single data base, which is comprised of logistic support analysis records (LSARs). The purpose of an LSAR is to provide a standardized method for compiling and storing logistics and logistics-related engineering data. This fulfills the fourth purpose of LSA, which is to maintain a single data base for all logistics data. The LSAR allows all relative supportability information to be used in an organized and uniform manner to identify and develop logistic support resource requirements.

DATA RECORDS

The LSAR data base is actually made up of a series of 14 different data records. MIL-STD 1388-2A, DoD Requirements for a Logistic Support Analysis Record, contains detailed descriptions and the data requirements of each data record. Each data record has a specific function, and requires definitive data that are produced as a result of analysis required by one or more of the LSA tasks contained in MIL-STD 1388-1A. Although each data record is significantly different in purpose and content, there are certain similarities among all. Each data record is a series of 80 card-column lines, with each line having a distinct three-digit identifier in the first three columns. This allows easy computer layout and entry and creates a standard data transfer medium so that information can be readily communicated between computerized data bases. Additionally, each data record contains one or more key data fields common throughout each set of data records for a particular item.

Data Record A: Operation and Maintenance Requirements

Data Record A (Fig. 17-1) is used to record information related to the planned operation and maintenance of the system. This data record is normally completed by the government and provided to the contractor in the ILSP, as a part of the contract or at the initial LSA guidance conference. The information contained on the data record should reflect the requirements of the product specification for operation and maintenance. Normally, only one Data Record A is required for an LSA program. In cases where different operational or maintenance requirements are placed on subsystems, such as on an aircraft, a separate Data Record A might be required for each subsystem.

Data Record B: Item Reliability and Maintainability Characteristics

Data Record B is prepared for each repairable or maintenance-significant item in the system. This includes system, subsystems, components, assemblies, subassemblies, etc. Data Record B describes the function of the item, the proposed maintenance concept for that item, and any design constraints placed on the item. Reliability and maintainability characteristics, summaries of the FMECA and RCM analysis, maintainability predictions, and other logistic evaluations are documented on the Data Record B, as illustrated in Fig. 17-2.

Data Record B1: Failure Modes and Effects Analysis

The results of the failure modes and effects analysis, performed in accordance with MIL-STD 1629, are recorded on Data Record B1. This data record (Fig. 17-3) contains three distinct sections for recording failure data. An entry in one section requires corresponding entries in the remaining sections. The failure modes are used to identify maintenance requirements. Data Record B1 is normally prepared as a companion to each Data Record B.

Data Record B2: Criticality and Maintainability Analysis

The results of the criticality analysis portion of the FMECA and also maintainability analyses on each item are recorded on the Data Record B2. This information (Fig. 17-4) is used to evaluate the need for maintenance tasks and to accumulate maintenance task times for future maintainability analyses.

Data Record C: Operation and Maintenance Task Summary

Data Record C is used to document the identification of all operation and maintenance tasks required to support an item. A separate data record is prepared for each repairable or maintenance-significant item and is paired with corresponding Data Records B, B1, and B2. Data Record C (Fig. 17-5) lists each required task and also identifies the need for special tools and support equipment, facilities, or training.

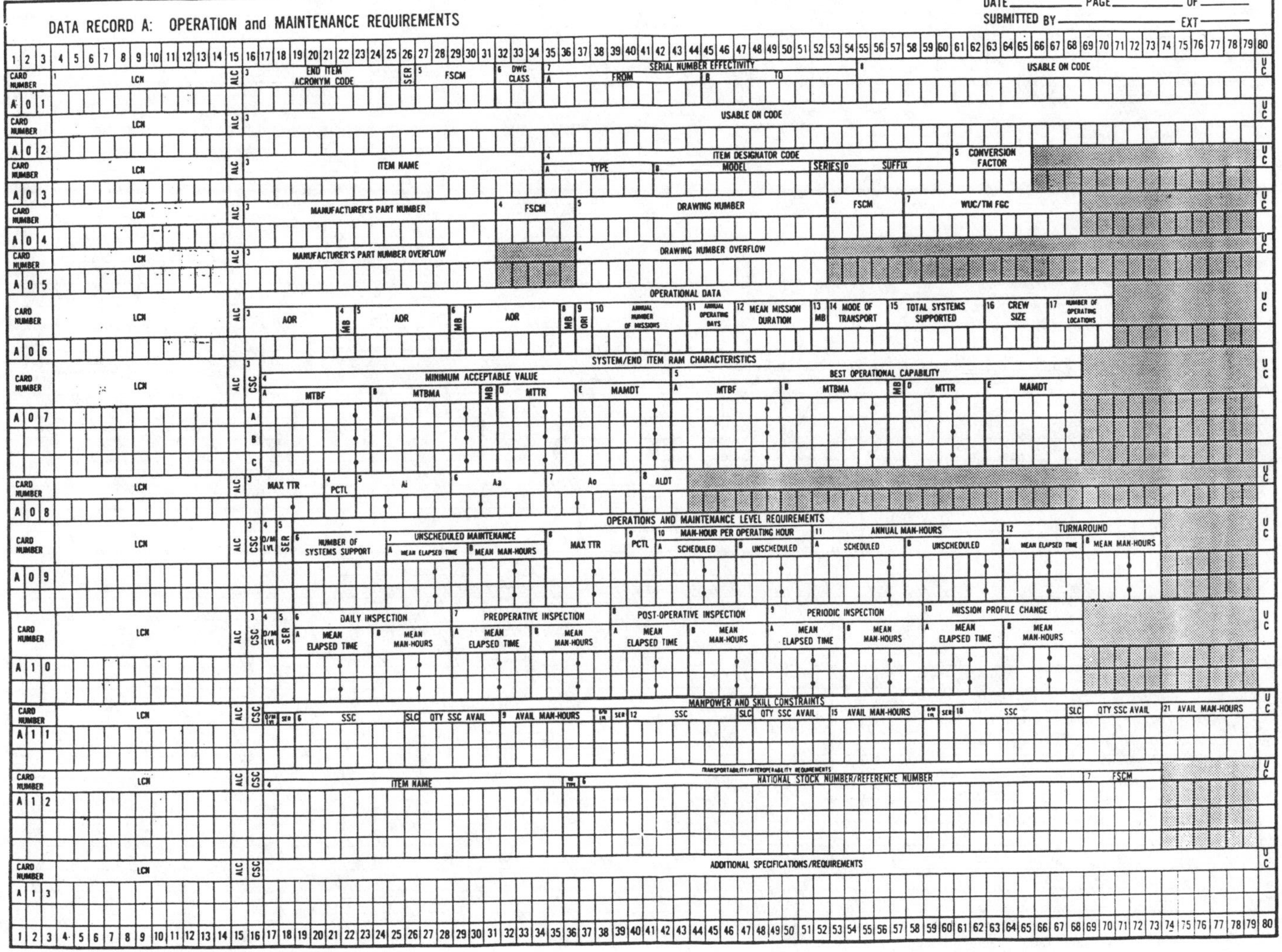
DATA RECORD A: OPERATION and MAINTENANCE REQUIREMENTS

DATE ______ PAGE ______ OF ______
SUBMITTED BY ______ EXT ______

Card columns 1–80

A01 — CARD NUMBER | 1 LCN | ALC | 3 END ITEM ACRONYM CODE | SER | 5 FSCM | 6 DWG CLASS | 7 SERIAL NUMBER EFFECTIVITY: A FROM, B TO | 8 USABLE ON CODE | UC

A02 — CARD NUMBER | LCN | ALC | 3 USABLE ON CODE | UC

A03 — CARD NUMBER | LCN | ALC | 3 ITEM NAME | 4 ITEM DESIGNATOR CODE: A TYPE, B MODEL, SERIES, D SUFFIX | 5 CONVERSION FACTOR | UC

A04 — CARD NUMBER | LCN | ALC | 3 MANUFACTURER'S PART NUMBER | 4 FSCM | 5 DRAWING NUMBER | 6 FSCM | 7 WUC/TM FGC | UC

A05 — CARD NUMBER | LCN | ALC | 3 MANUFACTURER'S PART NUMBER OVERFLOW | 4 DRAWING NUMBER OVERFLOW | UC

A06 — CARD NUMBER | LCN | ALC | OPERATIONAL DATA: 3 AOR | 4 MB | 5 AOR | 6 MB | 7 AOR | 8 MB | 9 ORI | 10 ANNUAL NUMBER OF MISSIONS | 11 ANNUAL OPERATING DAYS | 12 MEAN MISSION DURATION | 13 MB | 14 MODE OF TRANSPORT | 15 TOTAL SYSTEMS SUPPORTED | 16 CREW SIZE | 17 NUMBER OF OPERATING LOCATIONS | UC

A07 — CARD NUMBER | LCN | ALC | CSC | 3 SYSTEM/END ITEM RAM CHARACTERISTICS: 4 MINIMUM ACCEPTABLE VALUE: A MTBF, B MTBMA, MB, D MTTR, E MAMDT | 5 BEST OPERATIONAL CAPABILITY: A MTBF, B MTBMA, MB, D MTTR, E MAMDT | rows A, B, C | UC

A08 — CARD NUMBER | LCN | ALC | 3 MAX TTR | 4 PCTL | 5 Ai | 6 Aa | 7 Ao | 8 ALDT | UC

A09 — CARD NUMBER | LCN | ALC | OPERATIONS AND MAINTENANCE LEVEL REQUIREMENTS: 3 CSC | 4 D/M LVL | 5 SER | 6 NUMBER OF SYSTEMS SUPPORT | 7 UNSCHEDULED MAINTENANCE: A MEAN ELAPSED TIME, B MEAN MAN-HOURS | 8 MAX TTR | 9 PCTL | 10 MAN-HOUR PER OPERATING HOUR: A SCHEDULED, B UNSCHEDULED | 11 ANNUAL MAN-HOURS: A SCHEDULED, B UNSCHEDULED | 12 TURNAROUND: A MEAN ELAPSED TIME, B MEAN MAN-HOURS | UC

A10 — CARD NUMBER | LCN | ALC | 3 CSC | 4 D/M LVL | 5 SER | 6 DAILY INSPECTION: A MEAN ELAPSED TIME, B MEAN MAN-HOURS | 7 PREOPERATIVE INSPECTION: A MEAN ELAPSED TIME, B MEAN MAN-HOURS | 8 POST-OPERATIVE INSPECTION: A MEAN ELAPSED TIME, B MEAN MAN-HOURS | 9 PERIODIC INSPECTION: A MEAN ELAPSED TIME, B MEAN MAN-HOURS | 10 MISSION PROFILE CHANGE: A MEAN ELAPSED TIME, B MEAN MAN-HOURS | UC

A11 — CARD NUMBER | LCN | ALC | CSC | MANPOWER AND SKILL CONSTRAINTS: D/M LVL | SER | 6 SSC | SLC | QTY SSC AVAIL | 9 AVAIL MAN-HOURS | D/M LVL | SER | 12 SSC | SLC | QTY SSC AVAIL | 15 AVAIL MAN-HOURS | D/M LVL | SER | 18 SSC | SLC | QTY SSC AVAIL | 21 AVAIL MAN-HOURS | UC

A12 — CARD NUMBER | LCN | ALC | CSC | TRANSPORTABILITY/INTEROPERABILITY REQUIREMENTS: 4 ITEM NAME | NO TYPE | 6 NATIONAL STOCK NUMBER/REFERENCE NUMBER | 7 FSCM | UC

A13 — CARD NUMBER | LCN | ALC | CSC | ADDITIONAL SPECIFICATIONS/REQUIREMENTS | UC

Fig. 17-1. LSA data record A.

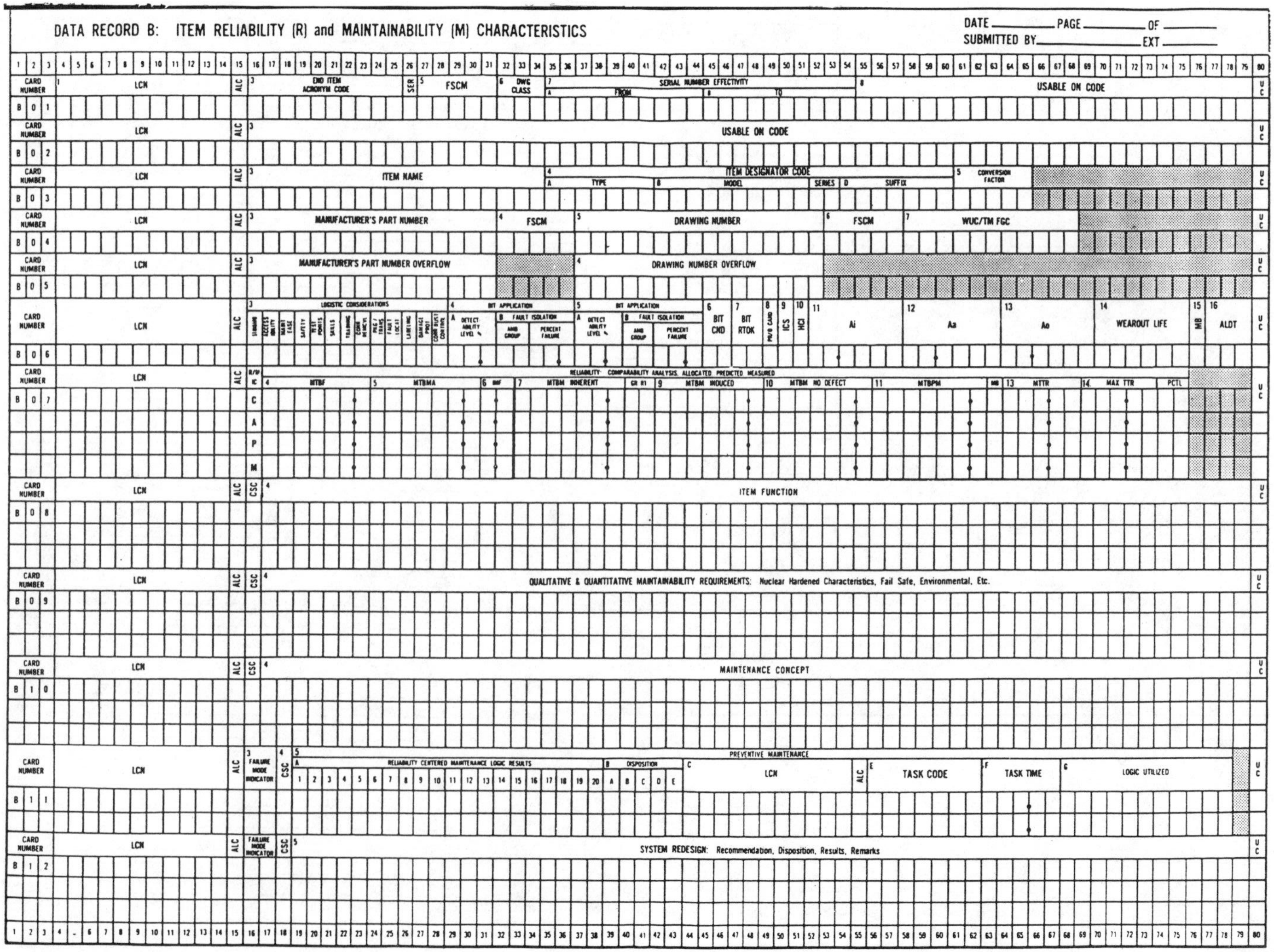

Fig. 17-2. LSA data record B.

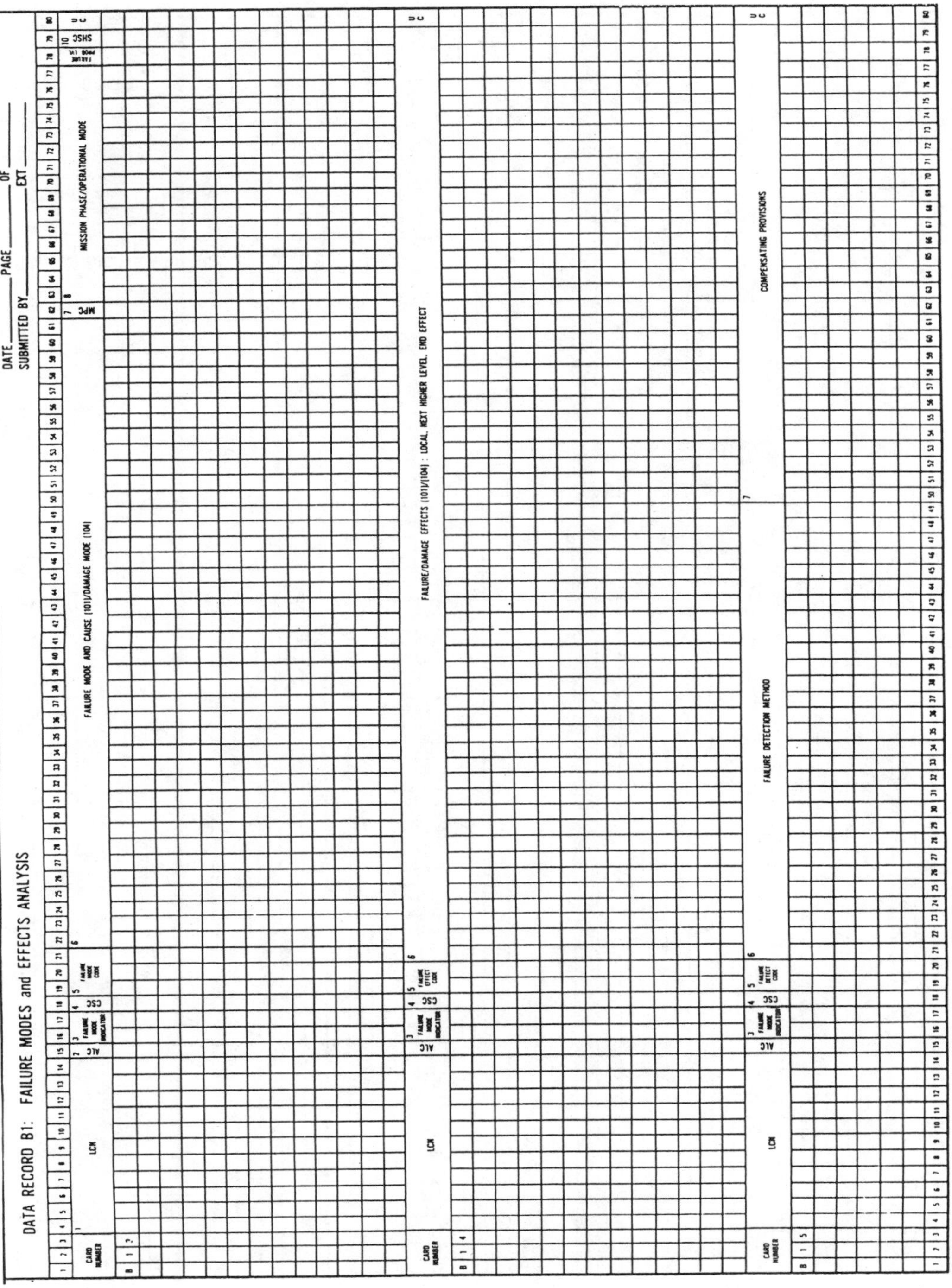

Fig. 17-3. LSA data record B1.

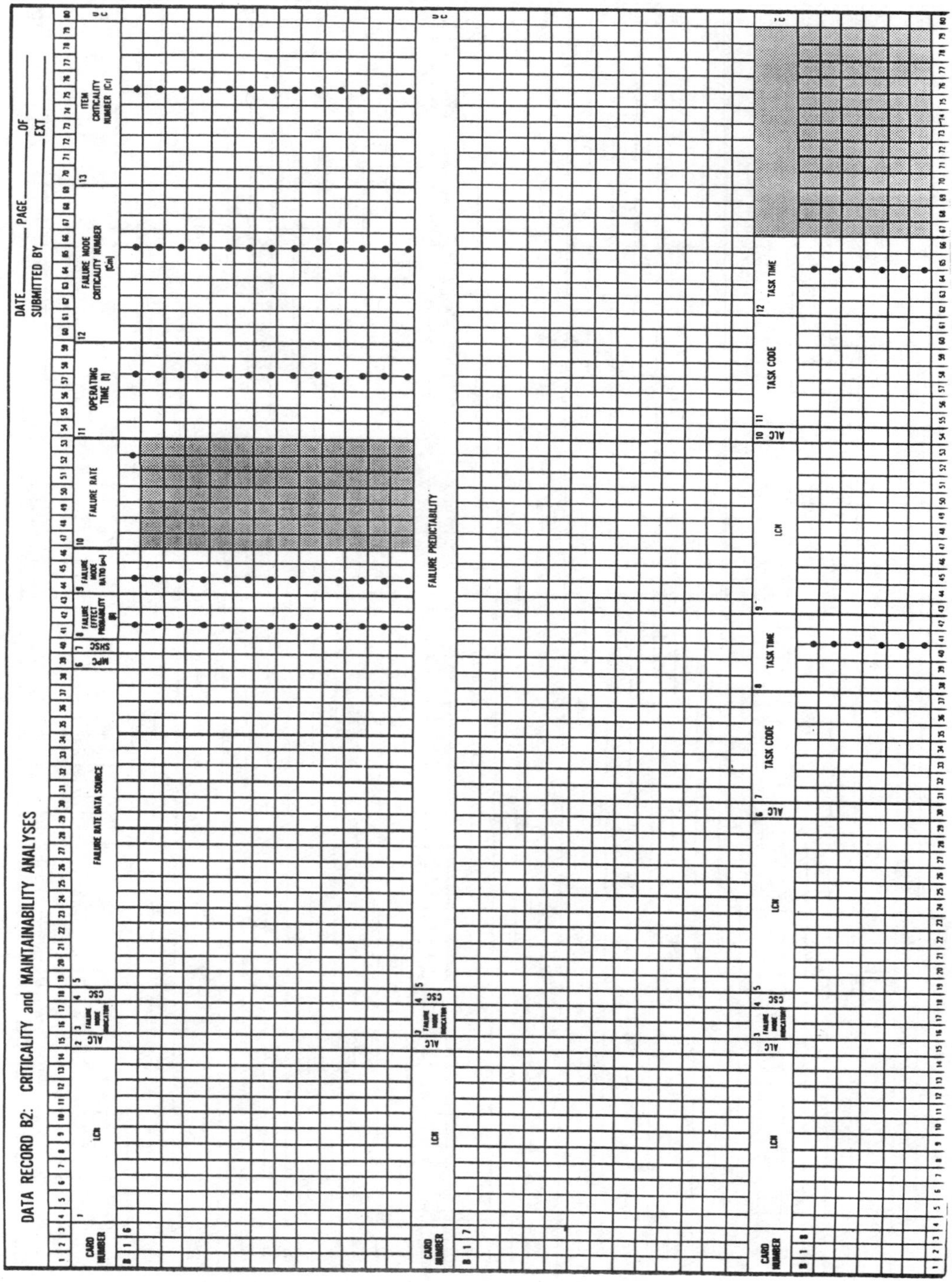

Fig. 17-4. LSA data record B2.

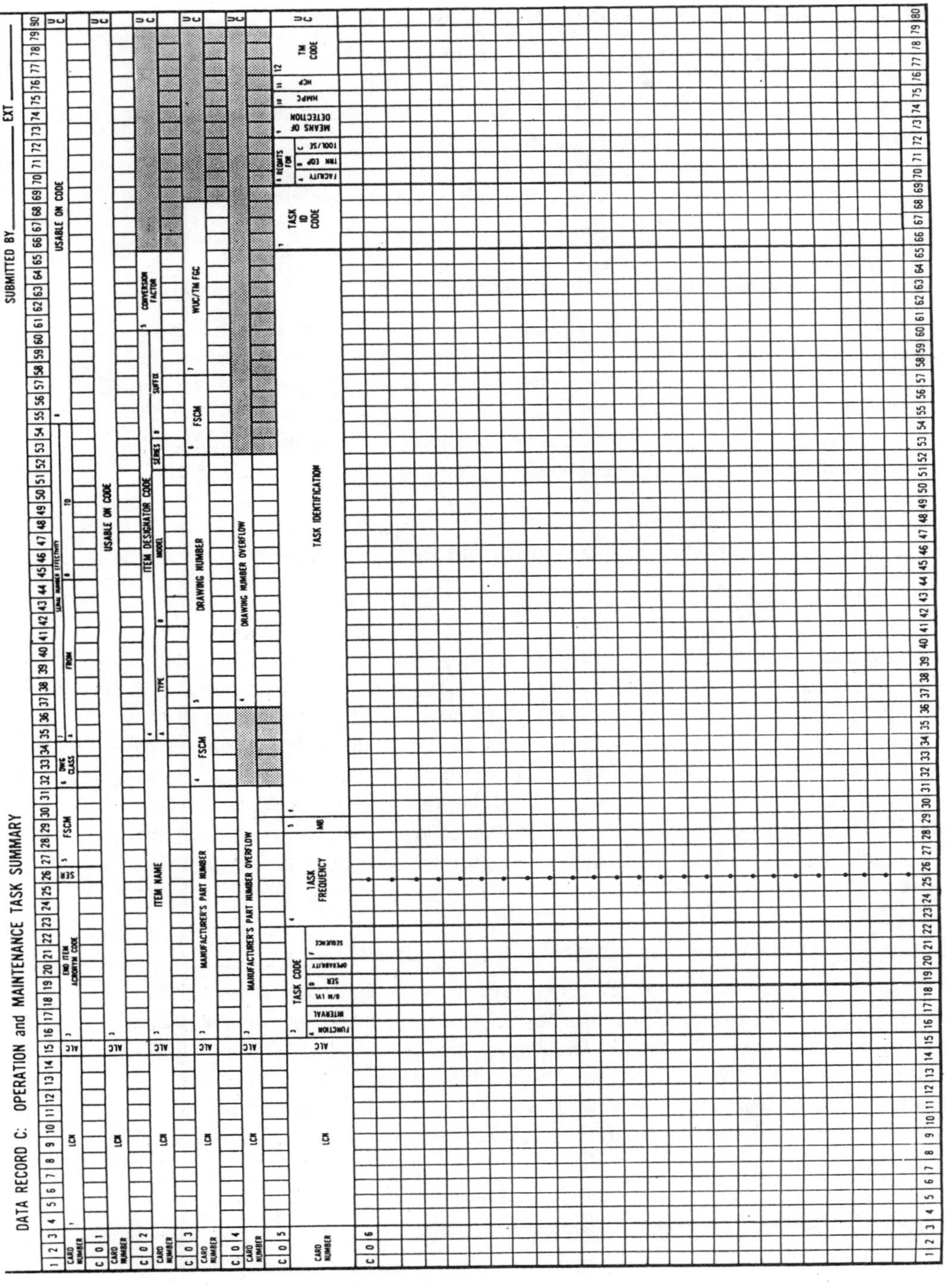

Fig. 17-5. LSA data record C.

Data Record D: Operation and Maintenance Task Analysis

A step-by-step, detailed narrative description of each operation or maintenance task identified on Data Record C is documented using Data Record D. If Data Record C identifies a requirement for 20 different operation or maintenance tasks for an item, then a separate Data Record D is prepared for each task. As illustrated in Fig. 17-6, the data record is used to document the task narrative, number and required skill specialty codes of personnel necessary to perform the task, and the elapsed and total times required to perform each sequential step of the task.

Data Record D1: Personnel and Support Requirements

The narrative identification of support requirements contained on Data Record D is recorded in tabular form on a companion Data Record D1. Also contained on this data record (Fig. 17-7) is a summary of the personnel and skill requirements to perform the task and maintenance task times. The information contained on Data Record D1 is combined with corresponding data for all other maintenance and operation task to identify and justify the logistic support resources required to support the system.

Data Records E and E1: Support
Equipment and Training Material Description and Justification

Data Records E and E1 are used to automate the preparation of support equipment recommendation data (SERD). Using the LSAR data base to prepare this document enables the logistics engineer to have complete supporting documentation available throughout the preparation process, in the form of detailed maintenance tasks that require the use of the item. Data Records E and E1 are used to consolidate all the pertinent information generated by the maintenance task analysis process relative to the use or need for a specific item of support equipment or training material. These data records list all the tasks that require the use of the item, and provide justification for procuring or authorizing use of the item. Data Records E and E1 (Fig. 17-8) are prepared for each nonstandard or special item of support or test equipment and training material needed to support operation or maintenance.

Data Record E2: Unit Under Test Description and Justification

Data Record E2 is used to record detailed information about testing assemblies and using automated test programs. The purpose of this data record is to consolidate all references in the maintenance task narrative on the Data Record D into a single requirement for a specific test program set (TPS) or other need for using automatic test equipment for fault isolation or testing. Data Record E2, illustrated in Fig. 17-9, identifies the UUT, the type of testing to be accomplished, test parameters, and TPS requirements.

Data Record F: Facility Description and Justification

Requirements for special facilities required to support operation, maintenance, or

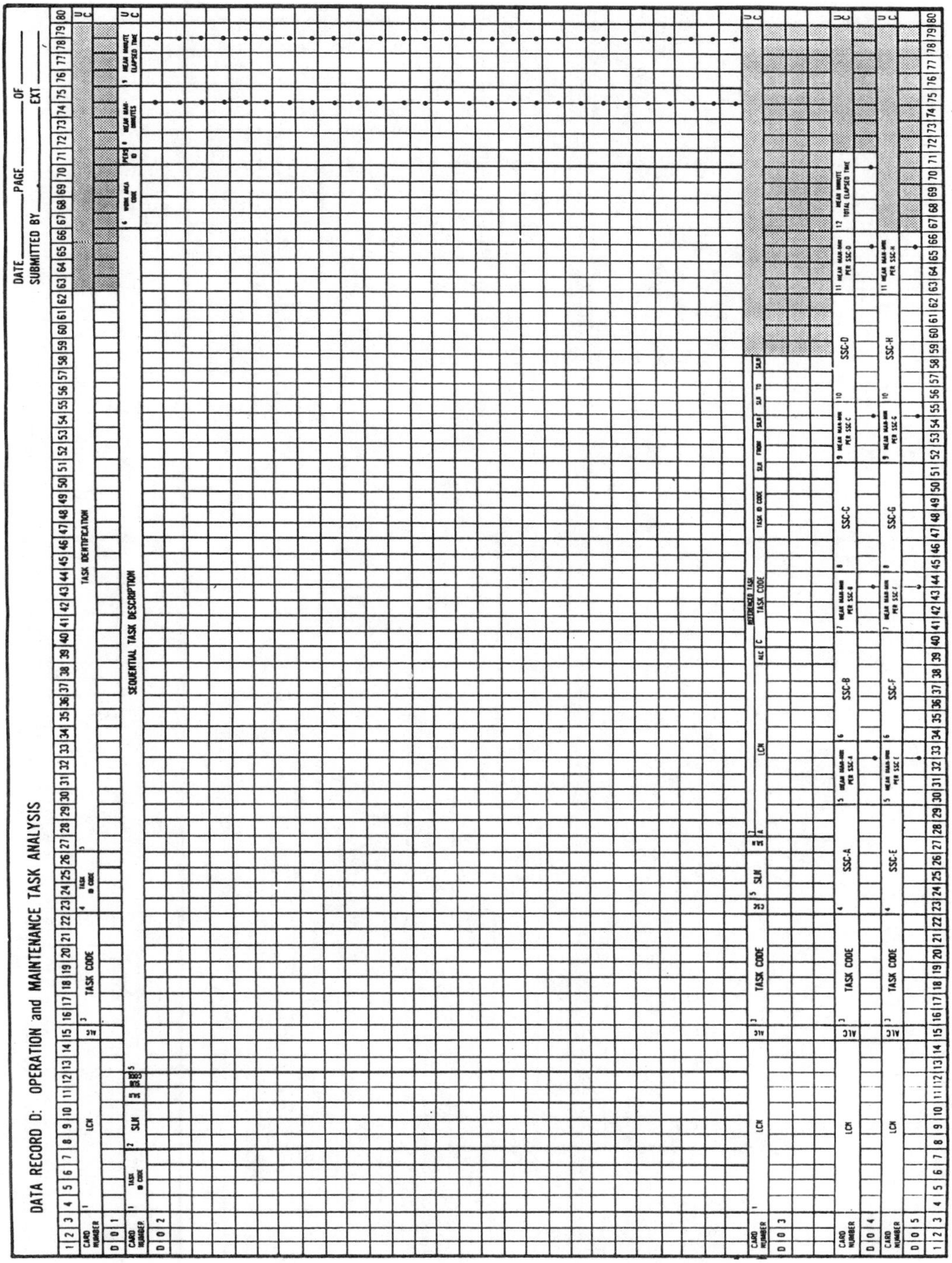

Fig. 17-6. LSA data record D.

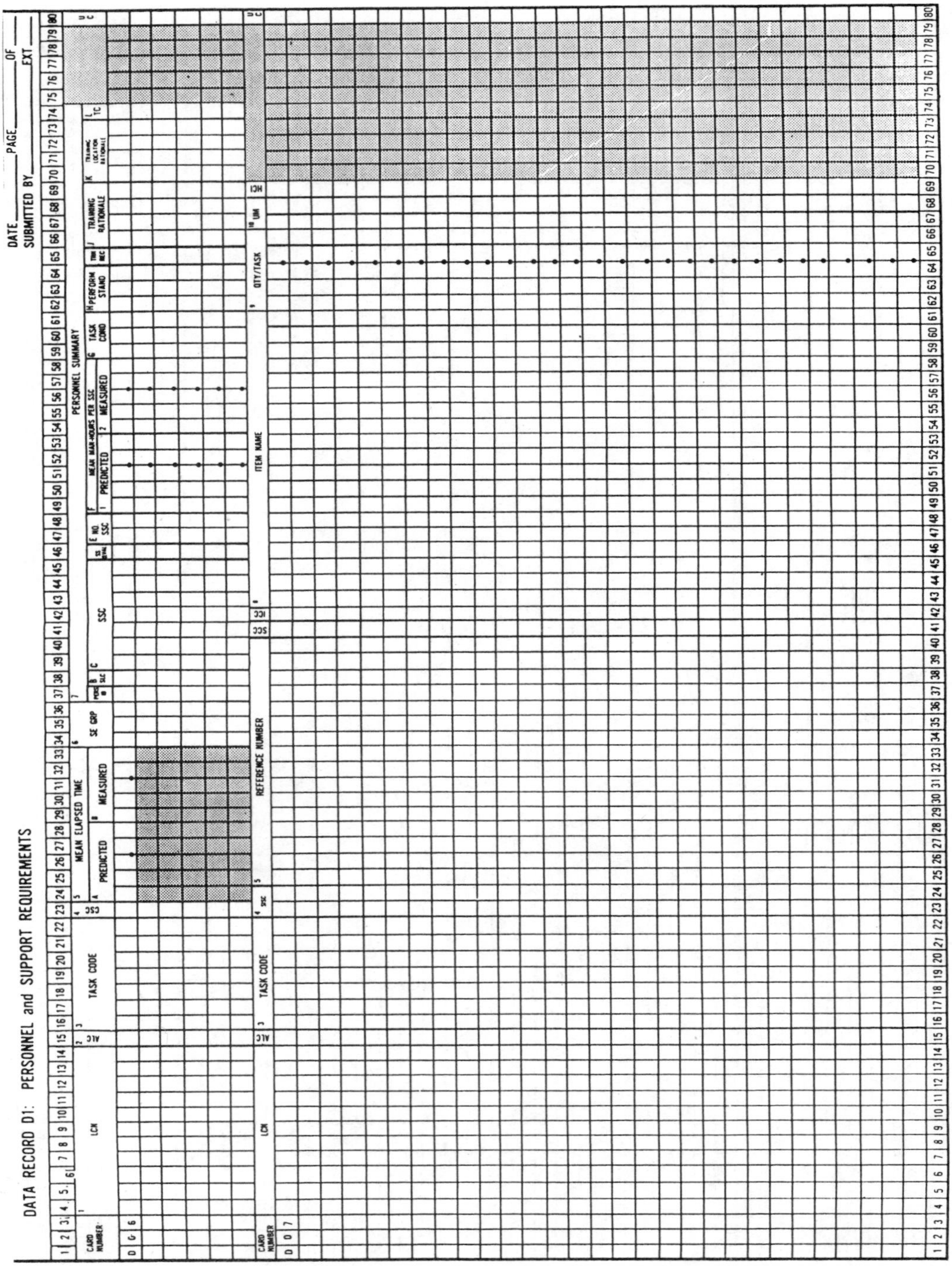

Fig. 17-7. LSA data record D1.

288

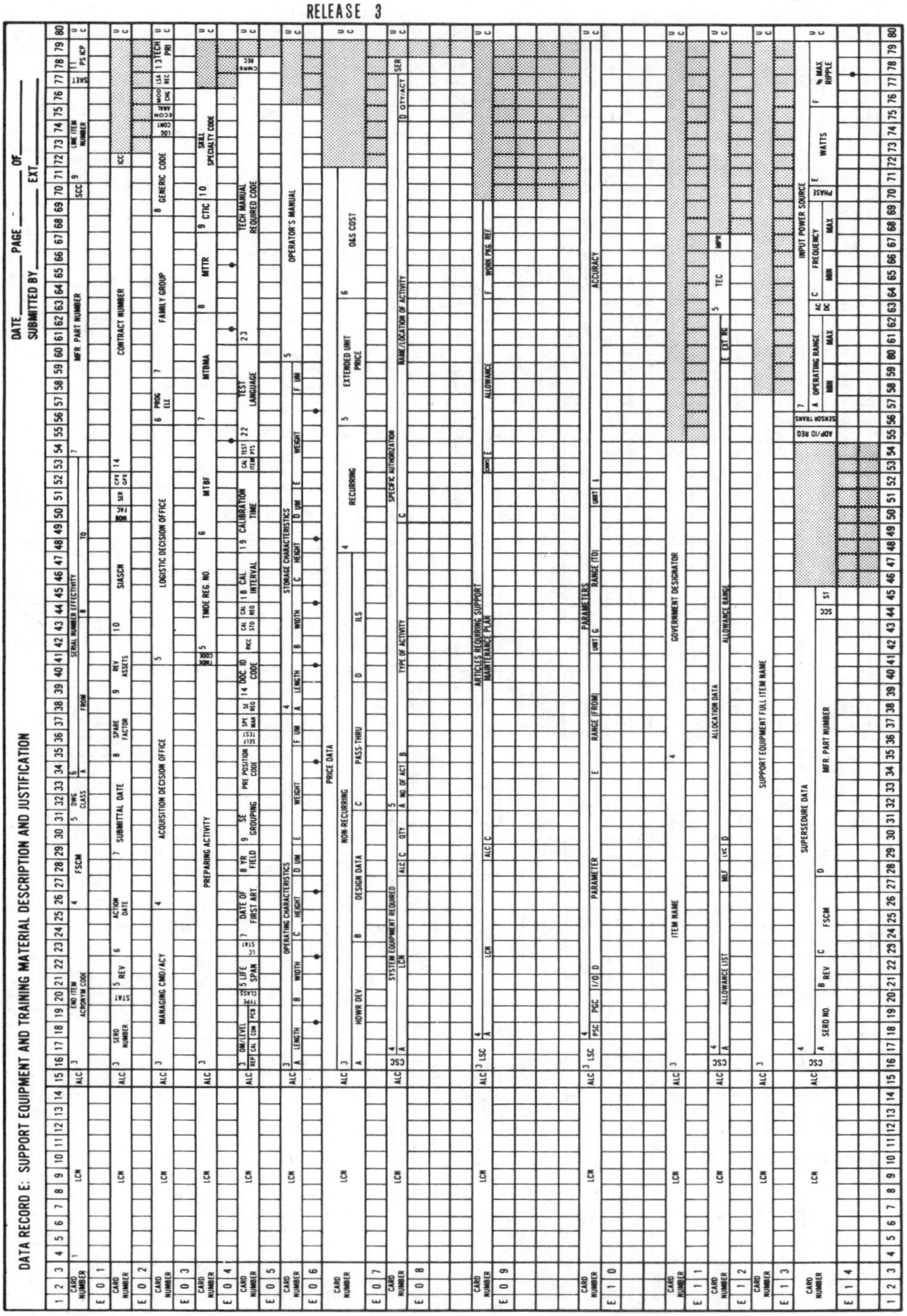

DATA RECORD E: SUPPORT EQUIPMENT AND TRAINING MATERIAL DESCRIPTION AND JUSTIFICATION

DATE_____ PAGE _____ OF_____
SUBMITTED BY _____ EXT._____

Fig. 17-8. LSA data record E/E1.

Fig. 17-8b. LSA data record E/E1. (Continued from page 289.)

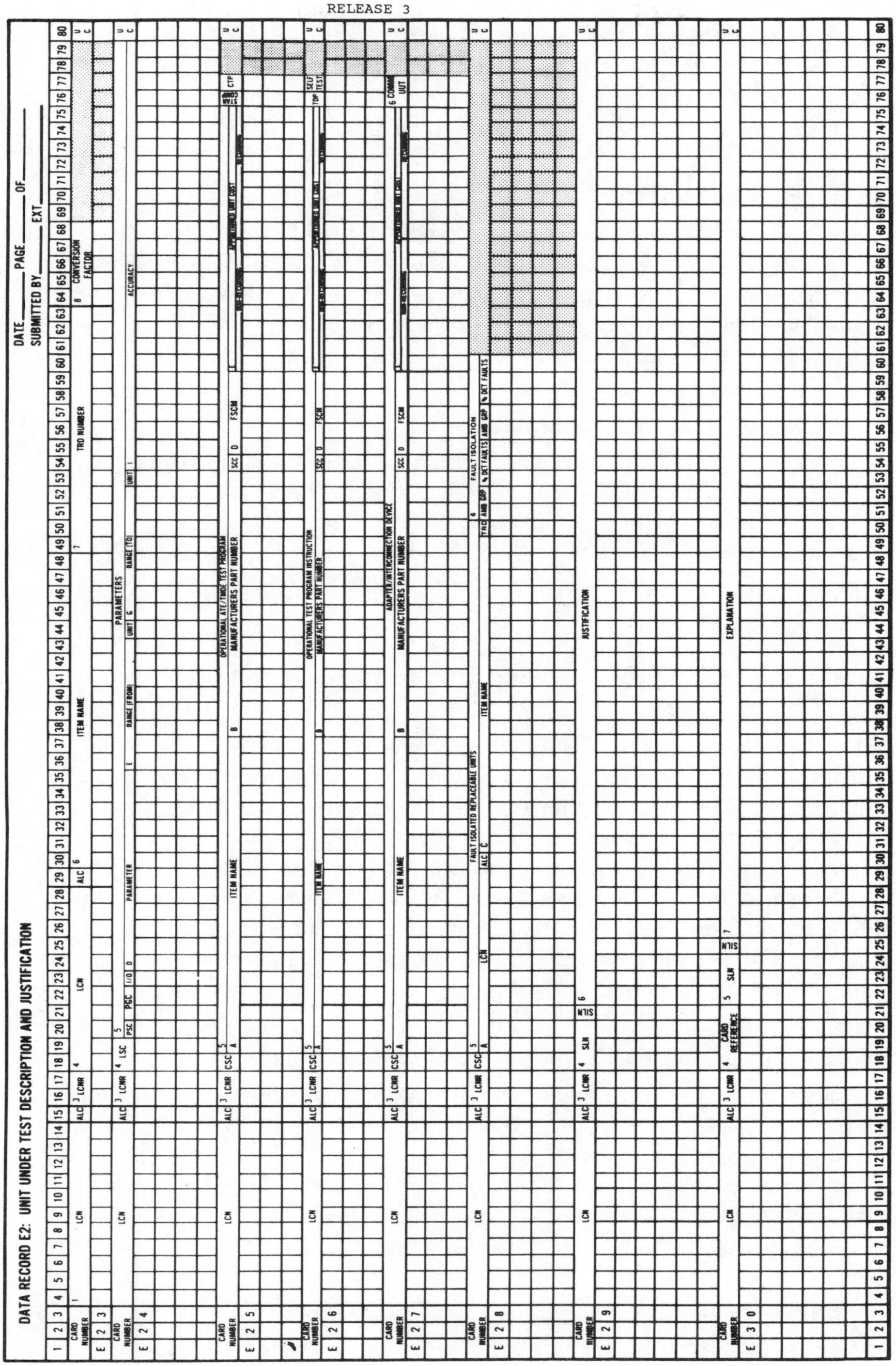

RELEASE 3

Fig. 17-9. LSA data record E2.

training identified on Data Record C are documented on Data Record F. The data record (Fig. 17-10) contains an identification of all tasks the facility is required to support, a technical description of the facility, and complete rationale and justification for using the facility. Data Record F is prepared for each new or modified facility.

Data Record G: Skill Evaluation and Justification

Data Record G is used to document the need for new or modified personnel skills to support maintenance of the new equipment. A data record is prepared for each new or modified skill. Figure 17-11 shows that the data record includes identification of tasks to be accomplished, personnel qualifications, training requirements, and a detailed justification for the new or modified skill.

Data Record H: Support Items Identification

Provisioning documentation is prepared using Data Records H and H1. Data Record H is prepared for each unique item or part used in a system. This data record (Fig. 17-12) contains cataloging information about the part, such as name, FSCM, price, and NSN, that does not change with different applications of the part. Only one Data Record H is prepared for each part, regardless of how many times the part is used in a system.

Data Record H1: Support Items Identification (Application Related)

All provisioning information related to the specific application of a part in a system is documented on Data Record H1. Also included on the data record is information pertaining to technical manuals and engineering changes. As shown in Fig. 17-13, Data Record H1 contains the SMR code, maintenance factors, quantities per assembly and end item, and other information about part usage.

Data Record H1 is prepared for each application or each part in a system. For example, if a valve is used ten times in a system, one Data Record H is prepared to record cataloging information about the valve, and ten Data Records H1 are prepared to document each use of the valve. A Data Record H1 should be prepared for each item listed on Data Records D1.

Data Record J: Transportability Engineering Characteristics

Information about the transportability characteristics of an item is documented on Data Record J. As shown in Fig. 17-14, the data record addresses the physical movement requirements of the system. Data Record J is prepared for each transportable portion of a system; so, in many cases, only one data record is required for a complete program. This information is used by the government to plan for transporting the system when delivered and for preparing mobility requirements for deployment.

Data Record Relationships

There are many interrelationships among LSA data records. Maintenance

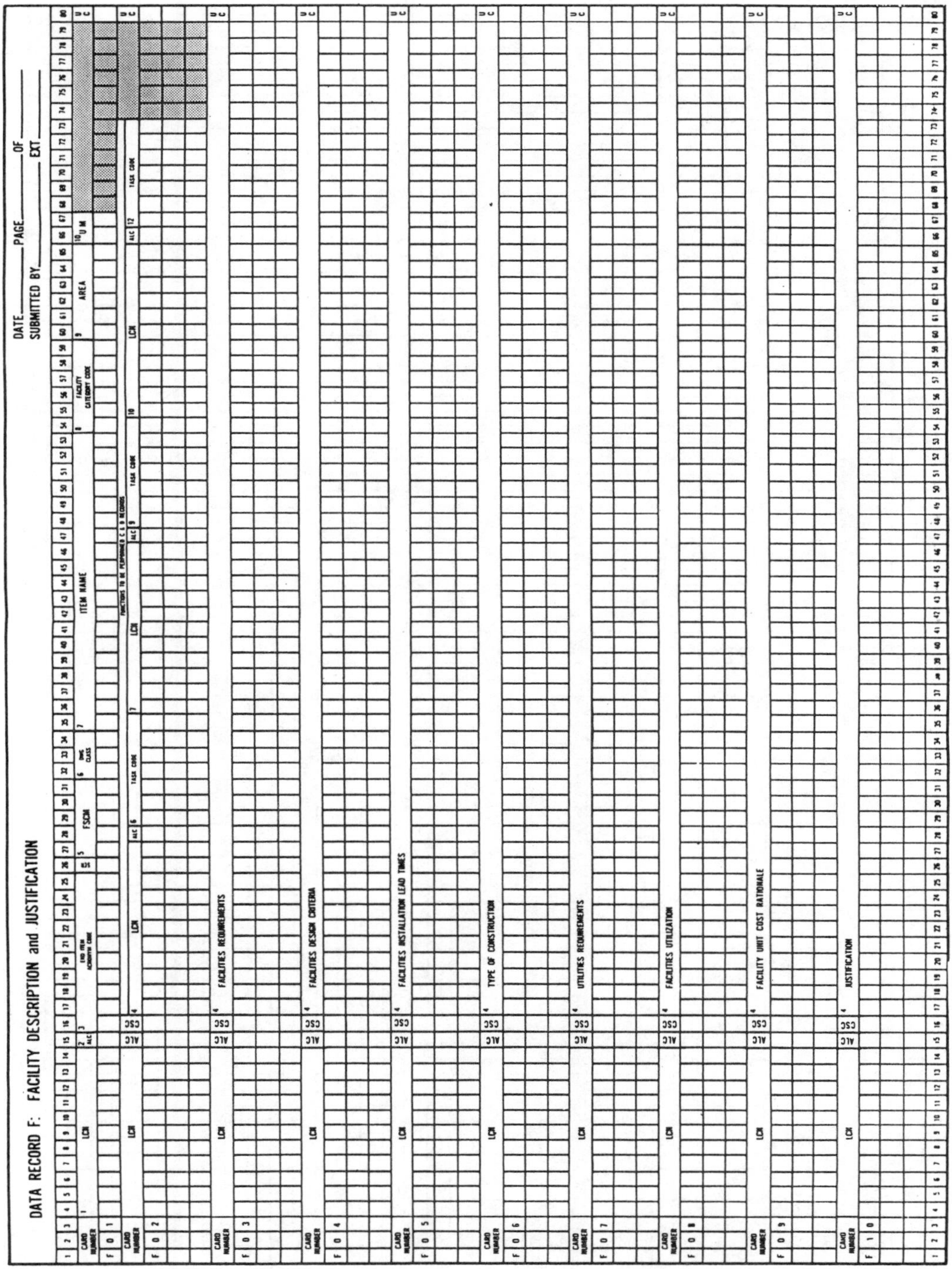

Fig. 17-10. LSA data record F.

293

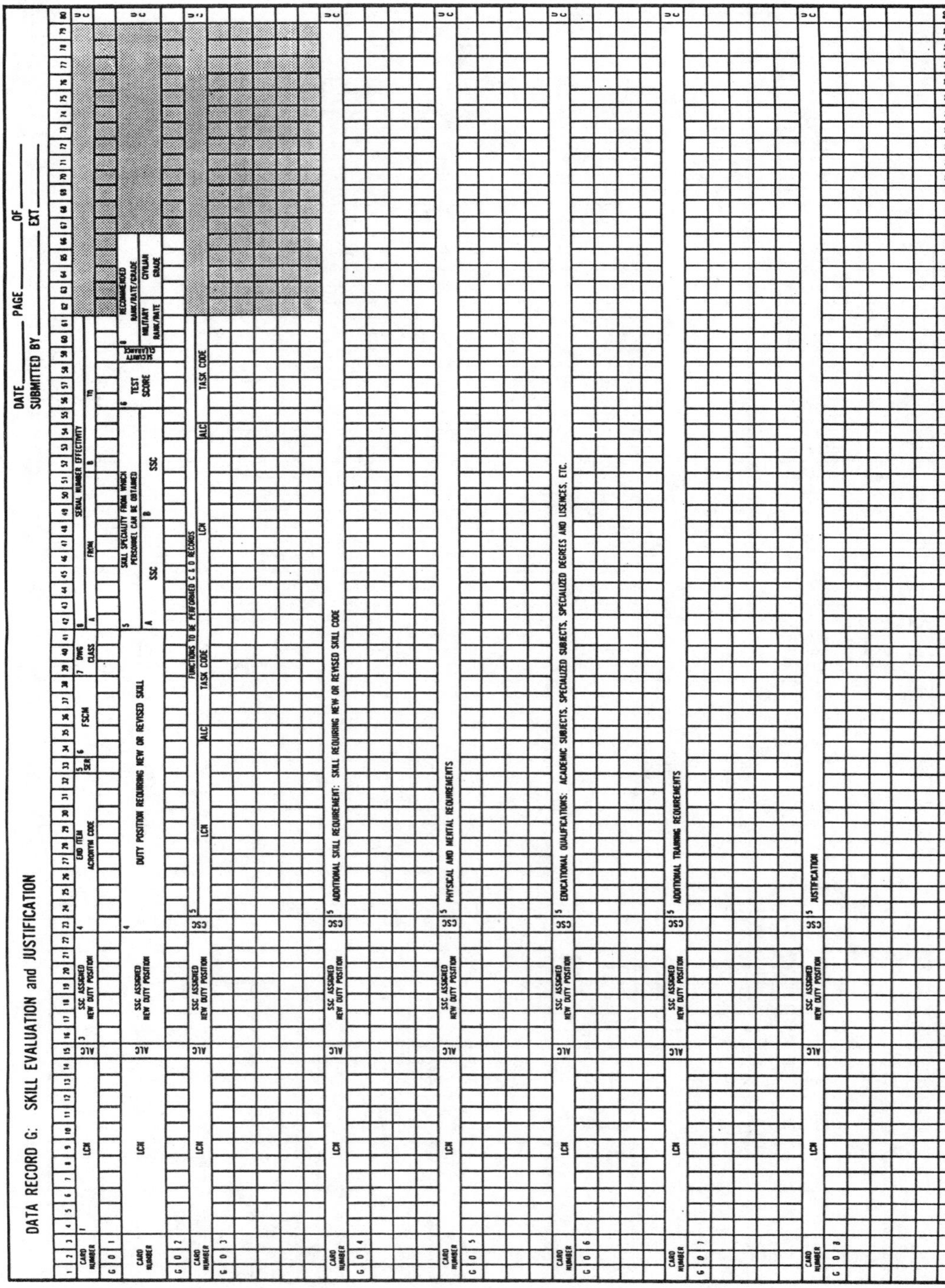

Fig. 17-11. LSA data record G.

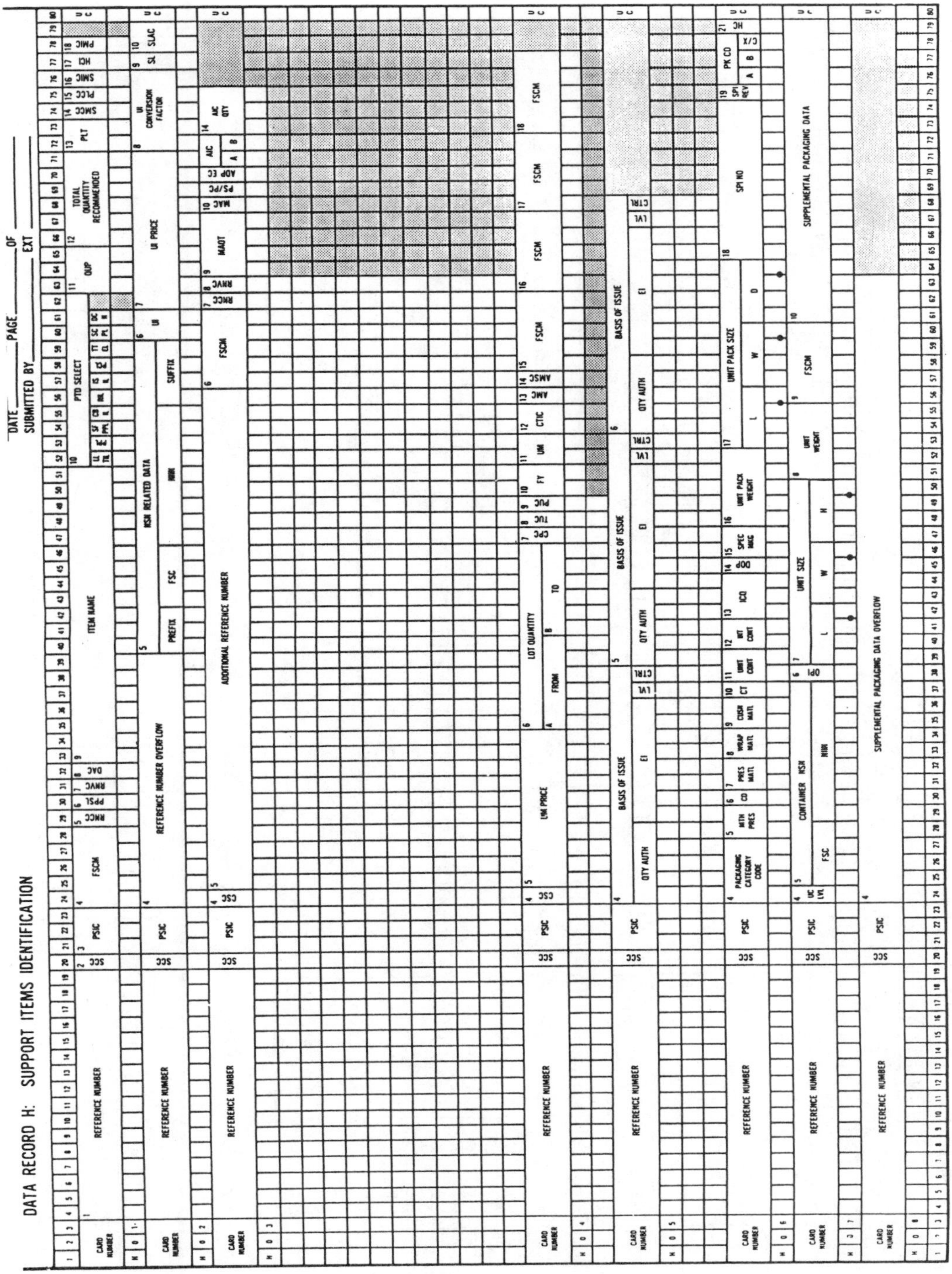

Fig. 17-12. LSA data record H.

DATA RECORD H1: SUPPORT ITEMS IDENTIFICATION (APPLICATION RELATED)

DATE ______ PAGE ______ OF ______

SUBMITTED BY ______ EXT ______

Card Number	Fields
H 0 9	REFERENCE NUMBER; SCC; LCN; ALC; PSIC; CSC; USABLE ON CODE; U C
H 1 0	REFERENCE NUMBER; SCC; LCN; ALC; PSIC; CSC; PCCN; PLISN; TOCC; QTY/EI; NHA PLISN; NHA IND; ORR; SAME AS PLISN; PRIOR ITEM PLISN; QTY/ASSY; STAT; SIC; U C
H 1 1	REFERENCE NUMBER; SCC; LCN; ALC; PSIC; ICC; SMR; DMIL; MRRI; MRRII; MRR MOD; MAINTENANCE TASK DISTRIBUTION (C E P S A W O F H SRA D CBD CAD); U C
H 1 2	REFERENCE NUMBER; SCC; LCN; ALC; PSIC; EC; LRU; NRTS; RSR; MRU; RISS BUY; RMSS LVL; RTLL; DESIGNATED REWORK POINT; U C
H 1 3	REFERENCE NUMBER; SCC; LCN; ALC; PSIC; REPAIR CYCLE TIME (O F H SRA D); CON; REPLACEMENT TASK DISTRIBUTION (O F H SRA D); U C
H 1 4	REFERENCE NUMBER; SCC; LCN; ALC; PSIC; CSC; REFERENCE DESIGNATION; RDCC; RDC; U C
H 1 5	REFERENCE NUMBER; SCC; LCN; ALC; PSIC; CSC; TM CODE; FIGURE NUMBER; ITEM NUMBER; TM CHG NO; TM IND; QUANTITY PER FIGURE; WUC/TM FGC; U C
H 1 6	REFERENCE NUMBER; SCC; LCN; ALC; PSIC; TMI; CSC; PROVISIONING NOMENCLATURE; U C
H 1 7	REFERENCE NUMBER; SCC; LCN; ALC; PSIC; CSC; CHANGE AUTHORITY NUMBER; IC; SERIAL NUMBER EFFECTIVITY (A FROM, B TO); REPLACED OR SUPERSEDING PLISN; RS/IND; U C
H 1 8	REFERENCE NUMBER; SCC; LCN; ALC; PSIC; CSC; CHANGE AUTHORITY NUMBER; TIC; QUANTITY SHIPPED; QUANTITY PROCURED; PRORATED ELIN; PRORATED QUANTITY; U C
H 1 9	REFERENCE NUMBER; SCC; LCN; ALC; PSIC; CSC; CHANGE AUTHORITY NUMBER; DCN UOC; U C
H 2 0	REFERENCE NUMBER; SCC; LCN; ALC; PSIC; CSC; REMARKS; U C

Fig. 17-13. LSA data record H1.

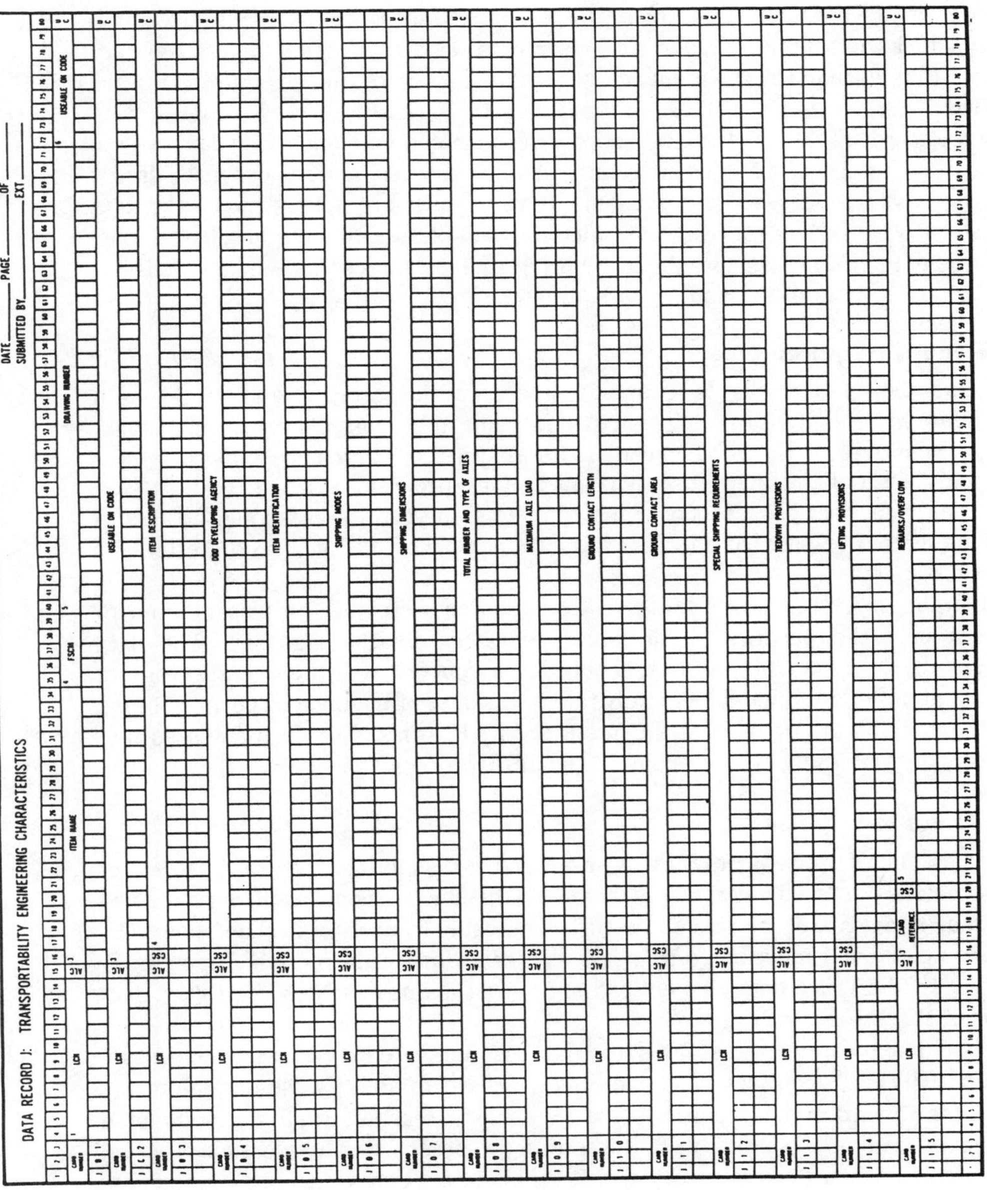

Fig. 17-14. LSA data record J.

engineering and the maintenance task analysis process deal with Data Records B, B1, B2, C, D, and D1 simultaneously. These data records must be considered as a single entity. The Data Record B series identifies the need for a maintenance task, Data Record C identifies what tasks are to be performed, and Data Records D and D1 contain the actual tasks and support resource requirements. This chain of actions is necessary to ensure that all required maintenance tasks and support resource requirements are identified.

Data Records E, E1, E2, F, and G support the maintenance engineering data records because they contain consolidated information and justification for decisions made during the task analysis process. The provisioning data records are dependent on the data in Data Record D1 for identification of all parts and other items required to support maintenance. Figure 17-15 shows the relationship among data records and how decisions made by one activity affect another.

Data Record Requirements

The number of data records required to document the results of an LSA program can be overwhelming. Determining exactly how many are required is not a simple task, but the magnitude of the project can be approximated. A general philosophy for estimating the number of data records required for a program is in Fig. 17-16. Using information from past programs or similar systems, the contractor should be able to determine, within reasonable tolerances, the scope of the LSA data record effort for a given program.

When an LSA program is tailored, the data requirements are also tailored to document only the pertinent information generated by the analyses. The data requirements for a specific program should be in concert with the overall program requirements. Generation of data merely for the purpose of having a large stack of paper at the end of a program is not the intent of LSAR. It should be remembered that the LSAR only documents the results of the LSA process.

Data Elements

Detailed data requirements must be prepared for each data record. A review of the data records discussed above shows that there are numerous data elements contained on each sheet. Several of the data elements are repeated from sheet to sheet; however, many are unique and must be generated specifically for a data record. Appendix F, MIL-STD 1388-2A, is a data element dictionary that defines and describes each data element in the LSAR system.

It should be pointed out that not every data element is required for each program. DD form 1949-1, LSAR Data Selection Sheet, is used to identify the data elements required for a program. DD form 1949-1 is a two-part, 16-page form that is completed by the government and is provided to the contractor either with a request for proposal or at the LSA guidance conference. Part I is for all data records, except Data Records H and H1, which are in Part II. This form addresses every data element block on each data record, one at a time, and identifies if the data element is required for the program. Part I, page 1 of DD form 1949-1, is shown in Fig. 17-17.

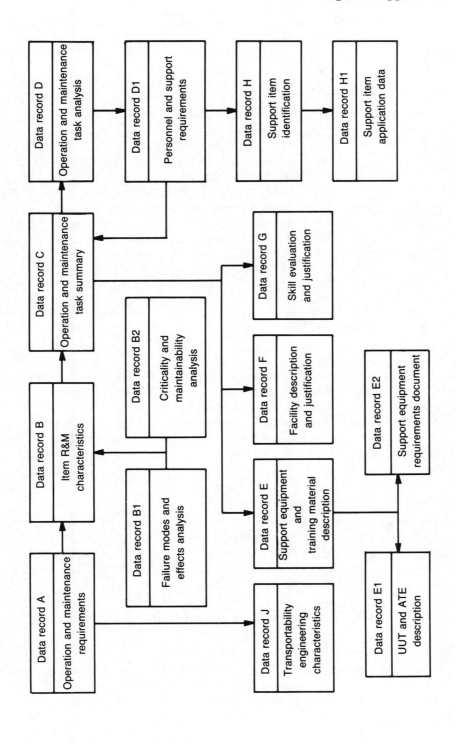

Fig. 17-15. LSA data record relationship.

Data record	Estimating rationale
A	One data record per program unless directed otherwise.
B, B1, B2, C	One data record required for each repairable or maintenance-significant item.
D, D1	One data record required for each maintenance task listed on the corresponding Data Record C. At a minimum plan for tasks for fault isolation, remove and replace, and repair for each item.
E, E1, E2, F, G	As required to document need and justification for support and test equipment, facilities, or personnel skills.
H	One data record for each unique part used in the system.
H1	One data record for every application of every part used in the system.
J	As required to document system transportability requirements. At least one per system.

Fig. 17-16. Estimating LSA data record requirements.

USING THE LSAR

The LSAR data base is designed to be used, not just created and filed away. The files should be used by all ILS disciplines throughout the program to develop the detailed support plans and requirements for the system. It is important to understand how the LSAR is to be used when a program starts so that the methods for use can be identified and understood by all concerned. Before any real benefit can be obtained from the LSAR, a basic understanding is necessary in three areas: LSA control numbers, LSA candidates, and LSA summary reports.

LSA Control Numbers

The LSAR process generates a tremendous amount of paper and data. The method for controlling data records is the LSA control number (LSACN). The LSACN is the key field in the 4th through 14th card columns of each data record. This number uniquely identifies each item in the system and correlates each data record with its corresponding item. Figure 17-18 shows how an LSACN is assigned for each item and how the indenture level dictates the number of digits used. Detailed directions for assignment of LSACNs is contained in Appendix D of MIL-STD 1388-2A.

LSA Candidates

Maintenance engineers begin the LSA process by identifying elements of the system that are candidates for an LSA. This breakdown of the system is normally referred to as an LSA candidate list. The candidate list identifies each part of the system that

PART I		LSAR DATA SELECTION SHEET	Form Approved OMB No 0704-0188 Exp. Date: June 30, 1986
CARD/ BLOCK NUMBER	DED NO	DATA ELEMENT NAME	REQUIRED
		LSAR DATA RECORD A	
01-1	197	LOGISTIC SUPPORT ANALYSIS CONTROL NUMBER *(Applies to complete A Record)*	▲
01-2	023	ALTERNATE LSA CONTROL NUMBER CODE *(Applies to complete A Record as needed)*	▲
01-3	106	END ITEM ACRONYM CODE	
01-4	414	SERVICE DESIGNATOR CODE	
01-5	139	FEDERAL SUPPLY CODE FOR MANUFACTURERS	
01-6	098	DRAWING CLASSIFICATION	
01-7	411	SERIAL NUMBER EFFECTIVITY	
01-8	536	USABLE ON CODE	▲
	536	OPTION 1	
	536	OPTION 2	
	536	OPTION 3	
01-9	535	UPDATE CODE *(Applies to complete A Record)*	▲
02-3	536	USABLE ON CODE	▲
03-3	181	ITEM NAME	
03-4	178	ITEM DESIGNATOR CODE	
03-5	069	CONVERSION FACTOR	
04-3	213	MANUFACTURER'S PART NUMBER	
04-4	139	FEDERAL SUPPLY CODE FOR MANUFACTURERS	
04-5	099	DRAWING NUMBER	
04-6	139	FEDERAL SUPPLY CODE FOR MANUFACTURERS	
04-7	545	WORK UNIT CODE /TECHNICAL MANUAL FUNCTIONAL GROUP CODE	
05-3	214	MANUFACTURER'S PART NUMBER OVERFLOW	
05-4	100	DRAWING NUMBER OVERFLOW	
06-3,5,7	029	ANNUAL OPERATING REQUIREMENTS	
06-4,6,8	244	MEASUREMENT BASE	
06-9	285	OPERATIONAL REQUIREMENT INDICATOR	
06-10	027	ANNUAL NUMBER OF MISSIONS	
06-11	028	ANNUAL OPERATING DAYS	
06-12	234	MEAN MISSION DURATION	
06-13	244	MEASUREMENT BASE	
06-14	254	MODE OF TRANSPORT	
06-15	499	TOTAL SYSTEMS SUPPORTED	
06-16	073	CREW SIZE	
06-17	268	NUMBER OF OPERATING LOCATIONS	
07-3	051	CARD SEQUENCING CODE	▲
07-4	248	MINIMUM ACCEPTABLE VALUE	
07-4A	235	MEAN TIME BETWEEN FAILURES	
07-4B	236	MEAN TIME BETWEEN MAINTENANCE ACTIONS	
07-4C	244	MEASUREMENT BASE	
07-4D	241	MEAN TIME TO REPAIR	
07-4E	219	MEAN ACTIVE MAINTENANCE DOWNTIME	
07-5	039	BEST OPERATIONAL CAPABILITY	
07-5A	235	MEAN TIME BETWEN FAILURES	
07-5B	236	MEAN TIME BETWEEN MAINTENANCE ACTIONS	
07-5C	244	MEASUREMENT BASE	
07-5D	241	MEAN TIME TO REPAIR	
07-5E	219	MEAN ACTIVE MAINTENANCE DOWNTIME	
08-3	218	MAXIMUM TIME TO REAPIR	
08-4	312	PERCENTILE	
08-5	158	INHERENT AVAILABILITY	
08-6	003	ACHIEVED AVAILABILITY	
08-7	283	OPERATIONAL AVAILABILITY	
08-8	015	ADMINISTRATIVE AND LOGISTIC DELAY TIME	

▲ Required for automated processing

DD Form 1949-1, 84 JUL   *Previous editions are obsolete.*   Part I, Page 1

Fig. 17-17. LSAR data selection sheet.

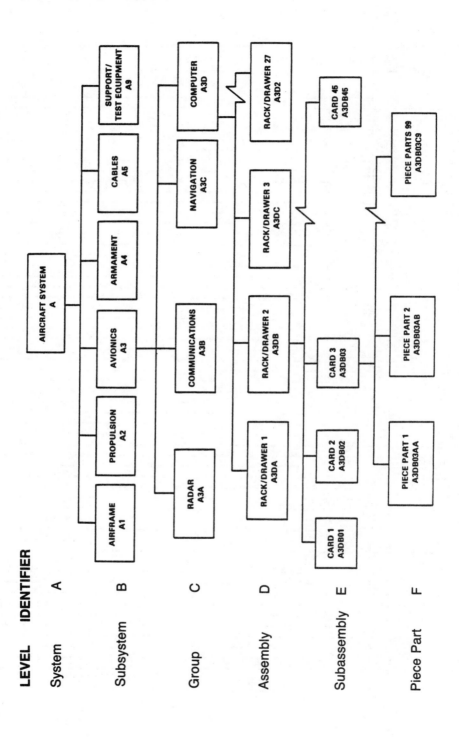

Fig. 17-18. LSA control number assignment.

LSACN	Item name	Part number
A	System	133-2
AA	Subsystem 1	133-2-1
AAA	Assembly 1	388-2
AAAA	Subassembly 1	388-2-1
AAAAA	Part 1	RCR07G100JS
AAAAB	Part 2	JANTX2N2222A
AAAAC	Part 3	MS1950-2
AAAB	Subassembly 2	399-294
AAABA	Part 4	399-294-1
AAABB	Part 5	399-294-2
AAB	Assembly 2	76487
AABA	Subassembly 3	5847
AABAA	Part 6	366628
AABAB	Part 7	14532
AABB	Subassembly 4	46632
AB	Subsystem 2	133-2-2
ABB	Assembly 3	24435-1
ABC	Assembly 4	24435-2
AC	Subsystem 3	133-2-3
AD	Subsystem 4	133-2-4
ADA	Assembly 5	74001
ADAA	Subassembly 5	AGR9756
ADAAA	Part 8	MS38510-155
ADAAB	Part 9	MS38510-2
ADAAC	Part 10	392006
AE	Subsystem 5	133-2-5

Fig. 17-19. LSA candidate list.

should be considered for further analysis through the LSA process. Figure 17-19 shows a candidate list. The government may require that a preliminary list be prepared and submitted as part of the contractor's proposal package.

LSA Summary Reports

The Data contained in the LSAR are used to determine logistic resources required to support the system and to identify needs for design changes to increase supportability. The government has prepared a set of computer-generated reports that summarize key information contained in the LSAR data base. These reports, referred to as LSA summary reports, provide detailed information from the LSAR data base in a usable and relevant format.

Reports can be generated that identify the most time-consuming or critical maintenance tasks, personnel and skill requirements, training requirements, maintenance allocation charts, and requirements for support and test equipment. LSA summary reports also generate all provisioning documentation for the system. A complete description of each report and the process for initiating each report is contained in Appendix B of MIL-STD 1388-2A.

Chapter 18

Logistics Plans

The key to the success of an integrated logistics support program is thorough, detailed planning followed by implementation of all aspects of the plan. All too often, plans are written with the intent of fulfilling a contractual requirement rather than creating a road map for success. Writing plans just for the sake of meeting a requirement is counterproductive and wasteful. It is incumbent on both the government and the contractor to produce plans that will be used.

Previous chapters have identified the need to generate plans that set the course for individual ILS disciplines in accomplishing the requirements of a government contract. In each case, the plans address the specific actions that must occur in order for the discipline to meet contract requirements.

There are two plans that guide the ILS program, which have not been discussed. They are an Integrated Logistics Support Plan (ILSP) and an Integrated Support Plan (ISP). The ILSP is prepared by the government organization that is responsible for planning, acquiring, monitoring, and implementing the ILS program. The ISP is prepared by the contractor responsible for planning and executing the ILS program. The ILSP and ISP must be in concert in order to accomplish the goals of the overall ILS program. The titles ILSP and ISP are used interchangeably by the different branches of the service, so a contractor, in some cases, may be required to prepare an ILSP that conforms to the ISP described below.

INTEGRATED LOGISTICS SUPPORT PLAN

The ILSP is a three-part document prepared by the government that describes in detail how the ILS program will be conducted to accomplish the program goals. Figure 18-1 provides an in-depth description of an ILSP. As shown, the ILSP addresses every aspect of the ILS program from the government's point of view. This document is initially prepared during the concept phase and updated throughout the life of the program.

Information contained in the plan must be coordinated with the efforts of other program activities and integrated into the overall program strategy to realize the full benefit of the ILS program. The ILSP is a dynamic document that is used through all phases of the acquisition cycle to guide ILS activities.

Part I: General

Section 1 - Introduction

Contained in the Introduction is a statement of the purpose of the plan and how it will be used throughout the ILS program, a summary of the program and background information, applicability of information used during the program, results of previous efforts in conjunction with the program, and references used as authority for the program. A list of acronyms and abbreviations may also be included.

Section 2 - Equipment Description

The equipment description provides a complete description of the equipment being developed along with any known supporting equipment that is being procured concurrently. Performance requirements of the equipment should be stated that reflect the contractual goals. Any resources that the government plans to provide as a part of the program, such as GFE, should be identified along with any logistics-related information available.

Section 3 - Program Management

The ILS program management concept should be described in detail. The ILS program manager and all ILS team members are identified by name, responsibility, and organization. A description of internal government interfaces and government-to-contractor interfaces should be provided. Additionally, the ILS review procedure, to include ILSMT activities, is discussed.

Part II: Plans, Goals, and Strategy

Section 1 - Operational and Organizational Plan

Describes the mission scenarios and requirements, operational environment, transportability requirements, employment concepts, deployment plans, and combat

Fig. 18-1. Integrated logistics support plan. (Continued to page 317.)

service support force structure. Provides detailed information for preparation of the LSAR Data Record A such as annual operating hours, mission times, sortie rates, and operation and maintenance requirements. A completed Data Record A may be attached to the plan as an appendix.

Section   2 - System Readiness Objectives

System readiness objectives are stated in terms of anticipated or required operational availability or mission capabilities. These figures represent the measured readiness objectives that the equipment must meet when operational. Both peacetime and wartime requirements should be stated.

Section   3 - Acquisition Strategy

Contained in this section is the government's strategy for acquiring ILS effort in conjunction with the equipment procurement. Discussions should address the scope and level of logistics required to support the new equipment. Particular attention should be given to what efforts will provide the most return for the investment. Any existing information or technologies that will reduce the program risks should be identified along with the applications.

Section   4 - Source Selection

The government's philosophy for selecting the source for ILS effort and how it complements or impacts the overall selection process is addressed in this section. Specific attention is given to identification of minimum criteria for qualifying potential contractors.

Section   5 - Elements of Support Acquisition

The method or methods that the government plans to use to acquire support for the new equipment is discussed in this section. Specific information should relate to requirements that are to be contained in solicitation documents and contracts. Use of existing support elements should also be discussed in the context reducing the overall cost.

Section   6 - Reliability, Availability, and Maintainability

Methods planned by the government to enhance the reliability, availability, and maintainability (RAM) of the new equipment are described along with any incentives that will be used to increase RAM. Initial RAM factors such as MTBF and MTTR should be included with indications of which factors are critical to meet system objectives.

Section 7 - Logistic Support Analysis

The strategy for acquiring LSA is discussed in detail. Selection criteria for LSA tasks to be performed by contractors are contained in this section. Included in the discussion are the controls that the government plans to use to ensure that duplication of effort is avoided. For programs that occur over several years where follow-on task phasing may also be discussed, interfaces between the government and contractors are described, including methods for review of progress and approval.

Section 8 - ILS Disciplines

The planned efforts of each ILS discipline are described in detail in this section. The required products of the overall ILS effort are identified to include the proposed method of accomplishing and coordinating each product. Particular attention is given to the interface between disciplines. At a minimum the following should be addressed:

Safety - Planned efforts to minimize hazards,
Provisioning - Planned provisioning process,
Supply Support - Proposed supply support concept,
Support Equipment - Planned use of support equipment,
Personnel - Anticipated personnel requirements,
Test Equipment - Planned use of test equipment,
Training - Planned training concepts,
Technical Data - Requirements for technical manuals,
PHS&T - Planned PHS&T concept and constraints, and
Facilities - Anticipated facilities requirements.

Section 9 - Design Influence

This section describes how ILS plans to influence design. The source selection process and acquisition decisions play a role in ILS's being able to have inputs to design and should be addressed as appropriate. A description of the process to be used to ensure that ILS personnel participate in design reviews and change proposal activities should be included. ILS factors that influence design, such as maintenance, manpower, personnel, and support equipment, should be addressed.

Section 10 - Constraints and Objectives

Support-related design constraints and objectives, such as limited resources, manpower, time, size, availability, etc., should be included in this section. Other factors that affect mission accomplishment should be identified as the program matures.

Section 11 - Maintenance Plans

> The maintenance plans for the equipment should be discussed in detail. Included in the discussion should be the overall equipment maintenance concept, specific plans for organizational and intermediate maintenance, depot maintenance and overhaul, use of level-of-repair analyses, and projected maintenance factors.

Section 12 - Funding Requirements

> This section contains the projected support funding requirements for the new equipment. Included are estimates for all phases of the acquisition process. The life cycle cost for support is addressed with emphasis on support requirements after the equipment is fielded. This information should be provided to the organizations that will be using the equipment in order to project funding requirements in time to have adequate operational funding. A statement should be provided identifying the impact of funding reductions or shortfalls.

Section 13 - Standardization and Interoperability Plan

> A description of the planned use of standard support equipment, test equipment, components, devices, or subsystems to reduce acquisition, training, operation, maintenance and support costs is provided in this section. Methods to use standard items to reduce risk are also identified.

Section 14 - Support Transition Planning

> Since the government organization that develops an item of equipment is normally not the end user, this section addresses the procedures that will be followed to transition the equipment program from the developer to the user. Specific attention should be placed on transferring support responsibilities between organizations, who is responsible for coordination, and accountability.

Section 15 - Modeling Techniques

> The plan should identify the proposed methods and procedures that will be used for modeling during the acquisition process. The range of models identified should include life cycle cost, reliability, maintainability, spares selection, repair-level analysis, etc.

Section 16 - Work Breakdown Structure

> A work breakdown structure (WBS) should be included in the plan. The WBS is used as a control for all equipment-related activities including cost accounting, support

planning, and data generated by all activities. As the design matures, the WBS should be expanded with the master structure being included in the plan.

Section 17 - Reporting

All reporting requirements to be invoked in support of the program should be identified in the plan. This allows dissemination of information to the necessary activities. ILS reporting requirements must be coordinated with other development efforts to eliminate redundancy.

Section 18 - Post-Fielding Assessment

An assessment of the ILS planning and execution should be planned after the equipment is fielded. This is normally done through an analysis of data generated by using organizations. The information gained through this process is invaluable for future programs or identification of needed product improvements and engineering changes.

Section 19 - Post-Production Support

The ILSP should contain a plan for supporting the equipment after the end of production. Included are identification of sources for spare and repair parts, engineering changes, and funding requirements. The post-production support of the equipment represents the largest percentage of life cycle costs due to operational and support costs. ILS efforts to reduce post-production costs should also be identified.

Part III: Schedule

Section   1 - Program Milestones

The program milestone schedule identifies all the events that must occur in order to meet the overall program objectives. The schedule must be detailed in nature and specifically identify every task that is to be accomplished. The schedule should include beginning and completion dates, dependency between tasks, and identification of task responsibility.

Section   2 - Coordination

The plan should identify the methods to be used to coordinate the ILS task schedule with other organizations and activities. Specific attention should be given to how changes to the schedule are to be assessed and coordinated.

Section 3 - Reporting Requirements

The procedures for reporting relative to schedule status should be provided in the plan. Reporting should be focused on obtaining significant and timely information pertaining to program progress and identification of schedule slippages or projected schedule impacts. This information should address the overall program schedule, not just the ILS program.

Fig. 18-1. Integrated logistics support plan. (Continued from page 310.)

INTEGRATED SUPPORT PLAN

The ISP is a contractor-prepared document that establishes the policies, procedures, and methods that will be used to implement the contractor's ILS program. It is the single planning and controlling document for all ILS activities and related efforts. Government requests for proposals may require that a preliminary ISP be submitted as part of a proposal package. Figure 18-2 addresses in detail the contents of an ISP. As shown in this figure, the ISP is comprehensive in nature and integrates the activities of all contractor ILS disciplines.

The separate plans written for each discipline are normally referenced in the ISP and attached as appendices to form a complete ILS planning package. The government approves the ISP to ensure that the plan supports the government-prepared ILSP and that the overall program concepts and strategies are implemented. Just like the ILSP, the ISP is a dynamic document that is maintained and updated by the contractor throughout the ILS program.

Section 1 - Introduction

1.1 Purpose and Scope

The purpose and scope provide a concise statement of the intent and use of the ISP. The section establishes the fact that the ISP is the controlling document for all ILS activities. The government ILS requirements are addressed with the intent of substantiating that the contractor's ILS program is in concert with the specified requirements.

1.2 Program Summary

This subsection provides an overview and summary of the contractor's ILS program. A brief description of the ISP is provided to establish a clear understanding of the content and organization of the document to show how the ISP controls the ILS program.

Fig. 18-2. Integrated support plan.

1.3   Updating Process

The process for periodic review and updating of the ISP is described. The methods for incorporating government or contractor-generated changes and the approval cycle for updates are identified.

Section   2 - System Characteristics

This section provides a brief synopsis of the significant system characteristics that are relevant to ILS and support planning. Supportability characteristics of the system, the design activity, and the intended operational environment are addressed in the context of establishing the base for all ILS activities. The procurement specification is the initial source document for this information.

2.1   System Description

A brief description of the functional and physical characteristics of the system is provided. Significant attributes of the system that affect ILS or that are considered drivers for supportability are identified.

2.2   Reliability Factors

System reliability factors, such as MTBF, MTBCF, and MMH/OH, are identified in this paragraph. The reliability program is summarized with emphasis on the FMECA, failure rate predictions, and effects of environmental conditions on operation. The use of reliability information by ILS disciplines is highlighted. The reliability program plan should be referenced.

2.3   Maintainability Factors

The required maintainability factors for the system such as MTTR and MTBMA are discussed. Emphasis is placed on using these factors as goals for the ILS program. The maintainability program plan should be referenced as to how the goals will be met.

2.4   Human Engineering Factors

The man-to-machine interface factors are identified in this paragraph. The efforts to address each area to optimize use of controls and indicators, ease of maintenance, and accessibility are summarized.

2.5   Standardization Factors

Requirements to minimize the number of different types of spares, repair parts, support equipment, test equipment, and training materials are summarized in this paragraph. Engineering efforts planned to address standardization are identified. If required, the standardization plan is referenced.

2.6   Safety Factors

This paragraph identifies the known and potential safety hazards that may occur with the operation and maintenance of the system. A summary of the system safety efforts and procedures for liaison, coordination, and exchange of information between safety engineering and other design and support activities is provided with emphasis on methods for identification and resolution of potential hazards. The system safety program plan should be referenced.

2.7   PHS&T Factors

Physical or material aspects of the system that pose problems in the areas of packaging, handling, storage, and transportability are identified. Specific requirements contained in the product specification that cause these problems should be highlighted along with a summary of the planned analyses or actions to negate the problems.

2.8   Other Significant Factors

Any other significant factors that affect design, support, or fielding of the system, such as funding constraints, extreme environmental conditions, multiservice applications, or specifications, should be identified. The impact of each factor along with corrective or alternative actions should be summarized.

Section   3 - Program Management

3.1   Objectives, Policies, and Procedures

The purpose of this paragraph is to summarize the contractor's ILS program management philosophy in general and state specifically how the overall program will be managed. It should describe the contractor's objectives, policies, and procedures that will be used to manage the ILS program. The objectives of the program must be in concert with the government objectives contained in the ILSP. Other internal objectives necessary to meet the overall program objectives should be identified. Existing policies and procedures that will be used should be tailored as appropriate to meet contractual requirements. New policies and procedures established specifically for a program should be discussed in detail.

3.2   Organization

The contractor's ILS organization and each ILS discipline are described in this paragraph. The identification of key ILS positions and personnel is normally a required entry. The relationship of the ILS organization to the overall company structure should be provided. The organizational structure should show how information is coordinated and shared between activities and within ILS. This structure should be reflected in all subsequent plans for individual ILS disciplines. If an integrated logistic support management team (ILSMT) is to be formed, the team organization should be addressed.

3.3   Subcontractor and Vendor Interface

When subcontractors and major vendors participate in a contract, the contractor must establish a formal methodology for the exchange of information. This paragraph should illustrate how interfaces will be established and monitored. Subsequent plans for individual disciplines that require formal accomplishment of this task should be referenced.

3.4   Government Interface

The interfaces between contractor and government ILS organizations are identified in this paragraph. Specific information should be provided relative to the individuals, by name, who are authorized and responsible for communicating information from the contractor to the government.

3.5   Review Procedures

This paragraph describes the procedures for conducting formal and informal reviews. Information provided should include identification of each review, review participants, procedures for preparing agendas and minutes for each review, procedures for handling action items, and approval authority of review results. If an integrated logistic support management review team (ILSMRT) is to be formed, the organization and function of the team should be discussed.

3.6   Management Control System

A summary description of the system to be used to control the cost, schedule, and technical performance of ILS disciplines is included in this paragraph. The control system should be described in sufficient detail to show how it is responsive to management needs. The relationship between this system and the system that will be used to control the overall contract should be established.

3.7   Configuration Management Program

The contractor's configuration management program should be summarized to provide a definition of the relationship between ILS and configuration management activities. ILS representation of the configuration control board and other procedures where ILS has direct influence on design status should be identified. The configuration management program plan should be referenced when applicable.

Section 4 - ILS Program

4.1 Support Concept

The overall logistic support concept for the system should be described in detail. Alternative support concepts to be considered should be identified as appropriate along with plans for evaluation. Describe the methods that will be used to resolve supportability problems.

4.2 Maintenance Planning

The overall system maintenance concept, as reflected in the ILSP, should be restated in detail. Describe the type and extent of maintenance planned to be performed by each level of maintenance. Explain how maintenance task analysis will be used to generate detailed maintenance and resource requirements. Special maintenance considerations or limiting factors should be addressed. If a separate system maintenance plan is required, the plan should be referenced.

(Note: Paragraphs 4.3 through 4.10 address the specific activities of the ILS disciplines. The efforts of each discipline must be explained thoroughly. The description should state explicitly what each discipline is to accomplish and how the effort is to be done. Included in the explanation should be individual goals for each discipline, interrelationships between disciplines, and processes and procedures that will be used to manage each activity. Separate discipline plans should be referenced when appropriate.)

4.3 Logistics Support Analysis

See note above.

4.4 Support and Test Equipment

See note above.

4.5 Provisioning and Supply Support

See note above.

4.6 PHS&T

See note above.

4.7   Technical Data

See note above.

4.8   Facilities

See note above.

4.9   Personnel

See note above.

4.10   Training

See note above.

4.11   .Test and Evaluation

Planned or contractually required test and evaluation of the ILS program results should be discussed in detail. The conduct of such activities as a maintainability or a logistics demonstration should be described, including goals, procedures, and evaluation requirements. If tests and evaluations are covered in other plans, the plans should be referenced.

4.12   Initial Contract Support

In many cases, a contractor is required to provide initial support to test systems or early fielded systems. When such a requirement exists, the contractor should include plans for accomplishment in the ISP.

Section   5 - ILS Schedule

The schedule section of the ISP contains detailed milestone schedules for three levels of management, overall program schedule, ILS program schedule, and ILS discipline schedules. The schedules must be compatible and support the accomplishment of all program goals.

5.1   Master Program Milestone Schedule

The master program schedule should be a reproduction of the schedule contained in the overall program management plan for the contract. The schedule should be highlighted to identify significant milestones that require ILS support for accomplishment.

5.2   ILS Milestone Schedule

The ILS milestone schedule should show all the milestones that the ILS organization must meet to complete the contract requirements. The schedule must be supportive of the overall program schedule. This schedule must be extremely detailed in order to provide sufficient management visibility of all ILS activities.

5.3   ILS Disciplines Milestone Schedules

A milestone schedule for each ILS discipline should be included in the ISP. The schedules should be in total agreement with the ILS milestone schedule. In most cases, these schedules should be reproductions of the schedules contained in individual discipline program plans.

Section   6 - Related Plans

A copy of each subordinate plan that is referenced in the ISP should be attached as an enclosure to the basic plan. This creates a single consolidated document that addresses every aspect of the ILS program.

Chapter 19

Contracts

Chapter 2 discussed the terminology and basic makeup of contracts, and throughout the preceding chapters references have been made to contractual requirements. A major consideration has not been addressed: how contractors get government contracts. Government contracts don't just appear on a contractor's doorstep. There is a formal process that must be followed. The contracting process is governed by the Federal Acquisition Regulations (FAR), which contain procedures that must be strictly adhered to by government contractors. This chapter describes the contracting process, different types of contracts, and proposal preparation and pricing.

THE CONTRACTING PROCESS

Contracts, with few exceptions, are awarded to contractors and completed through the series of events illustrated in Fig. 19-1. Each event has definitive requirements and procedures.

Request for Proposal

The contracting process begins when the government issues a solicitation, normally a Request for Proposal (RFP) or a Request for Quote (RFQ). The difference between an RFP and an RFQ is debatable and of no consequence for this discussion, so they will be treated as essentially the same. The government may issue a Request for Information (RFI) prior to issuing an RFP or RFQ. An RFI is a preliminary document that

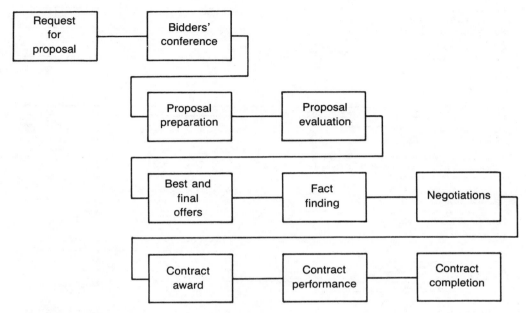

Fig. 19-1. The contracting process.

solicits information that may be used later in an RFP or RFQ.

The process begins with the government issuing, for the sake of simplicity, an RFP. An example of the cover sheet for an RFP is shown in Fig. 19-2. The RFP contains the product specification, Statement of Work (SOW), and Contract Data Requirements List (CDRL) that will later become a contract. It also contains other information that pertains to how potential contractors must prepare and submit proposals, how proposals will be evaluated, and how the contract will be administered.

There are two ways that contractors can receive an RFP. First, each government agency that issues contracts maintains a list of contractors that want to receive RFPs called a Bidders List. The Bidders List is divided into types of products and services. When an RFP is issued, it is automatically sent to every contractor on the Bidders List that claims to have the capability to fulfill the contract requirements. To be on the list, a contractor must submit a request to the agency. This request is renewed annually. Because there are many agencies that issue contracts, most contractors are on dozens of lists.

The second way to receive an RFP is to request it from the issuing agency. By law, every solicitation and contract award must be published in the *Commerce Business Daily* (CBD), a government publication available through the U.S. Government Printing Office. Most contractors subscribe to the CBD, which contains a brief description of each solicitation and gives information on how to obtain a copy of the RFP. Figure 19-3 is an example of a page of the CBD.

SOLICITATION, OFFER AND AWARD

3. CERTIFIED FOR NATIONAL DEFENSE UNDER DPS REG. 1 AND/OR DMS REG. 1 RATING	4. PAGE 1	OF

1. CONTRACT *(Proc. Inst. Ident.) NO.*	2. SOLICITATION NO ☐ ADVERTISED (IFB) ☐ NEGOTIATED (RFP)	5. DATE ISSUED	6. REQUISITION/PURCHASE REQUEST NO

7. ISSUED BY	CODE	8. ADDRESS OFFER TO *(if other than block 7)*

In advertised procurement "offer" and "offeror" shall be construed to mean "bid" and "bidder"

SOLICITATION

9. Sealed offers in original and _____ copies for furnishing the supplies or services in the Schedule will be received at the place specified in block 8, or if handcarried, in the depository located in _____ until _____ local time _____.
                                                                                (Hour)   (Date)

If this is an advertised solicitation, offers will be publicly opened at that time.
CAUTION – LATE OFFERS: See pars. 7 and 8 of Solicitation Instructions and Conditions.
All offers are subject to the following:

1. The Solicitation Instructions and Conditions, SF 33-A, _____ edition which is attached or incorporated herein by reference.

2. The General Provisions, SF 32, _____ edition, which is attached or incorporated herein by reference.

3. The Schedule included herein and/or attached hereto.

4. Such other provisions, representations, certifications, and specifications as are attached or incorporated herein by reference. *(Attachments are listed in schedule.)*

FOR INFORMATION CALL (Name & telephone no.) (No collect calls) ►

TABLE OF CONTENTS

(X)	SEC	PART I - THE SCHEDULE	PAGE	(X)	SEC	PART II - GENERAL PROVISIONS	PAGE
	A	Contract Form			I	General Provisions	
	B	Supplies, Services & Prices				PART III - LIST OF DOCUMENTS, EXHIBITS AND OTHER ATTACHMENTS	
	C	Description Specification			J	List of Documents, Exhibits and other Attachments	
	D	Packaging and Marking				PART IV - GENERAL INSTRUCTIONS	
	E	Inspection and Acceptance			K	Representations, Certifications and other Statements of Offeror	
	F	Deliveries or Performance			L	Instructions and Conditions and Notices to Offerors	
	G	Contract Administration Data			M	Evaluation Factors for Award	
	H	Special Provisions					

OFFER *(must be fully completed by offeror)*

In compliance with the above, the undersigned agrees, if this offer is accepted within _____ calendar days *(60 calendar days unless a different period is inserted by the offeror)* from the date for receipt of offers specified above, to furnish any or all items upon which prices are offered at the price set opposite each item, delivered at the designated point(s), within the time specified in the schedule.

16. DISCOUNT FOR PROMPT PAYMENT *(See par. 9, SF 33-A)*

% 10 CALENDAR DAYS	% 20 CALENDAR DAYS	% 30 CALENDAR DAYS	% _____ CALENDAR DAYS

17. OFFEROR	CODE	FACILITY CODE	18. NAME AND TITLE OF PERSON AUTHORIZED TO SIGN OFFER *(Type or print)*
NAME AND ADDRESS *(Street, city, county, State and ZIP code)*			
AREA CODE AND TELEPHONE NO. ►		19. SIGNATURE	20. OFFER DATE
☐ Check if remittance address is different from above – enter such address in Schedule			

AWARD *(To be completed by Government)*

21. ACCEPTED AS TO ITEMS NUMBERED	22. AMOUNT	23. ACCOUNTING AND APPROPRIATION DATA

24. SUBMIT INVOICES *(4 copies unless otherwise specified)* TO ADDRESS SHOWN IN BLOCK _____	25. NEGOTIATED PURSUANT TO	10 U.S.C. 2304(a) ( )
		41 U.S.C. 252(c) ( )

26. ADMINISTERED BY *(If other than block 7)*	CODE	27. PAYMENT WILL BE MADE BY	CODE

NAME OF CONTRACTING OFFICER *(Type or print)*	29. UNITED STATES OF AMERICA BY _____ *(Signature of contracting officer)*	30. AWARD DATE

Award will be made on this form, or on Standard Form 26, or by other official written notice

33–131

Standard Form 33 Page 1 (REV. 3-77)
Prescribed by GSA, FPR (41 CFR) 1-16.101

Fig. 19-2. Solicitation cover sheet.

Commerce Business Daily

Friday, September 26, 1986

Issue No. PSA-9183

A daily list of U.S. Government procurement invitations, contract awards, subcontracting leads, sales of surplus property and foreign business opportunities

U.S. GOVERNMENT PROCUREMENTS

Services

A Experimental, Developmental, Test and Research Work (research includes both basic and applied research)

Contr Section, ACT-51A, Fed Aviation Admin Technical Center, Atlantic City Airport, NJ 08405 Contract Specialist, Carmen P. Johnson, ACT-51A, 609/484-5384

A -- AIRFRAME SECTION (B707) LONGITUDINAL IMPACT TEST. Concerns having the ability to perform a test involving a assessment of the longitudinal braking strength between a large 10 ft B-707 fuselage section and seat loaded floor attachment structure when subject to a simulated servere crash load condition ar requested to give written notification (including) the tel no. to the procurement office listed in this notice within 30 cal.days from date of this synopsis. The test will require specialised facilitie to assure that the simulated crash load will be properly induced. And, because of th large test specimen size and high dynamic impact levels required, such facilities must b sufficient capacity. It is the govt intent to award this proposed sole source to Transportation Research Center of Ohio. This is not a formal sol; however, concerns that respond should furnish detailed data concerning their capabilities and may request to received a copy of the sol when it becomes avail. Further, it is requested that firms indicate in their response whether they.are interested in this procurement for submitting a RFP for for, info purpose. All requests must indicate no. DTFA03-87-R-70001. This notice may represent the only official notice of such a sol. (265)

H Expert and Consultant Services

Base Contr. Div, Bldg 861, 375 AAW/LGCA, Scott AFB, IL 62225-5320

H – TANDEM SYSTEMS ANALYST/PROGRAMMER SUPPORT Sol F11623-86-Q6135, RFQ closing 14 Oct 86. Contact Ellen Womack, 618/256-2672, Contr. Specialist: Carole J. Schneider, 618/256-6637, Contr. Officer. Intent to issue a del order against GSA contract to Tandem Computer, Inc., 788 Office Parkway, Creve Coeur, MO, for the period 01 Nov 86 thru 30 Sep 87. Analyst/programmer support shall be provided at Scott AFB, IL, 9 hrs per day, including a 1-hr lunch period, 5 days ea. week, excluding observed holidays, vacations and sick leave. Software analyst/programmer service to develop, implement and maintain software on Tandem computer systems. In addition, develop in-house computer systems analysis expertise thru training. Task orders will be issued that will define specific tasks and deliverables relating to the following categories of functions: a. Assist in using and maintaining Tandem supplied software for the Tandem computer systems. Make recommendations for Tandem software enhancement thru appropriate channels. b. Design, develop, program and document MAC-unique software to make optimum use of Tandem hardware and software. This includes the ability to develop Tandem 6100 communications subsystem software. c. Code, test, and document MAC-unique system and functional software. d. Participate in the testing of Tandem-supplied and MAC-unique system and functional software. e. Provide training to develop in-house computer system analysis expertise regarding Tandem computer hardware and software. Training will be of an informal nature through interaction and assistance. The Tandem systems analyst will explain concepts and procedures to ensure understanding by in-house personnel. f. Identify if a computer system problem is or is not related to a hardware malfunction. g. Analyze and resolve Tanden-developed and MAC-unique software problems. h. Assist and advise in the generation of the computer system environment. i. Assist, advise, and consult in coding and testing of MAC-unique software that interfaces directly with the operating system or other Tandem-supplied software. j. Assist in the evaluation of computer system performance. k. Attend management and technical level meetings as required and assist/advise as requested. The Govt will provide office furniture and space as normally supplied to assigned in-service personnel. Vendor shall submit a resume reflecting the analyst/programmer's qualifications (i.e., educational background, work experience, and training) for review and concurrence. The analyst/programmer must be trained, possess experience relating to Tandem Non-Stop II hardware and software, and have access with Tandem proprietary info. The analyst/programmer must be initmately familiar with Tandem executive level software, hardware, and related communication interfaces and protocols. Finally, the analyst/programmers must possess a TOP SECRET security clearance as much of the required work will be performed in controlled areas on Scott AFB. Vendors proposing compatible service must identify proposed prices and furnish sufficient data, including resumes, that will allow evaluation as to tech acceptability within 15 days after publ in the CBD. No contr award will be made on the basis of offers/proposals or other info received in response to this notice since the notice of intent to place a del order against a schedule contr cannot be considered a request for offers/proposals. Ref RFQ No. See Note 22. (265)

Warner Robins ALC Directorate of Contracting and Manufacturing, Robins AFB, GA 31098

H – PERFORM A FEASIBILITY STUDY TO DETERMINE THE BENEFITS OF ALUMINUM-LITHIUM (A-LI) AND POWDER METAL (PM) ALLOYS ON C-130 STRUCTURE WITH FREQ CORROSION OCCURENCE. This notice is for Sol FD2060-87-46704, which will be issued approx: 86 Oct 13. To: Lockheed-Georgia Co, Marietta GA with an aprox closing date of 86 Nov 11. For copy of sol contact: WR-ALC/PMXOA, include mfg code. For additional info contact: Bob Hawthorne, PMWCC, 912/926-3661 X2370. Authority: 10 USC 2304(C)(1). Justification: Supplies (or services) required are available from only one responsible source and no other type of supplies or services will satisfy agency requirements. Notes: 22, 27 & 40.

H – EVALUATE DAMAGE TOLERANCE ANALYSIS (DTA) UPDATE ON C-141 ACFT This notice is for Sol FD2060-87-46712 which will be issued approx 13 Oct 86 to Lockheed-Georgia Co, Marietta, GA with an approx closing date of 11 Nov 86. For copy of sol contact WR-ALC/PMXOA. Include mfg code, for addl info contact Bob Hawthorne/PMWCC, 912/926-3661 X2370. Authority: 10 U.S.C. 2304 (C) (1). Justification: Supplies (or services) required are avail from only one responsible source and no other type of supplies or services will satisfy agency requirements. Notes 22, 27 & 40.

H – ENSURE THE ORDERLY IMPLEMENTATION OF THE H-53 SERVICE LIFE EXTENSION PROGRAM This notice is for Sol FD2060-87-46716 which will be issued approx 13 Oct 86 to ARINC Research Corp, Annapolis, MD 21401 with an approx closing date of 11 Nov 86. For copy of sol contact WR-ALC/PMXOA. Include mfg code, for addl info contact Pat Anderson/PMWMA, 912/926-2972 X2242. Authority: 10 U.S.C. 2304 (C) (1). Justification: Supplies (or services) required are avail from only one responsible source and no other type of supplies or services will satisfy agency requirements. Notes 22, 27, 40 & 48.

H -- REQUIRED TO CONTINUE DEVELOPMENT OF C-130 ACFT INFO RETRIEVAL SYSTEM (AIRS) TO PROVIDE EFFECTIVE FRACTURE TRACKING FOR THE C-130 ACFT This notice is for Sol FD2060-87-46717 which will be issued approx 13 Oct 86 to Lockheed-Georgia Co, Marietta, GA with an approx closing date of 11 Nov 86. For copy of sol contact WR-ALC/PMXOA. Include mfg code, for addl info contact Bob Hawthorne/PMWCC, 912/926-3661 X2370. Authority: 10 U.S.C. 2304 (C) (1). Justification: Supplies (or services) required are avail from only one responsible source and no other type of supplies or services will satisfy agency requirements. Notes 22, 27 & 40. (266)

US General Accounting Office, Office of Acquisition Management, 441 G St, NW, Rm 2001/NC, Washington, DC 20548, Attn; Contracting Officer

H – TYPING SERVICES Generating hardcopy and floppy diskette output and with electronic data transmission capability. Using IBM PC, XT or AT with Wordperfect and Crosstalk Software. Date of award through 30 Sept 87 with two annual renewal options, 24 hrs, 48 hrs and 72 hrs turnaround time, indefinite qty. Sol OAM-86-N-0033 applies, to be issued o/a 15 Oct 86. Interested firms should submit a written request to the Contracting Officer. (265)

Naval Air Systems Command, Washington DC 20361, Attn: D. Rosenblatt, Code AIR-21614D, 202/692-1724

H – HARPOON INTEGRATION SERVICES AND DATA FOR THE F-27/PROGRAM PHASE II. This procurement will involve FMS. Negotiations will be conducted with McDonnell Douglas Astronautics Co, Box 516, St Louis MO 63166. McDonnell Douglas is the sole designer, developer and manufacturer of the AGM-84 system and is the only known firm possessing the requisite knowledge to furnish the services required to support the development and maintenance of this program. See Note 22. NAVAIR Synopsis 1000-86. (266)

Naval Air Systems Command, Code AIR-215B4, Washington, DC 20361-2010

H – INTEGRATED LOGISTICS SUPPORT (ILS) SERVICES AND DATA FOR THE SH-60B Under an existing contract. The Naval Air Systems Command intends to negotiate with International Business Machines, Route 17C, Owego, NY 13827 the only known firm possessing the unique knowledge, specialized skills and qualified personnel to perform ILS services and produce the required ILS data within the required time frame. See Note 22. NAVAIR synopsis 1,033-86. POC M. Anelli, (1,033-86), 202/692-1272. (266)

Naval Air Systems Command, Code AIR-21535, Washington DC 20361-2010

H – DESIGN STUDY ON THE SH-2F HELICOPTER'S MAIN GEARBOX FOR COMPONENT IMPROVEMENT PROGRAM To include analysis on wear effects and gear life and vibration analysis. The Naval Air Systems Command intends to negotiate with Kaman Aerospace Corp, PO Box 2, Bloomfield, CT 06002-0002 the designer, developer and sole manufacturer of the SH-2F helicopter and the H-2 gearbox. See Note 22. NAVAIR synopsis 1,035-86. POC P. Saunders (1,035-86), 202/692-3506. (266)

J Maintenance and/or Repair of Equipment

US Dept of Energy, Western Area Power Administration, POB 3402, 1627 Cole Blvd, Golden CO 80401

J -- MAINTENANCE OF EASTMAN KODAK COPIERS Control Data Corp Electric Dispatcher Training Simulator, and Xerox Telecopier Machines, Aperture Card Readers, and Memory Writers. Sol 7-11-A6-70001. Opening and closing dates: 9-29-86/11-15-86. Contact: Linda Kleihege, 303/231-7287, Contracting officer, Richard Steele, 303/231-7965. Maintenance of the following equipment: Three Kodak Ektaprint 150AF copier/duplicators, including approx 100,000 copies per machine per month, maintenance of a dispatcher training simulator for the Electric Power Training Center located at 1627 Cole Blvd, Golden CO 80401, including parts, labor and other costs, from 7 am thru 6 pm, Mon thru Fri, maintenance of Xerox telecopying machines, aperture card readers, memory writers, and photo copiers. The Government intends to award these maintenance agreements to Eastman Kodak, POB 819015, Dallas TX 75381-9015, Control Data Corp, 7995 E Prentice Ave, Englewood CO 80111 and Xerox Corp, 5700 S Quebec St, Englewood CO, 80111 rexpectively. These firms are the manufacturers of the foregoing equipment and are the only firms known to the Government that have the capability of maintaining their respective systems (See Note 22). (266)

USPS, Procurement Services Office, Attn: Mfg POB 6040, Denver CO 80206-0040

J -- VEHICLE REPAIR AND MAINTENANCE Agreement for approx 36 postal

CBD ON-LINE

An electronic edition of the COMMERCE BUSINESS DAILY is available from the following. Interested parties may contact them for full details. CBD SEARCH SERVICES, INC 7777 Leesburg Pike, Suite 300, South Falls Church, VA 22043 703/883-2510 or toll free 1-800/CBD-4750. UNITED COMMUNICATIONS GROUP, 4550 Montgomery Ave, Suite 700, North Bethesda, MD 20814 301/656-6666; SOFTSHARE, a Div. of MCR Technology Inc., 55 Depot Rd., Goleta, CA 93117, Sue Ernst 805/683-3841 (Collect); DATA RESOURCES INC. (DRI), 24 Hartwell Ave., Lexington, MA 02173, 617/863-5100; DIALOG INFORMATION SERVICES, INC., 3460 Hillview Ave., Palo Alto, CA 94304, toll free 800/334-2564. SHAKED - ELNET VIDOETEX, Int'l. Bids, 1121-12th St. NW Washington, DC 20005 Jeanine Maclary 202/682-2100

Content

Fig. 19-3. Commerce business daily.

Bidders Conference

Shortly after an RFP is issued, the government normally holds a bidders conference. The purpose of the conference is to answer technical and administrative questions that interested contractors have about the RFP. Contractors might be asked to submit a list of questions prior to the conference so that the government is prepared to give all the information possible. Minutes of the conference are distributed to each contractor that received the RFP, whether or not the contractor attended the conference. FAR requires that all interested contractors have complete information so that everyone has an equal opportunity to respond to the RFP.

Proposal Preparation

The next step in the contracting process is proposal or bid preparation by the contractor. A proposal is the contractor's response to the RFP that describes why the contractor is qualified to do the work, how the required tasks will be accomplished, and the proposed price for the effort. The RFP normally contains specific instructions for proposal preparation such as content, format, and maximum length. It might also require contractors to respond to a detailed set of technical questions to determine who has the best capabilities to complete the contract. Proposal preparation is discussed further in a subsequent paragraph.

Proposal Submittal

An RFP contains specific instructions concerning how, when, and where a proposal is to be submitted. It will normally tell how many copies are required and how the copies should be bound. The RFP also tells the place, date, and time when the proposal is to be submitted. This might seem like an insignificant point, but it is very important that proposals be submitted on time. If the RFP says that proposals are due at 3:00 P.M., it means 3:00 P.M., not some time around three o'clock.

Proposals that are submitted late, even as little as one minute, will not be accepted unless the contractor can prove extraordinary circumstances. Late planes or broken copy machines don't qualify as excuses. Many contractors, even major defense contractors, have had proposals rejected because they did not make the deadline.

Proposal Evaluation

Contractors' proposals are screened to identify those that are responsive to the RFP requirements. Those deemed nonresponsive are rejected. Reasons for rejection include lack of technical capability, unreasonable price, or failure to meet the minimum criteria stated in the RFP. The remaining proposals are distributed within the agency for evaluation. Based on this evaluation, the government may choose the contractor that will be awarded the contract or, if two or more are essentially equal, issue a request for best and final offers.

Best and Final Offers

Many times the government will allow the most qualified contractors to revise, or fine tune, their proposals by issuing a request for best and final offers (BAFOs). The purpose of a BAFO is to encourage contractors to review the technical and, especially, the costing sections of proposals in order to get the best technical approach and most reasonable price possible. A contractor's BAFO is used by the government to make a final decision on who will be awarded a contract.

Fact-finding

The government might visit contractors that are candidates for the award of a contract. These fact-finding visits are conducted by auditors and technical personnel. During fact-finding, a contractor's proposal is scrutinized to verify its contents. Areas of interest include pricing backup data, determination of what was actually bid in the way of labor and material required, and validation of the contractor's capability to perform the required tasks.

Negotiation

After the contractor has been selected, the government negotiates a contract on the final terms and requirements of the contract. Negotiations can affect price, quantities, scope of effort, and schedule. This is like any other type of give-and-take negotiating, where both sides ultimately reach a mutually acceptable position.

Contract Award

After negotiations are concluded, the successful bidder is awarded the contract. The results of the negotiations provide a definite contract that states exactly what the contractor is legally bound to provide to the government. It is important to remember that there might be a significant difference between what the contractor proposed and the actual contract that was awarded. This difference can be not only in price, but also in the technical requirements of the product specification and the SOW. Always refer to the contract, not the original proposal on BAFO, to determine what is required.

Contract Performance

Contract performance is where the contractor does all the things required by the contract, and the logisticians do the things discussed in this text. The period of contract performance can vary from a few months to several years, depending on the scope of work.

Contract Completion

When all contract requirements have been met, the contract is complete. To receive payment, contractors must submit a DD form 250, Material Inspection and Receiving Report, to the government. An example of this form is in Fig. 19-4. In most cases, the final DD form 250 is not for the total contract value. Contractors receive progress payments over the course of a contract based on key events or in increments agreed on during contract negotiations.

MATERIAL INSPECTION AND RECEIVING REPORT	1. PROC. INSTRUMENT IDEN. (CONTRACT)	(ORDER) NO.	6. INVOICE	7. PAGE	OF
				8. ACCEPTANCE POINT	

2. SHIPMENT NO.	3. DATE SHIPPED	4. B/L TCN		5. DISCOUNT TERMS

9. PRIME CONTRACTOR	CODE		10. ADMINISTERED BY	CODE

11. SHIPPED FROM (If other than 9)	CODE	FOB:	12. PAYMENT WILL BE MADE BY	CODE

13. SHIPPED TO	CODE		14. MARKED FOR	CODE

15. ITEM NO.	16. STOCK/PART NO. DESCRIPTION (Indicate number of shipping containers - type of container - container number.)	17. QUANTITY SHIP / REC'D *	18. UNIT	19. UNIT PRICE	20. AMOUNT

21. PROCUREMENT QUALITY ASSURANCE

A. ORIGIN

☐ PQA ☐ ACCEPTANCE of listed items has been made by me or under my supervision and they conform to contract, except as noted herein or on supporting documents.

DATE   SIGNATURE OF AUTH GOVT REP

TYPED NAME AND OFFICE

B. DESTINATION

☐ PQA ☐ ACCEPTANCE of listed items has been made by me or under my supervision and they conform to contract, except as noted herein or on supporting documents.

DATE   SIGNATURE OF AUTH GOVT REP

TYPED NAME AND TITLE

22. RECEIVER'S USE

Quantities shown in column 17 were received in apparent good condition except as noted.

DATE RECEIVED   SIGNATURE OF AUTH GOVT REP

TYPED NAME AND OFFICE

* If quantity received by the Government is the same as quantity shipped, indicate by (✓) mark, if different, enter actual quantity received below quantity shipped and encircle.

23. CONTRACTOR USE ONLY

DD Form 250, JUN 86   Previous editions are obsolete.   Form Approved / OMB No. 0704-0248 / Expires Apr 30, 1989

★ U.S.G.P.O.: 1986–491–133/52563

Fig. 19-4. Material inspection and receiving report.

TYPES OF CONTRACTS

Contracts can be categorized in different ways. The most common are by (1) who may bid, (2) how many contractors are deemed capable of bidding, or (3) how the government pays for the contract. Each method can be combined with the others to produce many distinct ways that the government can issue contracts.

Unrestricted Contracts

Any contractor can bid on an unrestricted contract. The only requirements are that the contractor must have the capability to fulfill the contract requirement. Most major government contracts are unrestricted.

Restricted Contracts

The government sometimes places restrictions on the types of contractors who can bid on a contract. The most common restriction relates to the size of the contractor's business and are called "small business set- asides." This means that only contractors who qualify as a small business can bid on the contract. There are varying definitions for a small business, but it is normally a company with fewer than 500 employees and gross sales of less than $10 million per year.

Competitive Contracts

The contracting process described above is typical for competitive contracts. This type of contract is the result of competitive bidding where all qualified contractors have the opportunity to submit proposals to the government. The majority of contracts awarded by the government are competitive.

Sole-Source Contracts

The government might determine that there is only one contractor capable of responding to the requirements of a contract. In these cases, a sole-source contract may be issued. The contracting process for a sole-source contract is much like that for a competitive contract except the steps shown above that are used to identify the receiving contractor are bypassed. An RFP is issued, the contractor must submit a proposal, and fact-finding and negotiations are still conducted.

The next five categories of contracts are based on the type of effort requested by the government and the financial risk assumed by the contractor. Each type has definite advantages, disadvantages, and risks for both the government and the contractor.

Firm Fixed Price Contract

The most rigid type of contract is firm fixed price (FFP). This type of contract requires the contractor to complete the total contracted effort for a set price. As long as the requirements of the contract don't change, the contractor cannot expect to receive an increase in the negotiated value. An FFP contract represents the least risk to the government and the greatest risk to the contractor. The government knows exactly what it will be required to pay and exactly what the contractor must deliver. On the other hand, the contractor is at risk because there is no relief for contract overruns or allowances for underbidding the contracted effort.

If a contractor is awarded an FFP contract with a value of $1,000, and it actually takes $1,500 worth of effort to complete, the contractor loses $500. The contract must still be completed even though the contractor loses money. Conversely, if the contractor successfully manages the effort and spends less than anticipated, the difference is retained as increased profit. This is the most common type of government contract in this category.

Cost Plus Fixed Fee Contract

When the effort required by a contract is difficult to quantify or is dependent on other activities for completion, the government may award a cost plus fixed fee (CPFF) contract. On a CPFF contract, the government agrees to reimburse the contractor for all costs incurred plus pay a fixed fee, or profit. There is a big difference between a CPFF and a FFP contract. A contractor has virtually no risk on a CPFF contract, and the government doesn't know the final price of the contract until the effort is completed. The government normally establishes a ceiling amount that the contractor's expenses cannot exceed, which is called a not-to-exceed (NTE) price. This type of contract is normally used for research and development efforts where the scope of the effort is unknown.

Time and Materials Contract

A time and materials (T&M) contract is similar to a CPFF but has limited risk for both the government and the contractor. This type of contract is normally used for service-type efforts and is administered through a basic ordering agreement (BOA) that contains set prices for different types of labor and a set reimbursement rate for materials.

The government issues task orders against the BOA that request a contractor to accomplish a specified task. The amount of labor and materials required to perform the task is estimated, an NTE is established, and the contractor performs the task. The amount of profit the contractor receives is included in the labor rates in the BOA. Labor rates are discussed later in this chapter.

Service Contract

The government uses service contracts just like other businesses that desire support for operations. A common service contract is for on-call repairs, such as a service contract for a computer center. The contractor agrees to maintain the capability to service and repair the computers for a set annual price. This is similar to an FFP contract in that there is a degree of risk to the contractor because the amount of effort required to support the contract is somewhat undefined.

Incentive Contract

Any of the above contracts can also be incentive contracts. In these cases, the government includes provisions for paying an incentive, or premium, for a contractor's performance for such things as accelerated delivery of required equipment, better reliability, or completion of effort without reaching a NTE price. The most common incentive contracts are firm fixed incentive price (FFIP) and cost plus incentive fee (CPIF). Possible incentives are identified in the RFP and are agreed on during negotiations.

Each government contract is different. The type depends on what the government needs. The categories discussed above can be combined to describe most types of contracts, for example, unrestricted competitive FFP; restricted sole-source CPFF; and restricted, small-business, set-aside competitive FFP; etc.

PROPOSAL PREPARATION

When a contractor receives an RFP, a sequence of events begins that results in a completed proposal. Each company has a standard procedure for proposal preparation; and although the exact actions taken by an individual company may vary, there is an almost universal approach to this task. Figure 19-5 depicts the typical actions taken to prepare a proposal.

Bid/No-Bid Decision

The first action that any contractor should take upon receipt of an RFP is to make a bid/no-bid decision. There are many reasons why a contractor decides to not submit a proposal in response to an RFP; the contractor might not be technically qualified, the requirements might be unrealistic, or the company might feel that it cannot submit a competitive bid. Proposal preparation is not cheap. It takes time and money. Contractors can't arbitrarily respond to every RFP that the government issues. The purpose of the bid/no-bid decision is to verify that the company has the capability to accomplish the work and that management feels the company has a good possibility of being awarded a contract.

Proposal Manager

A proposal manager is assigned to manage all activities related to the proposal either upon receipt of the RFP or upon reaching a decision to bid for the contract. This is a key element in ensuring that the proposal is prepared in time and is responsive to the government's requirements. The proposal manager is normally a senior staff member or an individual with experience in the type of work requested by the RFP.

Larger contractors might have individuals whose sole responsibility is to act as proposal managers, while smaller contractors might assign this responsibility to functional managers. In either case, the proposal manager must be experienced in preparing proposals since this person is responsible for forming the proposal team, establishing a preparation schedule, determining the detailed requirements and scope of effort requested, and orchestrating the total effort necessary to complete the proposal.

Proposal Team

The proposal manager forms a proposal team that will normally do the actual pro-

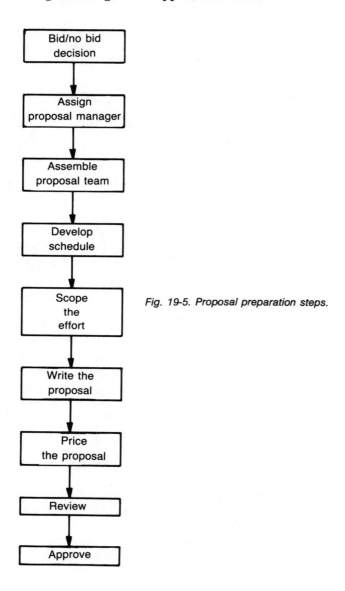

Fig. 19-5. Proposal preparation steps.

posal writing. Team composition will vary depending on the RFP, but each functional discipline or organization that will be involved with the contract, if awarded, must participate in preparing the proposal. The team members are normally the experts from each group and have the experience necessary to determine how tasks will be accomplished and how much effort and material will be required to complete the contract requirements.

Proposal Schedule

The proposal manager must establish a schedule for completing the proposal in a timely manner. Making the schedule is easy; keeping to it is not so easy. The schedule must be realistic and provide as much time as possible for writing a response to the RFP, but it must also contain sufficient time for administrative processing, review, changes, and final approval. Inevitably, most proposals end up as last minute "fire drills." This is a fact of life for government contractors; however, with good scheduling, this can be avoided.

Scoping the Effort

The next major step is determining the actual scope of the effort requested by the RFP. This is not as easy as it sounds. In order to prepare a proposal that responds to the RFP and accurately identify everything that the contractor must do to complete the contract, the team must first determine exactly what the RFP requires. As stated previously, an RFP contains a product specification, a SOW, and a CDRL. Each of these items must be addressed in the proposal. The specification describes what the contractor must produce. The SOW tells what actions the contractor will be required to accomplish to produce the item described by the specification. The CDRL identifies all data items that must be submitted to the government as part of the contract.

For example, the SOW might state that "the contractor will accomplish all the engineering and manufacturing tasks required to successfully produce the item described in the product specification." That short statement can encompass thousands of individual tasks and require several million dollars worth of labor and material. The proposal team must identify each task, labor hour, and all materials. That is scoping the effort. Figure 19-6 is a typical SOW excerpt that requires an ILS effort, in this case

---

Paragraph 4.5   Integrated Logistics Support

The contractor shall prepare an Integrated Logistics Support Plan (ILSP). Upon approval of the plan by the government, the contractor shall conduct an ILS program in accordance with the approved plan. As part of the ILS program, the contractor shall prepare and submit the following:

1. Logistic Support Analysis Plan (LSAP) (CDRL D001)
2. Logistic Support Analysis (LSA) documentation (CDRL D002)
3. Logistic Support Analysis Record (LSAR) data (CDRL D003)
4. Provisioning and Other Preprocurement Screening (CDRL D004)
5. Support Equipment Recommendations Data (SERD) (CDRL D005)
6. Maintenance Plan (CDRL D001)
7. Spare Parts List (CDRL D006)
8. Level of Repair Program Plan (CDRL D001)
9. Level of Repair Analysis (LORA) (CDRL D007)

Fig. 19-6. Statement of work (ILS excerpt).

preparation of logistic support analysis record (LSAR) data. Note that the SOW provides a reference to the CDRL shown in Fig. 19-7.

The CDRL is prepared on DD form 1423, Contract Data Requirements List, which is the standard method for contractually identifying data requirements. Each block of DD form 1423 provides specific information about how the data item will be prepared, frequency of submittal, copies required, and other pertinent information. Block 4 of DD form 1423 provides key information—reference to a data item description (DID).

DIDs are standard government documents that provide detailed guidance on the preparation of data items. Figure 19-8 is the DID referenced in Fig. 19-7. This illustrates the complete chain of documentation for data items. The SOW tells the contractor to produce a data item; the CDRL provides information about the requirements for submittal; and the DID tells how the data item will be prepared. Because a majority of ILS tasks result, either directly or indirectly, in preparation of data items, it is important that each task be traced completely through the SOW, CDRL, and appropriate DID.

Proposal Layout

After the effort has been completely scoped, actual proposal preparation can begin. Preparation is divided into two distinct activities; proposal writing and proposal pricing. An RFP will normally provide guidance as to the format that a contractor must follow when preparing the proposal. The most common sections, or volumes, of a proposal are management, technical, and cost. The management section describes the overall capabilities of the company and tells how the company will use these capabilities to accomplish the contract requirements. Relevant experience may also be included in the management section. The technical section specifically addresses the SOW and tells how the contractor proposes to complete the contract requirements. It usually follows the outline of the RFP to allow easy cross- referencing between the two documents. The cost section is the only part of the proposal that contains information or reference to the price of the bid.

Proposal Writing

Proposal writing is an art. When a proposal is written, the writer conveys to the government the capability of the contractor to perform the effort required by the RFP. There is no set style or method for writing proposals, but it is of the utmost importance that the final proposal adequately address each point contained in the RFP. Proposals should be readable, as brief as possible, and factual. Too many times proposals merely repeat the words contained in the RFP, and this is unacceptable. Figure 19-9 illustrates two responses—acceptable and unacceptable—to the SOW excerpt in Fig. 19-6.

The technical section of the proposal is made up of responses like this to every task required by the SOW. The contractor must tell the government what will and will not be done. If assumptions are made about the SOW requirements, then they must also be provided. If information or materials are expected to be provided by the government, the contractor must identify in detail exactly what is needed and when it must be received to support the contract delivery schedule.

CONTRACT DATA REQUIREMENTS LIST

ATCH NR _____ TO EXHIBIT ___D___

CATEGORY _____ILS_____

SYSTEM ITEM ___BXB360-1___

CONTRACTOR ___LMA___

TO CONTRACT PR

1. SEQUENCE NUMBER	2. TITLE OR DESCRIPTION OF DATA / 3. SUBTITLE	4. AUTHORITY (Data Item Number) / CONTRACT REFERENCE	4. TECHNICAL OFFICE	5. APP CODE A / 6. INPUT TO IAC (A)	10. FREQUENCY	11. AS OF DATE	12. DATE OF 1ST SUBMISSION	13. DATE OF SUBSEQUENT SUBS EVENT KD	14. DISTRIBUTION AND ADDRESSEES (Addressee - Regular Copies Repro Copies)	15.
D001	Integrated Logistics Support Plan (ILSP)	SOW 4.5		DD A	ONE/R		30DAC		AAFCXX	1/0
	REMARKS								TOTAL	1/0
D002	Documentation, Logistic Support Analysis (LSA)	SOW 4.5		DD A	ONE/R		12MAC		AAFCXX	2/0
	REMARKS								TOTAL	2/0
D003	Logistic Support Analysis Record (LSAR) Data	DI-ILSS-80114 SOW 4.5		DD A	QRTLY		6MAC		AAFCXX	1/0
	REMARKS TO BE REVIEWED AT QUARTERLY LSAR REVIEWS. HARD-COPY ACCEPTABLE. FINAL DELIVERY BY MAGNETIC TAPE IN MIL-STD-1388-2A FORM AT 30 DAYS PRIOR TO CONTRACT COMPLETION.									
D004	Provisioning and Other Pre-procurement Screening Data	SOW 4.5		DD A	ONE/R		SEE BLOCK 16,		DLSC	1/0
	REMARKS TO BE SUBMITTED TO DLSC 60 DAYS PRIOR TO EACH PROVISIONING CONFERENCE.								TOTAL	1/0

PREPARED _(signature)_ Catherine E. Jones   DATE 10/0/86

APPROVED BY _(signature)_   DATE 10/3/86

DD FORM 1423, 1 JAN 75   S/N 0102 LF 014-2300   REPLACES EDITION OF 1 JUN 69, WHICH IS USABLE.   PAGE ___ OF ___ PAGES

Fig. 19-7. Contract data requirements list.

	DATA ITEM DESCRIPTION	*Form Approved* *OMB No. 0704-0188* *Exp. Date: Jun 30, 1986*

1. TITLE	2. IDENTIFICATION NUMBER
Logistic Support Analysis Record (LSAR) Data	DI-ILSS-80114

3. DESCRIPTION/PURPOSE

3.1 This Data Item Description (DID) identifies deliverable LSAR data and describes the hardcopy media for their delivery. These data identify logistic support resource requirements in a correlated and integrated fashion and provide a basis for support system development activities and subsequent procurement actions and decisions.

3.2 LSAR data are used as source data in the preparation of equipment (continued on Page 2)

4. APPROVAL DATE (YYMMDD) 860221	5. OFFICE OF PRIMARY RESPONSIBILITY (OPR) TM	6a. DTIC REQUIRED	6b. GIDEP REQUIRED

7. APPLICATION / INTERRELATIONSHIP

7.1 This DID contains the format and content preparation instructions for LSAR data required by 205.2, 301.2 and 401.2 of MIL-STD-1388-1A, and appendices A and F of MIL-STD-1388-2A.

7.2 This DID is applicable to the acquisition of military systems and equipment.

7.3 DI-ILSS-80115, DI-ILSS-80116 and DI-ILSS-80117 are related to, and may be used in lieu of this DID to obtain the three automated LSAR Master files.

(continued on Page 2)

8. APPROVAL LIMITATION	9a. APPLICABLE FORMS	9b. AMSC NUMBER A3780

10. PREPARATION INSTRUCTIONS

10.1 <u>Source Document</u>. The applicable issue of the documents cited herein, including their approval date and dates of any applicable amendments and revisions, shall be as reflected in the contract.

10.2 <u>Data Record Preparation</u>. LSAR data element definitions, data field lengths, and formats for recording and reporting LSAR data shall be in accordance with appendix A of MIL-STD-1388-2A. Data requirements shall be as specified on DD Form 1949-1, LSAR Data Selection Sheet, contained in the Statement of Work. Supplemental forms, worksheets, block diagrams, and other data that are specified in the logistic support analysis plan or other plan as forming part of the LSAR data shall be deliverable to the extent specified in the Contract Data Requirements List (CDRL), DD Form 1423.

10.3 <u>Data Delivery Media</u>. LSAR data shall be deliverable by one or a combination of the following media, as specified by the requiring authority on DD Form 1423.

    a. Hardcopy (original, reproduced copies or computer generated copies).

    b. Microcopy (microfiche/microfilm).

10.4 <u>Data Records</u>. The following LSAR data records, as described in MIL-STD-1388-2A, are deliverable data under this DID as specified by the requiring authority on DD Form 1423.

    a. Data Record A, Operations and Maintenance Requirements (paragraph 20, appendix A of MIL-STD-1388-2A).

    b. Data Record B, Item Reliability and Maintainability (R&M) Characteristics (paragraph 30, appendix A of MIL-STD-1388-2A).

    c. Data Record B1, Failure Mode and Effects Analysis (paragraph 40, appendix A of

(continued on Page 2)

DD Form 1664, FEB 85   *Previous edition is obsolete.*   PAGE __1__ OF __2__ PAGES

Fig. 19-8. Data item description.

DI-ILSS-80114

3. DESCRIPTION/PURPOSE (continued)

publications; maintenance procedures; manpower and personnel requirements; training requirements; tool, support, test, measurement and
diagnostic equipment requirements; and other Integrated Logistic Support
element documentation.

7. APPLICATION/INTERRELATIONSHIP (continued)

7.4 This DID supersedes DI-L-7145A.

10. PREPARATION INSTRUCTIONS (continued)

MIL-STD-1388-2A).

    d. Data Record B2, Criticality and Maintainability Analysis
(paragraph 50, appendix A of MIL-STD-1388-2A).

    e. Data Record C, Operations and Maintenance Task Summary
(paragraph 60, appendix A of MIL-STD-1388-2A).

    f. Data Record D, Operations and Maintenance Task Analysis
(paragraph 70, appendix A of MIL-STD-1388-2A).

    g. Data Record D1, Personnel and Support Requirements (paragraph
80, appendix A of MIL-STD-1388-2A).

    h. Data Record E, Support Equipment and Training Material
Description and Justification (paragraph 90, appendix A of
MIL-STD-1388-2A).

    i. Data Record E1, Support Equipment and Training Material
Description and Justification continued (paragraph 95, appendix A
of MIL-STD-1388-2A).

    j. Data Record E2, Unit Under Test and Automatic Test Program
(paragraph 100, appendix A of MIL-STD-1388-2A).

    k. Data Record F, Facility Description and Justification
(paragraph 110, appendix A of MIL-STD-1388-2A).

    l. Data Record G, Skill Evaluation and Justification (paragraph
120, appendix A of MIL-STD-1388-2A).

    m. Data Record H, Support Items Identification (paragraph 130,
appendix A of MIL-STD-1388-2A).

    n. Data Record H1, Support Items Identification (Application
Related) (paragraph 140, appendix A of MIL-STD-1388-2A).

    o. Data Record J, Transportability Engineering Characteristics
(paragraph 150, appendix A of MIL-STD-1388-2A).

☆U.S. GOVERNMENT PRINTING OFFICE: 1986 — 605-035/80056

Fig. 19-8. Data item description. (Continued from page 332.)

Unacceptable response

LSAR data will be submitted as required by the Statement of Work. The data will be the results of the required LSA effort.

Acceptable response

LSAR data will be prepared as required by paragraph 4.5 of the Statement of Work. The data will be produced as an output of Task 401 of MIL-STD 1388-1A and will be recorded using LSAR data records contained in MIL-STD 1388-2A. Data elements for applicable LSAR data records will be provided as required by DD form 1949-1 contained in Appendix A of the Statement of Work. The data will be managed through the use of a single computerized data base currently installed in the ILS activity facility. Through the use of this computerized system, all logistics personnel will have access to the information as it is generated for use in developing support resource requirements for the new equipment design. The LSAR data base will be delivered, as required by DI-ILSS-80114, using magnetic tape media. LSAR summary reports will be provided in hardcopy media.

Fig. 19-9. Example proposal responses to LSAR portion of Fig. 19-6.

Proposal Pricing

Contractors must not only be able to perform the work required by the RFP, they must be able to complete the contract for a reasonable price. Pricing of proposals is as important as writing. In many cases, the bottom line price is the determining factor as to which contractor will receive a contract. When a contractor submits a price to the government, it must be fully justified, and the cost for performing each task must be identifiable. Proposal pricing is accomplished systematically by first determining the labor and materials required to perform each task. After the labor and materials are identified, the total proposal price is computed using a standard method based on the following:

- **Direct Labor**. Direct labor is required to specifically accomplish tasks that are required by the SOW. Examples of persons who normally perform direct labor are design engineers, logistics engineers, assembly line personnel, direct supervision personnel, and technicians. Direct labor is always quantified in terms of the hours required to perform a task.
- **Overhead**. Expenses incurred as part of the overall effort to complete a contract that cannot be attributed to a single task are considered overhead. They normally fall into three categories: indirect labor, employee benefits, and resources. Examples of indirect labor include secretaries, shipping clerks, and computer operators. These individuals are required to support the contract, their daily activities are neces-

sary to complete the contract, but their effort normally cannot be linked to a specific task. Employee benefits typically include health and life insurance, vacation and sick pay, pension plans, and any other benefits the company provides to employees. Resources can include any expense required to support the contract. Examples of this expense include rent, utilities, office supplies, and equipment. Overhead is expressed as a percentage of direct labor and is based on historical expenditures. A company's overhead rate must be approved by the government. Overhead rates vary among companies. A company that does not require a large investment in equipment or facilities may have an overhead rate of 75%, while a heavy manufacturer might have a rate as high as 250%.

- **Material**. The cost of materials required to build or produce an item of equipment or accomplish tasks that result in deliverable items must be specifically identified. Examples of material costs are raw steel, resistors, nuts and bolts, or any other items that can be traced through the manufacturing process. Actual quotes from the sources of materials or reasonable historical data are used when preparing a proposal price.
- **Other Direct Costs (ODC)**. These are expenses required to fulfill the contract but are not categorized above. ODC can include contract labor, consultants, and travel expenses if directly connected with the contract. Actual anticipated costs must be used.
- **General and Administrative Expenses (G&A)**. These are expenses incurred by the company as a part of doing business but that cannot be attributed directly to any contract. Examples of G&A include salaries for upper management, contracts and marketing personnel, legal expenses, and expenses for facilities that are not directly related to any contract. G&A is also expressed as a percentage based on historical data and must be approved by the government.
- **Fee or Profit**. The amount of money a company expects to earn by completing the proposed effort is its fee or profit. Fees are expressed as a percentage and normally range between 7% and 15%, depending on the type of effort being performed.

The standard method for determining the price for a proposal is illustrated in Fig. 19-10. In this example, direct labor hours have been estimated for the SOW task from Fig. 19-6. The best way to estimate the number of hours required to accomplish a task is to use historical information from completion of similar tasks. After the direct labor hours have been estimated, they are multiplied by the labor rate for each type of labor required.

Labor rates are maintained by contractors and reflect the median wage for different categories of personnel. In the example, the labor rates are for a senior logistics engineer, a logistics engineer, and a clerk. The direct labor dollars are then multiplied by the overhead rate to determine the anticipated cost for employees and other resources to support the contract.

The next step is to add the cost of materials and ODC. This figure is then multiplied by the G&A rate. The resulting figure is the projected cost for the company to complete the contract. The final step is to multiply the total cost by the fee rate, which results in the final proposal price. Although each company has its own rates, the method for using the rates to price a proposal is the same.

Step 1. Compute direct labor
Task: Prepare LSAR data

Labor grade	Hours required	Labor rate	Total
Senior logistician	800	$22.00	$17,600.00
Logistician	2100	16.00	33,600.00
Data entry	200	9.50	1,900.00
Total direct labor cost			$53,100.00

Step 2. Compute overhead costs
(Direct labor cost × Overhead rate)
$53,100.00 × 75% =

	39,825.00
Total direct labor and overhead costs	$92,925.00

Step 3. Compute material and other direct costs
Material costs = none
Other direct costs

Computer charge (400 hrs @ $55/hr) =	22,000.00
LSAR software	20,000.00
Total costs	$134,925.00

Step 4. Compute G&A expenses

(Total costs × G&A rate)
$134,925.00 × 15% =

	20,238.75
Total cost plus G&A	155,163.75

Step 5. Compute fee

(Total cost plus G&A × Fee rate)
$155,163.75 × 15% =

	$23,274.56
Total proposal price	$178,438.31

Fig. 19-10. Pricing methodology.

Review and Approval

Proposals normally go through several reviews prior to approval and submittal. The contractor might assemble a review team (sometimes called a red team or tiger team) of individuals who have not been involved in preparing the proposal to review it. This method has proven extremely successful in producing winning proposals. Sufficient time must be allowed for the review team to complete the review and for the proposal team to make required changes if this method is used.

After reviews and updates have been completed, the final action is approval by senior management. Adequate time must be planned to allow review and approval. Remember, senior management is ultimately responsible for ensuring that the company can fulfill the SOW requirements, if awarded the contract, and make a profit.

Chapter 20

Organization

Every company organizes in a manner that is supportive of its products, services, and business goals. No two companies are organized exactly the same; however, there are similarities among the organizations of most companies. Organizational methods have been developed over the years that have proven conducive to efficient management of resources and activities.

Most organizations can be divided into two areas; activities that handle the overall business operation and activities that are directly related to the products or services the company provides. The business-related activities consist of things that a company must do in order to accomplish any goal. The organization of these activities are relatively standardized throughout industry. Product- or service-related activities are tailored to meet specific requirements and, therefore, differ in each company. The logistician should understand the basic organizational structure and charters for activities contained within this structure in order to effectively interface with the other disciplines that are involved with the design and production of military equipment. The brief descriptions provided in this chapter are to explain, in general terms, the functions and responsibilities of the typical government contractor organization.

BUSINESS ORGANIZATION

The business organization of most companies includes those activities required to run the business, regardless of the product or service that the company provides to its customers. Figure 20-1 illustrates the typical business organization of defense contractors. The organization includes contracts, marketing, finance, and personnel. The

338

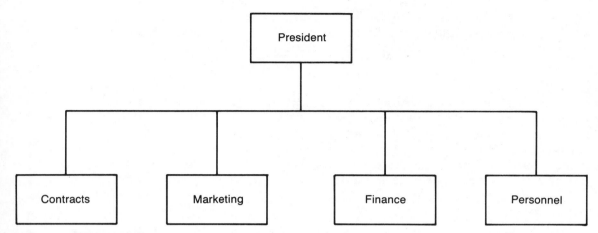

Fig. 20-1. Business organization.

functions performed by these activities are necessary for any company doing business with the U.S. government and are normally classified as general and administrative (G&A) expenses of the company.

Contracts

The purpose of the contracts organization is to act for the company in all matters related to the administration of government contracts. This includes preparing the final proposal submitted to the government in response to an RFP; negotiating contract terms, conditions, and scope; and representing the interests of the company with all customers. The contracts organization is the only entity of the company that can legally deal with customers on contractually related matters.

It is important to understand that agreements with customers can only be made by the contracts organization. A common problem that plagues contractors is when company personnel outside the contracts organization make agreements or promises to customers. When this occurs, the company is at risk due to legal considerations. This must be avoided at all costs. ILS should always interface with the contracts organization when dealing with definition of contractual requirements, changes in scope of work, or when contractual problems arise in dealing with a customer.

Marketing

The marketing organization is charged with the responsibility of identification and acquisition of new and follow-on business for the company. The marketing staff maintains contact with all potential customers of the company and leads all proposal preparation efforts. This organization is comparable to a sales organization in the commercial industry. Marketing leads all efforts related to business development and receipt of new orders and contracts. ILS should receive information from marketing on prospects for new business, which is used for long-range planning and determination of future staffing requirements. Additionally, ILS assists marketing in identification of logistics-related business and interpretation of the logistics portions of RFPs.

Finance

Responsibility for all financial and accounting matters of the company rests with the finance organization. Finance keeps the books on all company business. In the process of accumulating cost information related to each contract, finance computes the standard rates used to prepare proposals based on actual costs. This includes determination of standard rates for labor, overhead, and G&A. Finance also provides information to managers on actual expenditures for labor and material for each contract.

Personnel

The personnel organization is charged with the responsibility of all aspects of personnel administration for the company. This includes training programs, recruiting, hiring, terminations, and maintenance of personnel records. Additionally, personnel oversees the company benefits provided to employees such as health and life insurance, stock plans, vacation, sick leave, etc.

PRODUCT ORGANIZATION

The typical product organization of a government contractor is very difficult to identify. Each contractor establishes an organization that best fits the product or service that is provided to the government. Therefore, no two contractors have the same product organization. In many cases, the names of the organizations are different even though the activities are the same. Figure 20-2 shows an organization that contains most of the activities performed by a contractor. The activities normally found in a contractor's product organization include engineering, operations, product support, program management, and quality assurance.

Engineering

The engineering organization is responsible for developing the product that the company will build. This entails taking an idea or a specification and creating an equipment design that meets all stated and inherent product requirements. The design progresses through the activities that normally comprise an engineering organization, including advanced engineering, systems engineering, hardware and software design, and test. The final design must be fully documented to allow production by the operations organization. ILS disciplines normally work very closely with elements of the engineering organization throughout the design process.

Advanced Engineering. Research and development activities are performed by the advanced engineering organization to develop new products or explore new concepts. The efforts of this organization may be supported through internal company R&D funds or through contracts for such activities from the government. The activities of

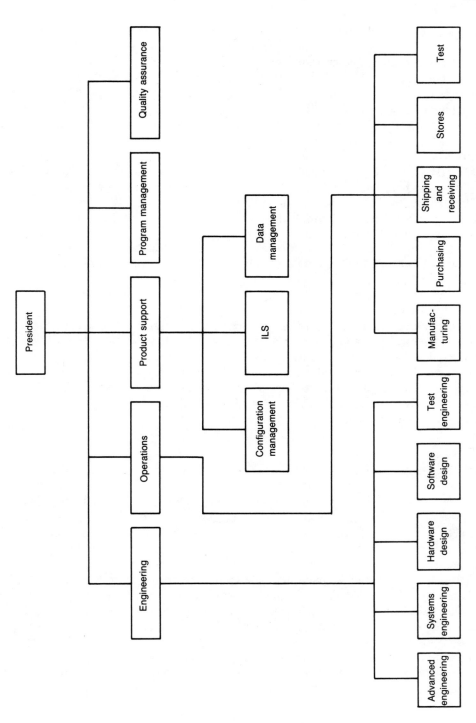

Fig. 20-2. Product organization.

advanced engineering are normally closely linked with those of marketing, where a company is developing a new product to meet the requirements of a marketing target.

Systems Engineering. The systems engineering activity is charged with the responsibility to determine the overall requirements for a product and ensure that the total package is integrated in a manner that meets specification requirements. Specifically, systems engineers are concerned with integrating the components or portions of an item that comprise the total system. Definition of interface requirements among interconnecting assemblies, hardware, and software, and testability are all addressed by this organization.

Hardware Design. The organization responsible for the detailed design of equipment may be composed of several different engineering disciplines such as electrical, mechanical, hydraulic, etc. The actual title and staffing of this activity depends on the contractor and the products being designed. Drafting and creation of detailed engineering documentation normally occur within the hardware design activity.

Software Design. Generating the software necessary to operate and test products is normally accomplished by a software design activity located within engineering. Because of the different types of software that might be required and the multitude of software languages that could be used, the composition of this activity is very product-dependent.

Test Engineering. The purpose of the test engineering activity is to evaluate the designs developed within the engineering organization. This activity may also include the ability to build engineering models or simulate actual equipment used to test the final equipment designs. Most products must pass the tests performed by this activity before being released for manufacturing. Many of the tests performed by test engineering are directly related to ILS activities, such as reliability growth test, reliability development test, and maintainability demonstration.

Operations

The operations organization is responsible for building and delivering the products designed by engineering or other sources. The activities contained in operations are tailored to fit the appropriate needs of the products being built, but normally include manufacturing, purchasing, shipping and receiving, stores, and testing. The functions of these activities are directly related to the production of equipment for a customer.

Manufacturing. The actual work of fabrication, assembly, and integration is done by the manufacturing activity. All efforts of manufacturing are controlled by the design documentation generated by engineering. The manufacturing activity does not have the authority to change engineering documentation without approval through the change control process. The most significant changes that do occur are related to producibility where changes allow more efficient production of the equipment.

Purchasing. Purchasing is normally the only activity in the company that is authorized to procure materials used in the manufacturing process. This is accomplished through issuance of purchase orders to suppliers or vendors for materials listed on engineering documentation. The actual list used for procurement is the bill of materials

(BOM) that is generated from all the parts lists of the designed equipment. Additionally, purchasing might be responsible for ordering office supplies, furniture, equipment, and other materials used by the company.

Shipping and Receiving. All equipment, materials, supplies, or other items sent to organizations outside the company and items that come from outside the company must pass through shipping and receiving. The shipping and receiving activity is responsible for sending all items out of the company, accepting all deliveries from suppliers and vendors, and ensuring that the necessary documentation is prepared to record such activity. Items shipped to the government must be accompanied by standard DoD shipping documents that are prepared by shipping. Materials that are ordered using a purchase order must be received and documented by receiving for the supplier to receive payment.

Stores. Materials purchased for use in the manufacturing process are secured in stores until required. Additionally, kitting for assembly may occur in stores to add efficiency, and finished equipment awaiting shipment may be located in stores. Government property in the possession of the contractor might be stored in a segregated, secure, bonded stores area if required by government regulations.

Test. Finished equipment must be tested to ensure that it meets the requirements of the product specification. This is accomplished by the test activity. Typical efforts include burn-in prior to test, acceptance test, qualification test, and environmental testing. In some companies, this activity is actually subcontracted to a vendor who performs this type of service.

Product Support

The product support activity is charged with the responsibility of providing "cradle-to-grave" support to the products delivered to customers. All the efforts of this activity are product oriented and can include configuration management, integrated logistics support, and data management.

Configuration Management. The task of identifying and controlling the complete configuration of a product is the responsibility of configuration management (CM). This activity must maintain accurate records of exactly what is used to make or build a product. These records are invaluable to ILS in determining logistic support resource requirements. Additionally, configuration management maintains records of each serially numbered item that is built. The as-built records are necessary to determine the impact of recommended design changes on items that have been delivered to the customer and in planning repairs of failed items and spares.

Data Management. The data management (DM) activity is responsible for scheduling, reviewing, and managing all data that must be submitted to a customer in response to a CDRL. All submittals of data to a customer by ILS must be accomplished through data management. The records maintained by this activity are useful in planning staffing requirements and coordinating workloads commensurate with customer data requirements.

Program Management

The program management organization is responsible for overseeing effort to complete specific programs. This organization is discussed in detail in Chapter 21.

Quality Assurance

The quality assurance (QA) organization is the internal audit and inspection agency for the company that is chartered with the responsibility to review all processes and materials used to design and manufacture its products. QA has the authority to reject any substandard parts, processes, or procedures that do not meet established specifications. This organization has authority to direct any changes necessary to ensure that the company produces equipment that performs as required by the government.

QA works closely with representatives of the Defense Contract Administration Service (DCAS) when contracts require government source inspection (GSI). Additionally, QA might perform contractor source inspection (CSI) at vendor facilities to ensure that items procured from vendors meet the required standards. QA establishes and maintains a qualified vendor list (QVL) that identifies vendors that meet the company's standing requirements for product quality. Internal in- process inspections and overview of manufacturing procedures are accomplished by quality control (QC), which is an element of QA.

ILS ORGANIZATIONS

ILS organizations are composed of the functional disciplines that have been de-

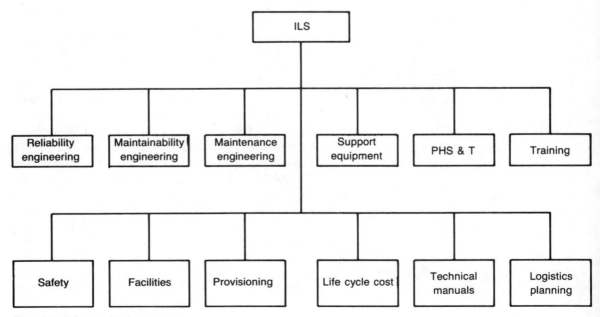

Fig. 20-3. ILS organization concept.

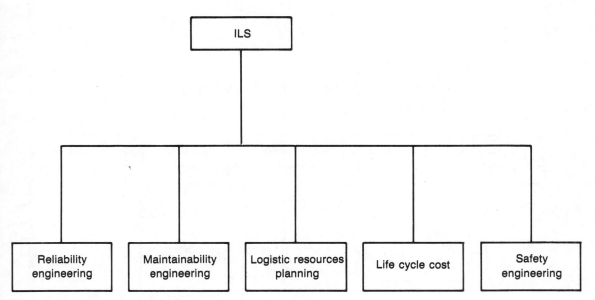

Fig. 20-4. ILS organization (concept phase).

scribed in previous chapters. The manner in which the disciplines are organized to form a specific ILS organization may depend on the acquisition phase of the specific product, company size, and type of products supported. Figure 20-3 shows a conceptual organization of ILS disciplines based solely on function. This concept, with slight variations, is used by many large government contractors that have on-going logistics programs for many different products. Other companies have tailored ILS organizations to fit specific program requirements.

Acquisition Phase

Figure 20-4 illustrates an ILS organization that supports up- front logistics planning during the concept phase of an acquisition program. Figure 20-5 is an organization for the FSD phase that emphasizes identification of logistic support resources. Figure 20-6 provides support to a product during the operational phase. Notice that, in each of these illustrations, emphasis is placed on disciplines that can have the most impact on the program at the point where support is provided. In actuality, all these organizations could be used on the same program, being changed as necessary as the product progresses through the acquisition process.

Company Size

The size of a company has a direct bearing on the ILS organization. Larger companies normally have organizations similar to those described above, while smaller companies tend to have ILS organizations composed of a few logisticians who perform all

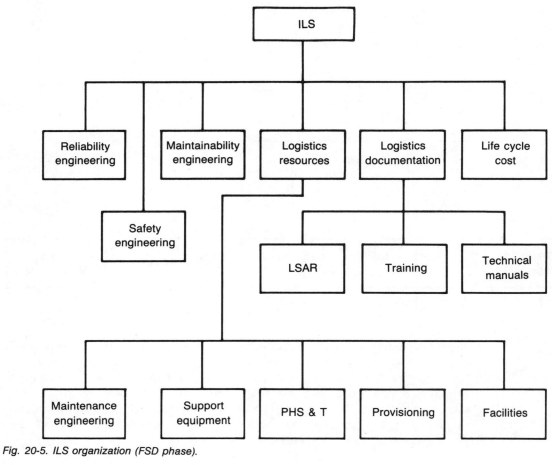

Fig. 20-5. ILS organization (FSD phase).

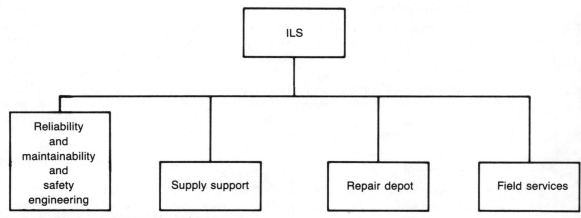

Fig. 20-6. ILS organization (Post-production phase).

346

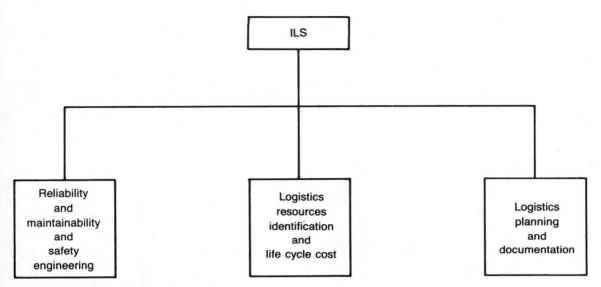

Fig. 20-7. Reduced ILS organization.

the functions required to design a support package. Figure 20-7 shows a typical ILS organization for a small government contractor. This organization does not normally change as products go through the acquisition cycle. Some companies have only one logistician who does it all.

Product Type

All products require ILS activities; however, the product itself might dictate changes to the ILS organization. For instance, if a contractor is a subcontractor that makes only parts that are integrated by a prime contractor, then the subcontractor may provide source information to the prime who does the detailed system planning. If a contractor makes a product that is nonrepairable and requires no maintenance, then certain ILS functions are not required. The ILS organization of a contractor that is designing and producing a tank or aircraft is drastically different from the organization of a contractor that designs and builds rifles because the logistic support requirements of the products are different.

The ILS organization must be responsive to the needs of the company and its customers. It should be organized to realize the greatest benefit possible from the efforts of the appropriate ILS disciplines as they relate to the company's products. The position of ILS within the overall company organization must provide visibility to the supportability requirements of the products that the contractor provides to the government as the products are being designed and produced.

Chapter 21

Program Management

ILS activities are normally coordinated and controlled through the program management process. The concept of program management is centered around a person or persons responsible for seeing that contractual requirements are met. The program management process consists of many different aspects that must be continually addressed in order to be successful. These aspects include assignment of responsibilities, changing actions as the program moves through different phases, using program management tools such as schedule and budget, and considering the unique requirements of DoD programs. The success or failure of a company to fulfill contractual requirements depends on how well a program is managed.

RESPONSIBILITY

The key to successful program management is a clear definition of who is responsible for what. Most contractors use the program management process to coordinate and control all activities related to a specific government contract. This process starts with identifying the person who has overall responsibility for administering the program. In addition to this program manager, there is normally a supporting organization called a program management office (PMO) that is tailored for each contract. It is important that the responsibilities of the program manager and the PMO be clearly defined at the beginning of each program, and that changes in responsibility be identified as they occur, to ensure that programs are successfully completed.

At this point, it might be appropriate to define ''success'' with regard to government contract completion. On the government side, success is normally defined as re-

348

ceipt of equipment from a contractor that meets the contractual procurement specifications, is delivered on schedule, and at no additional cost above the contract award price. From the contractor's perspective, success is defined as delivering a product to the government that meets the minimum procurement specification requirements and results in a reasonable profit. These definitions are different, but complementary. If the contractor meets the government requirements and properly manages the program, a profit should be realized. That is the ultimate responsibility of the program management process; to satisfy both the government's and the contractor's definitions of success.

The Program Manager

The person who bears the total responsibility for program success or failure is the program manager. Therefore, the person selected to be a program manager should have abilities commensurate with the requirements of the position. Figure 21-1 briefly lists general qualifications for a program manager. If a program manager is to be successful, authority must come with responsibility. Without the authority to make decisions and implement those courses of action necessary to complete the program, a program manager can never fulfill the responsibilities of the position.

A program manager does not have to be a technical expert in the product being designed or possess extraordinary talents. The basic role of a program manager is that of an orchestrator. Figure 21-2 shows the typical activities of a program manager. A successful program manager is able to continually oversee the entire scope of a program, judiciously acquire and use the resources necessary to complete the program, and identify and correct problem areas before they affect the program. The position of program manager requires dedication and hard work.

Program Management Office

The program manager cannot function without adequate support. The program management office (PMO) provides the administrative and management support to the program manager necessary to coordinate and control program activities. Figure 21-3

1. Technical understanding of the program.
2. Understanding of government contracts.
3. Ability to manage people.
4. Ability to motivate people.
5. Communication skills.
6. Understanding of scheduling.
7. Understanding of budgets.

Fig. 21-1. Program manager qualifications.

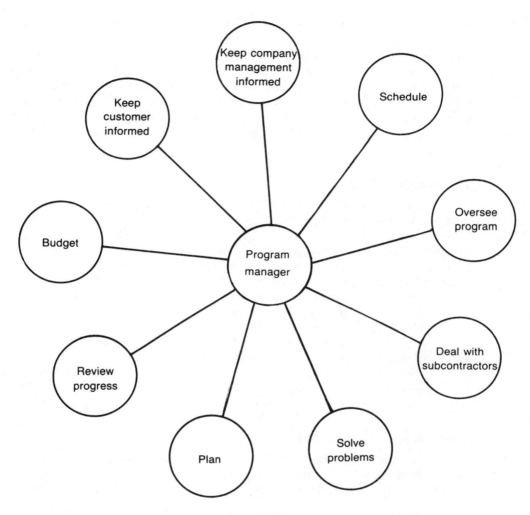

Fig. 21-2. Typical program manager activities.

illustrates typical PMO elements. The normal responsibilities of a PMO include budgeting, cost accounting, scheduling, and data management for the program.

The PMO staffing level is dependent on the size of the program. A large weapon system program might require a PMO consisting of as many as 100 people, while a small program might have only three or four. Although the staffing level might vary, the basic responsibilities of the PMO are the same regardless of the size of the program.

PROGRAM PHASES

Upon receipt of a contract, the contractor cannot start work without systematically planning how to complete the required effort. This is normally done by establish-

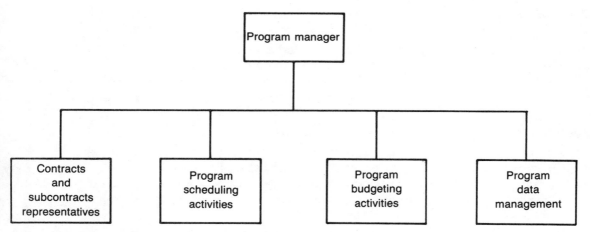

Fig. 21-3. Program management office organization.

ing a program for the contract. There are four distinct phases of a program: (1) definition; (2) planning and organization; (3) performance; and (4) post-performance. Each phase requires specific actions in order to ensure that the program will be completed successfully.

It is important to realize what each phase requires and how the results of one phase affects subsequent phases and other programs. These phases should not be confused with the phases of the acquisition process. However, the program phases apply to how a program is conducted to fulfill acquisition process requirements regardless of the phases.

Definition Phase

The purpose of the definition phase is to delineate exactly what the program is to accomplish. During this phase, the contractor identifies the program manager, establishes requirements for a PMO, publishes a charter for the program, and authorizes the program manager to proceed to the next program phase. The results of this phase are a clear understanding of why the program is necessary, the goals of the program, and the schedule and milestones for program completion.

Planning and Organization Phase

The definition phase of the program establishes requirements. The purpose of the planning and organization phase is to identify the resources that will be required to complete the program, plan how the program will be accomplished, and establish the organization that will be necessary to carry out the plan. The planning done during this phase does not generate plans required by the contract (e.g., reliability plan or ILS plan), but formalizes the plans pertaining to how the program goals will be accomplished. The outputs of this phase should be a program budget, detailed schedule, resource definition and allocation, and final charter for the program manager and the PMO.

Performance Phase

The performance phase is where the actual contract work is done. All activity of this phase is directed toward program completion. During the performance phase, the program manager functions as the focal point for all direction on program task completion. The PMO performs the supporting role of providing the management oversight and information necessary to enable the program manager to manage the program.

Post-Performance Phase

The last phase of the program cycle is the post-performance phase. This phase is often overlooked or given only token effort. The purpose of this phase is to close the program in an orderly manner by documenting the lessons learned, completing data requirements, closing books on the accounts of the program, and performing other tasks necessary to complete all program actions.

PROGRAM DEFINITION

The first step to having a successful program is completely defining exactly what the program is to accomplish. The basis for defining the intent and direction of the program is the government contract. All contracts state explicitly the goals of the program; whether delivery of equipment, services, or other effort. Any activity, effort, or resource required to reach the contract goals should be considered part of the program.

Remember that the contract SOW identifies the effort that the government requires of a contractor. This, in fact, is the program definition. The product specification is used to describe the equipment to be delivered, not the contractor's program.

The second part of the program definition is determining the actual resources that will be required to fulfill the contractual requirements. The source for this part of the definition is the negotiated contract. Contained in the contract is the detailed justification for the bottom line price that the government will pay the contractor. This justification is the labor and materials required to complete the work. Using this as a baseline, the program manager and the PMO should be able to develop a budget for the program.

The final part of the definition is the schedule for completion. The government's required schedule is contained in the contract. It is important to understand that the contractual schedule is not sufficient for program management purposes. The contract schedule normally identifies only key events that must occur during the program. This is not a program management schedule; however, the contract schedule is necessary to define the boundaries of the program. All these items are required to adequately define the program, which must be done before any detailed planning or organizing can be initiated.

PROGRAM PLANNING AND ORGANIZING

Once the program has been defined, planning and organization begin. This is not to say that the program definition never changes. However, the basic program direction, once set, should not change without a redefinition of the program goals. Using the definition as the guide for planning and organizing, the program manager and the PMO develop detailed plans for accomplishing all defined program goals. These plans are internal to the company and do not necessarily respond to any of the government contract reporting requirements.

A key process that occurs during the planning phase is identification of the resources, e.g., time, money, labor, materials, and facilities, that will be required to complete the program. It is important that the program manager receive approval from the company to commit the required resources to the program. If the program manager does not receive this approval and backing, then the program will not succeed. The next step in the planning process is to determine when and how the resources will be applied to the program. If this is not adequately addressed, the resources will be wasted due to lack of proper utilization.

The final step is organizing the program to best control the use of resources and provide the necessary management overview and interfaces for the program tasks. This is not necessarily the creation of a new organization. It might be as simple as assigning program-related charters to an existing organization, defining the responsibilities of each portion of the organization for completing the program goals. The significant documents prepared as part of the planning process are the program schedule and budget. These documents must be of sufficient detail to allow the program manager to effectively manage the day-to-day activities of the program. Without these documents, the program cannot be managed.

PROGRAM PERFORMANCE

Actual program performance includes all the actions necessary to complete the program as required by the government contract. The results of the definition and planning and organizing phases determine how smoothly the program is accomplished. If the previous phases were done haphazardly, then the performance phase will be continually interrupted with periods of redefinition, reorganizing, and crisis management to fix problems. Too often a company will jump into the performance phase of a contract with little or no planning. When this occurs, the program will be plagued with problems throughout the performance phase.

PROGRAM POST-PERFORMANCE

The post-performance phase of the program must not be overlooked. Many times a company considers the program completed when the last item has been accepted by the government or the final documents have been submitted to the government. This is not true. At the close of each program, the contractor must perform several tasks to complete the program after the government requirements have been satisfied.

One of the most important actions is to ensure that payment is received for all work performed. This could take several months to accomplish depending on the reporting and documentation requirements of the specific contract. Another important action that should be taken is to document the lessons learned during the program for use on future programs and for use in preparing proposals. There is no better justification for a proposal submitted to the government than documented results on previous programs.

The final step in closing a program is to dispose of the documentation that was generated in accomplishing the program goals. In most cases, there is a tremendous volume of paperwork that must be reviewed, consolidated, and stored or discarded when a program is completed. Much of this paper can be discarded; however, significant materials should be stored for use on future programs as planning documents, sources for research information, or for modification to fulfill future program requirements.

PROGRAM MANAGEMENT TOOLS

Several tools are used to effectively manage a program. The most common are the schedule and budget. The schedule describes what is to be done when, and the budget is used to control the expenditure of funds and other assets.

Before a creditable schedule or budget can be developed, the work to be accomplished must be identified. This is done through a process known as a work breakdown, which defines each action that must be accomplished to reach the program goals. Additionally, the amount of time required to complete each action must be determined through some type of work measurement process. These can then be used to develop a usable schedule and budget for the program.

Work Breakdown Structure

The most effective method of identifying the detailed tasks required to complete a program is to prepare a work breakdown structure (WBS). The purpose of a WBS is to break the program into separate manageable tasks and subtasks that are used to allocate funds and resources toward program accomplishment. MIL-STD 881A, Work Breakdown Structures for Defense Materiel Items, provides a detailed description of the WBS preparation process. The key to the WBS process is to divide a program down to the lowest manageable task level by time and resources required to complete the task. The resulting tasks should be detailed to the point where a task can be assigned to one or more persons for completion with complete understanding of what the task is, the resources allocated to task completion, and the required completion schedule. If possible, the tasks should be no greater than 40 hours in duration.

On a large program, this entails a tremendous amount of planning and resource allocation. Figure 21-4 shows the first three levels of a WBS for a typical program. As shown, the first levels simply segregate the work into general topical areas. Most WBSs require many levels of breakdown before the discrete tasks are identified. However, once the WBS is completed, the entire program is defined. Whenever possible, the WBS should be developed when a contractor is preparing a response to an RFP; otherwise, the rationale for the proposed contract is not adequately supported and justified.

The WBS provides the contractor with a structured method for identification of the time and other resources necessary for contract completion. This is necessary for realistic program scheduling and budgeting. If the detailed requirements for program completion are not known, then any schedule or budget prepared for the program will not be adequate for accurate program management.

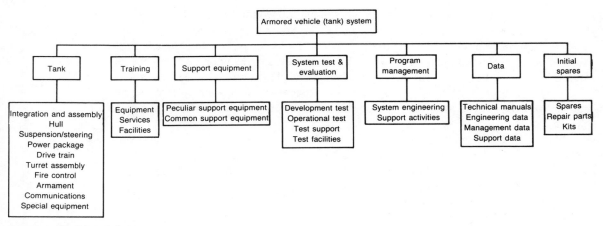

Fig. 21-4. Work breakdown structure.

Work Measurement

The preparation of a WBS is the start of the complete identification of the information necessary for adequate program management. After the detailed tasks to be accomplished have been identified in the WBS, the time required to perform each task must also be identified. Most companies have established standard times for completing a task based on actual historical data. Where historical data do not exist, the contractor must estimate the time based on similarity or best engineering judgment. In either case, the time must be reasonable and supportable.

MIL-STD 1567A, Work Measurement, contains the definitions and parameters to be used when developing task times. The use of standard times allows the contractor to estimate the overall time that will be required for program completion by applying the times to the tasks identified in the WBS. The results of this process can then be used to develop a valid schedule and budget for the program.

Schedule

The program schedule is probably the most overrated document with which the program manager must deal. Everyone places so much importance on the schedule without knowing what it represents. Too often a schedule is a document that shows where a program has been rather than where it is going. Schedules that have not been developed using a detailed WBS and work measurement procedure cannot be relied on to be accurate or usable.

Many times a program management schedule will be a one-page document showing events that are to occur during the program. This is not a program management schedule. A program management schedule must be detailed to the point where each task required for program completion is identified in relation to other dependent and interrelated tasks so that the program manager can manage the resources required to complete all tasks.

Two common methods for program scheduling are program evaluation and review technique (PERT) and critical path method (CPM). Both methods provide a detailed network approach to identification of the tasks that must be performed and their relationships in order to complete a program. The advantage to using one or both of these methods is that they allow the program manager visibility over the critical tasks of the program. With this visibility, the program manager can allocate resources according to the needs of the program and identify problem areas before the program is affected.

Figure 21-5 is format of a PERT network. Each point on the network represents a program task and shows the relationship between tasks. This management tool is extremely effective in identifying which tasks must be done when and establishing priorities for allocation of resources.

Budget

The budget reflects the resources to be expended to complete the tasks of the program as identified by the program schedule. These two documents, the schedule and the budget, must be developed and used in concert if the program is to be successful. The schedule tells what is to be done when, and the budget tells what resources will

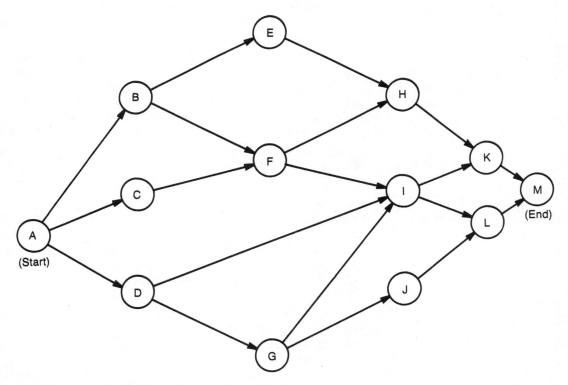

Fig. 21-5. Program evaluation and review technique network.

356

be used to support the schedule. If the schedule changes, then the budget must also change. Even if the tasks do not change, a change in time represents a change in the utilization of allocated resources. All too often, the budget is not developed at the same time as the schedule, so one will never complement the other. When this occurs, the program manager will not be able to effectively manage the program.

Another consistent problem with budgets is that they are used to report what has been spent on a program rather than to plan for controlling expenditures in relation to the program schedule. The typical budget is composed of three figures: estimate at completion (EAC), actual expenditures to date, and estimate to complete (ETC). The actual expenditures plus the ETC equals the EAC. The EAC should be the total anticipated expenditures for the program. If a contractor is continually generating ETCs and adjusting the EAC, then the budget is being used as a report, not as a management tool.

Another common problem with budgets is that the tasks completed according to the schedule do not match the budgeted expenses. For example, if 50% of the budget has been spent, then some managers assume that the program is 50% complete. This cannot be a valid assumption unless 50% of the effort, according to the WBS, has been completed. A program cannot be managed on the basis of expenditures. The budget and the schedule must match, otherwise, neither is a useful tool for program management.

DOD PROGRAMS

Several program aspects are unique to DoD contracts. These include specific program milestones, technical reviews and audits, and definitive interfaces. The successful program manager must understand the significance of each of these events in order to be effective.

Program Milestones

Milestones are specific events that occur during a program. There are several different types of milestones, and the program manager must understand the significance of each. There are DoD milestones that refer to the point in time when significant decisions or reviews occur; e.g., program initiation, Defense System Acquisition Review Council (DSARC) approval for program continuation, and weapon system delivery. For the contractor, the most significant milestones are contract award, preliminary design review, critical design review, production go-ahead, and weapon system qualification and acceptance. Significant ILS milestones include guidance conferences, LSA reviews, data approvals, and fielding of the support package.

The program manager must be able to correlate each and all of these milestones into the overall program objectives. Failure to meet any program milestone will have an impact on both the program schedule and budget. Figure 21-6 shows a typical program milestone schedule.

Reviews and Audits

All DoD programs require the contractor to hold or participate in a series of reviews

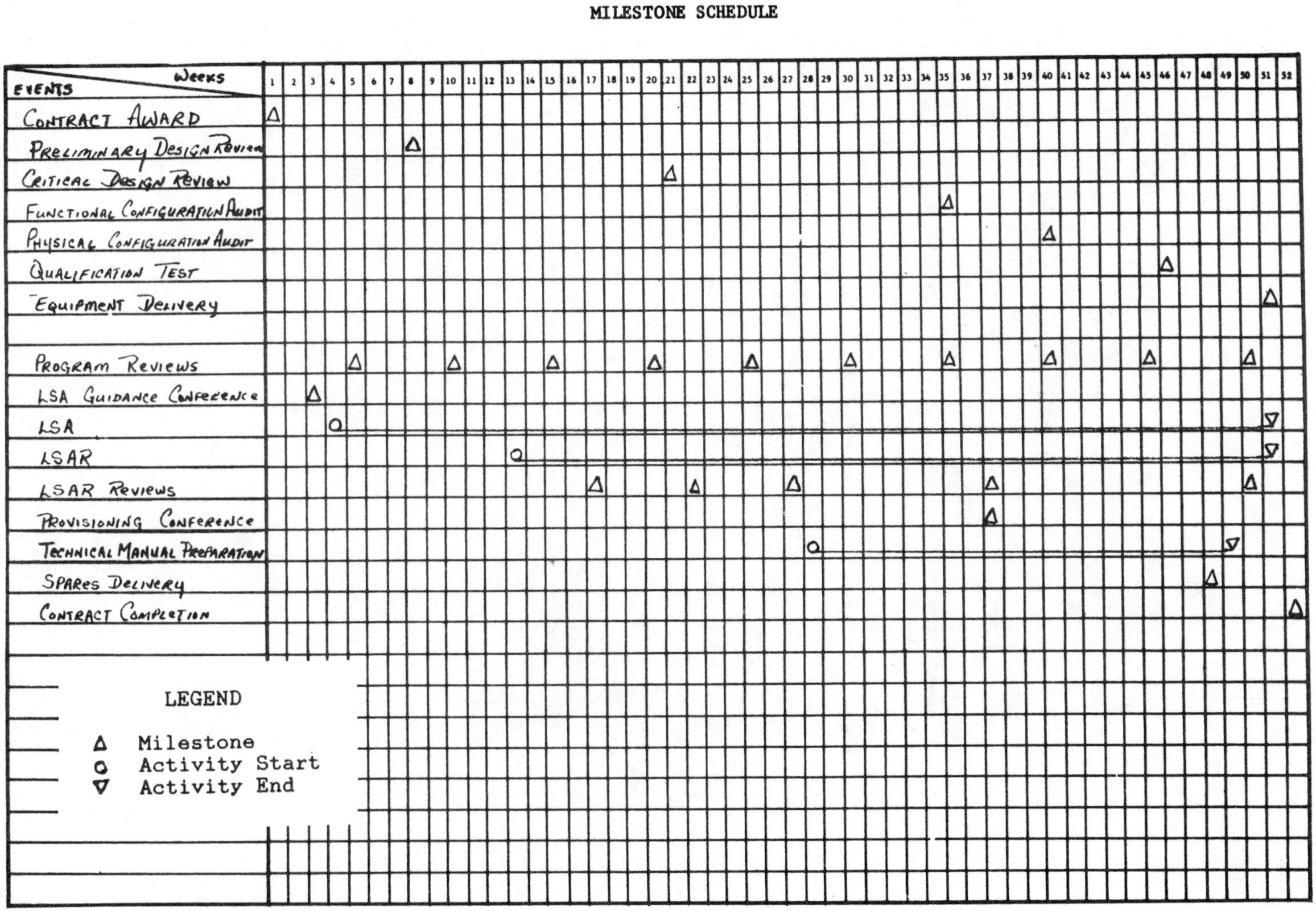

Fig. 21-6. Program milestone schedule.

and audits that are designed to give the government continual insight into the progress the contractor is making toward program completion. MIL-STD 1521B, Technical Reviews and Audits for Systems, Equipments, and Computer Software, contains the standard requirements for conducting these meetings. Normally, the contractor is required to provide the government with a recommended agenda for the meeting and also submit minutes of the meeting at its conclusion. Figure 21-7 lists the typical reviews and audits conducted on a DoD program. Successful completion of these meetings can be milestones for the contract.

Interfaces

The program manager, as the focal point for the program, is required to interface with many internal and external organizations in order to effectively manage the program. In effect, the PMO becomes a facilitator for the flow of information between all concerned activities and organizations. There are both formal and informal interfaces required to manage a program.

The key to effective program interfaces is for all activities to understand their respective roles in the interface process. The program manager must control this process at all times or the program will lose direction, with individual activities making agreements or decisions without proper authority or review. Such events lead to chaos and ineffective program management.

Figure 21-8 illustrates common program interfaces. It is important that the program manager be the key formal interface between the contractor and the government program office, which also has a program manager assigned to oversee the program for the government. The contractor's counterparts in the government program office are normally the procurement contracting officer (PCO) and the administrative contracting officer (ACO). Communications between these individuals and their organizations must be documented to ensure compliance with the contract and to record agreements that affect the program.

---

1. System requirements review
2. System design review
3. Software specification review
4. Preliminary design review
5. Critical design review
6. Test readiness review
7. Functional configuration audit
8. Physical configuration audit
9. Formal qualification review
10. Production readiness review

Fig. 21-7. Technical reviews.

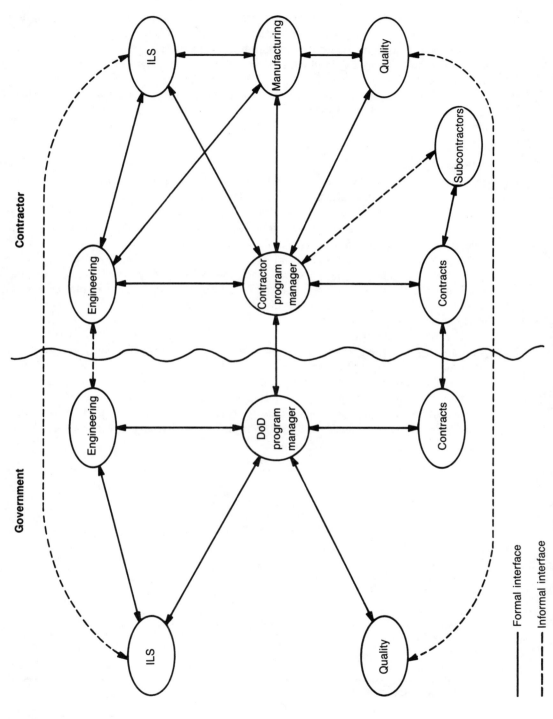

Fig. 21-8. Program interfaces.

— Formal interface

‐ ‐ ‐ Informal interface

ILS Management Team

A unique program management organization is the ILS management team (ILSMT). This organization is composed of representatives from all ILS disciplines of both the government and the contractor. The purpose of the ILSMT is to encourage communications between the logistics activities of the government and the contractor, identify and address ILS problems early in the program, and develop alternatives or solutions that can be implemented as quickly as possible. This process has proven to be very beneficial to both the government and the contractor in identifying and resolving logistics-related concerns. The ILSMT does not bypass the program management process; it enhances the process by providing an additional method for solving program-related problems.

Chapter 22

Software ILS

ILS for software is a difficult area to discuss. The previous references to software have been in the context of applications in conjunction with the hardware being developed. It seems that software is a forgotten or ignored subject when discussing support requirements. Most ILS activities are directed to supporting hardware. Hardware support planning is easier to visualize because the logistician can look at drawings and see what maintenance tasks must be performed and determine the support resources that will be required. Software is not that simple.

Once software has been delivered, it does not require maintenance to keep it operational in the same manner as hardware. Maintenance personnel who perform maintenance on hardware are not equipped to maintain the software. Either the software works or it doesn't. Because of this, ILS has, for the most part, disregarded software support planning. However, the majority of new weapon systems being developed for the military now make extensive use of software as an integral part of the total system. Therefore, it is important that ILS disciplines consider software when planning life cycle support.

DEFINITIONS

The following definitions of common software-related terms will aid in understanding the ILS requirements for software.

- **Software** is a combination of coded computer instructions and associated procedural data that enables computer hardware to perform computations or control functions.

- **Firmware** is software that is installed on a hardware device as a read-only computer instruction. This type of software cannot be readily changed once installed on the hardware device.
- **Computer software configuration item (CSCI)** is a functionally oriented or logically distinct segment of software that is controlled by configuration management in the same manner as an item of hardware.
- **Computer software component (CSC)** is a functional or logical segment of a CSCI. It may be at any level of the software structure. A typical software hierarchical structure for a system may include top-level computer software components (TLCSC), lower-level computer software components (LLCSC), and units. Units are the smallest logical segments of software. Figure 22-1 shows this structure.
- **Computer software documentation (CSD)** is the technical documentation that describes the capabilities and limitations of a CSCI or provides operating or maintenance instructions for the software.
- **Hardware configuration item (HWCI)** is a functionally oriented or logically distinct segment of firmware that is controlled in the same manner as a CSCI.
- **Software development library (SDL)** is a controlled set of software, documentation, records, and procedures used to facilitate the orderly development and subsequent support of software.

SOFTWARE DEVELOPMENT

Software does not magically appear. It is developed through an orderly and systematic process to ensure that the final software meets contractual specifications and provides the necessary instructions for computers to perform required operations. Requirements for software are contained in the procurement specification, and contractor responsibilities with regard to software development are contained in the contract statement of work. The basic requirements for software development that must be followed by a contractor are contained in DOD-STD 2167, Defense System Software Development. This standard provides guidelines for overall software development and tailoring guidance for specific contract applications.

Software Procurement Specification

The purpose of the Software Procurement Specification (SPS) is to provide detailed requirements for the development of software for a specific application. It is normally provided to contractors as an attachment to the overall procurement specification for a weapon system or as a stand-alone document. The reason for this segregation of software and hardware requirements is that there are two different vocabularies and languages used when describing these products.

The hardware specification addresses equipment operational performance, dimensions, materials, etc., while the software specification addresses computer languages, format, conventions, etc. However, it is extremely important for a contractor to ensure that the two are compatible since the software provides instructions to the hardware for operation. This point must be continually reviewed throughout the development

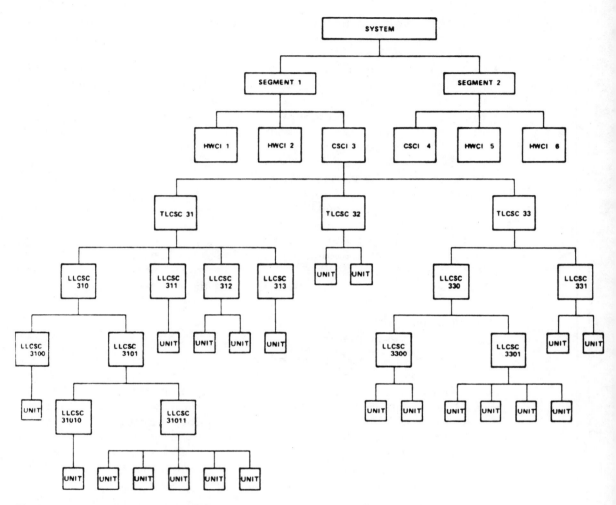

Fig. 22-1. Software structure (DOD-STD 2167).

process since the engineers who will actually design the hardware and software work semi-independently to accomplish a common goal of a total integrated system.

Statement of Work

The contract statement of work (SOW) normally describes the efforts that are required to be performed to develop operational software. These requirements include preparation of applicable planning documents, configuration management processes and procedures, software testing, and software design reviews. The software development activities must closely parallel hardware development.

Development Cycle

The software development cycle is composed of a step-by-step process for orderly production of software to meet the requirements of the SPS. There are three generic complexities of planning software development: functional baseline, allocated baseline, and product baseline. The functional baseline identifies the functions that the final software must perform. The allocated baseline identifies all the software segments that will be generated to perform the functions identified by the functional baseline. The product baseline is the final software configuration that is delivered to the government as part of the total system. The process for developing these baselines is shown in Fig. 22-2.

System Software Requirements Analysis. The initial analysis of system software requirements is accomplished to define the complete functional, performance, interface, and qualification requirements of each CSCI in the system. The results of this analysis form the functional baseline for system software, which is used to determine the overall hierarchy of the software, plan further software effort, and establish design criteria, standards, and constraints to be followed during subsequent software development.

Software Requirements Analysis. After establishment of the functional baseline, a software requirements analysis is conducted to identify the CSCIs and CSCs that will be required to meet the software requirements of the procurement specifica-

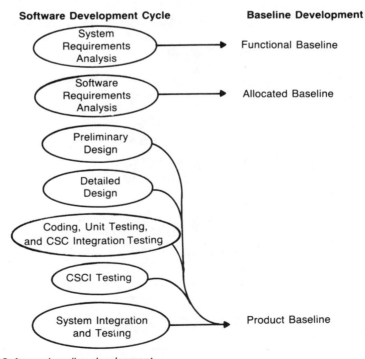

Fig. 22-2. Software baseline development.

tion. The analysis is used to develop the segments of the software structure and document the detailed requirements of each segment. The result of this effort is the allocated baseline. The allocated baseline defines each CSC, describes the functions that each must perform, and delineates the interfaces between each CSC.

Preliminary Design. The preliminary design activity consists of development of detailed requirements for each CSCI and CSC. Using the allocated baseline, software engineers begin to develop the requirements for TLCSCs and begin detailed documentation of required software functions. This development includes applicable inputs and outputs of each TLCSC, timing, interrupts, sequencing, and interface among TLCSCs. The preliminary design phase also provides an initial projection of the support resources that will be required to support the software after delivery. This projection should address the areas shown in Fig. 22-3 and provides ILS disciplines with

---

1. Support environment

   Support software
   Support equipment
   Facilities
   Personnel

2. Support operations

   Operation instructions
   Administration
   Software modification
   Software integration and testing
   System and software generation
   Software quality evaluation
   Corrective action system
   Configuration management
   Simulation
   Emulation
   Reproduction
   Operational distribution

3. Training plans and provisions

4. Predicted level of change after delivery

Fig. 22-3. Software support resource planning information.

essential information that can be used to develop detailed support resource requirements for the software.

Detailed Design. The results of the preliminary design phase are used during detailed design to develop specific functional requirements for each CSC down to and including each unit of the software architecture. The results of the detailed design phase are a "road map" of the functions that will be performed by each CSC, inputs and outputs, and interface requirements. The final software design can then be translated into computer-readable code for operation.

Coding and Unit Testing. The coding and unit testing phase consists of actual translation of the detailed software design of each CSC into computer-readable code. This is a critical phase of the overall software engineering process because here is where the commands for computer computations and functions are translated from human to computer-usable form.

The software code for each unit is tested thoroughly to ensure that the proper coding has been accomplished. Reliability, safety, and human engineering disciplines have a vested interest in this phase of software development. The proper coding of software determines how reliably the total system will operate in the field. Safety is concerned with the operation of the system in a hazard-free manner, and software performance can either increase or reduce hazards depending on how the coding is accomplished. Proper software coding that minimizes the occurrence of human-induced error increases overall system performance.

CSC Integration Testing. The next phase in software development is CSC integration and testing. During this phase, coded software developed during the previous phase is combined and tested in integrated segments. The purpose of this phase is to allow an orderly process to build the total software package. It is easier to find errors that occur as the individual units are combined rather than trying to put it all together at one time then having to find errors at the system level.

CSCI Testing. The final phase in software development is CSCI testing, which is the final step in the software development cycle. CSCI testing is accomplished by actual loading of the software on the system and testing as a part of the overall system acceptance test effort. This activity produces and validates the final software product baseline.

Operational Test and Evaluation. After product baseline, the software is combined with the production hardware to produce the total system. The system is then subjected to various tests (ESS, qualification, acceptance, etc.) to demonstrate that it meets the requirements of the procurement specification. Tests may be conducted by the contractor, the government, or a combination of both. Software errors identified during these tests must be corrected prior to delivery to the government.

Deployment and Distribution. Acceptance by the government of the total system, including hardware and software, begins the deployment phase. This phase ends when the last system is removed from service, which may be many years. During this phase, the software may be modified or changed due to hardware changes, to remove latent errors, or to enhance software capabilities.

Throughout the phase, it is extremely important that the configuration of the software be adequately controlled to eliminate safety or operational problems through unauthorized software changes. As authorized changes occur, ILS disciplines should participate in the change process to ensure that software performance, supportability, reliability, and maintainability are not impaired.

Reviews. During the software development cycle, a series of periodic internal and external reviews are conducted to ensure that the software being developed meets the requirements of the procurement specification. Figure 22-4 shows when these reviews occur in relationship to the cycle phases. The normal reviews include the following.

- **System Requirements Review (SRR)** is conducted to ensure that the software requirements are completely identified prior to the start of the system requirements analysis.
- **System Design Review (SDR)** is conducted at the completion of the system requirements review to determine if the established functional baseline meets the requirements of the procurement specification.
- **Software Specification Review (SSR)** is conducted at the completion of the software requirements analysis to review the adequacy of the allocated baseline.
- **Preliminary Design Review (PDR)** occurs when the proposed approach to hardware and software design is approved by the government. Detailed design of the software should not begin until this approval is received.

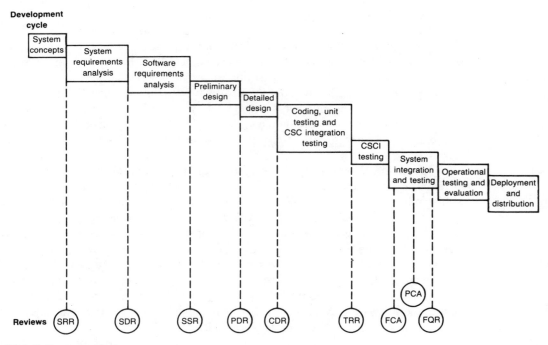

Fig. 22-4. Software reviews.

- **Critical Design Review (CDR)** occurs when the government approves the design approach to be used by the contractor for final design of both hardware and software. Actual translation of software instructions to computer-readable code should not be initiated until this approval is received.
- **Test Readiness Review (TRR)** is conducted prior to the start of CSCI testing. The review should address the readiness of all software codes for testing and the methods to be used when the tests are conducted.
- **Functional Configuration Audit (FCA)** is when both hardware and software are audited by the government to confirm that the design meets the functional requirement of the procurement specification.
- **Physical Configuration Audit (PCA)** is when both hardware and software are audited with relationship to the contractor's documentation to ensure that design documents accurately reflect the products to be delivered to the government. Approval of the contractor's product baseline occurs as a result of this audit.
- **Formal Qualification Review (FQR)** occurs when the final software package is demonstrated to the satisfaction of the government for qualification as an acceptable system that is ready for operational test and evaluation.

SUPPORTABILITY

ILS planning for software supportability is different from planning for support of hardware, but many of the basic ILS concepts apply to software planning. Design criteria should be established for software just as it is established for hardware. Reliability and maintainability should also be addressed in detail. A system safety task specifically addresses the safety aspects of software.

Other ILS disciplines (Fig. 22-5) that have a direct input to software supportability are human engineering, maintenance planning, support equipment, personnel, facilities, training, and supply support. Each ILS discipline must be involved with the software development process if the total system is to meet the operational and supportability requirements of the procurement specification.

Design Criteria

The design criteria for software should address the ILS areas of concern that would produce a deliverable product that can be supported in the field. These criteria consist of program structure, notations, standard formats, standard error identification, and modularity of segments. Establishment of standard methods for software engineers to use during the development process will aid in the subsequent maintenance of the delivered software. A key criterion is the modular approach to designing the software that allows isolation of problems and implementation of changes to a single segment of the software package and avoids having to hunt for cause-and-effect problems.

Reliability

Software reliability is based on the premise that once proven to work, the software should always function identically each time it is used. This might or might not

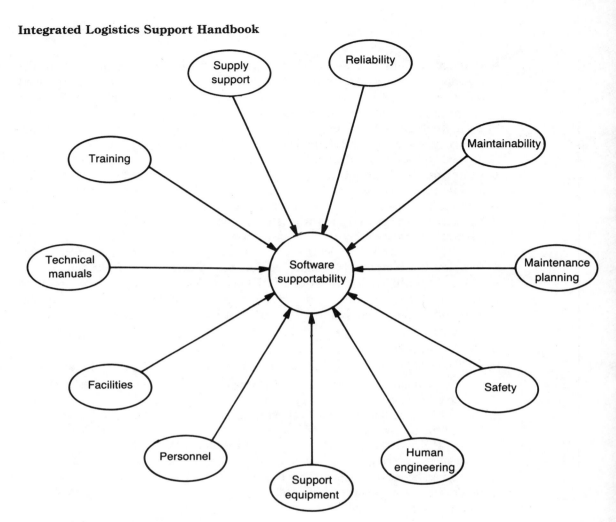

Fig. 22-5. Software supportability.

be true, depending on the hidden operations and peripheral effects of other portions of the software. Extreme care should be exercised during the design phase to identify latent defects in the software code that may cause later failures. This is comparable to performing a sneak circuit analysis on the software to identify unwanted functions.

Maintainability

The ease of maintenance of software is not as easy to determine as that for hardware. However, the design and format chosen for the software package will greatly affect how maintainable it is. Standardization of notations, modularity of segments, and use of comments to describe functions aid in performing maintenance. Maintainability engineers should participate throughout the software development cycle to ensure that the final software package can be maintained as easily as possible.

Safety

Task 212 of MIL-STD 882B specifically addresses the safety aspects of software. The software design must consider the hazards that might be caused due to functions performed by the computer system. Safety engineers should participate during the design and testing of software to identify areas of concern for analysis to determine and eliminate possible hazards.

Human Engineering

The man-machine interface required to operate software is just as important as that for hardware operation. The desired result is to require as little human intervention in software operations as possible, thereby reducing the possibility of human error. When human involvement is necessary, the number and types of responses possible should be kept to a minimum. The location and relationship between controls and displays play an important role in reducing human-induced software errors.

Maintenance Planning

Maintenance of software is not accomplished using wrenches and screwdrivers. Such maintenance normally requires sophisticated equipment, special facilities, and detailed documentation. The planning sequence for software maintenance is identical to that for hardware maintenance, except that maintenance of delivered software is normally accomplished by a single facility responsible for all changes and control of the configuration of the operational software versions in use by fielded systems. There is, in fact, a single level of maintenance for software where the operator's only responsibilities are to install the software, identify problems, and report them to the depot that provides maintenance.

Support Equipment

The equipment required to support software maintenance is distinctly different from that needed to support hardware. Planning for the quantities and types of support equipment required to support the maintenance of software must be accomplished early in the development stages of a program to ensure that the equipment is available when the system is fielded. Whenever possible, the support equipment used to support the software when fielded should be identical to that used when the software was developed. This enhances the capabilities of the maintenance personnel to use the same processes and procedures to maintain the software as were used to initially create the computer code.

Personnel

The number and types of personnel required to support maintenance of software is dependent on the anticipated number of changes that will be required over the life of the system. Programmers and other system engineers will be needed to implement changes due to hardware changes or latent software defects identified through field use. The technological changes that occur over the life of a system must be considered. The methods and procedures that were used to generate the software might be outmoded in just a few years. This might reduce the number of personnel knowledgeable and available to work on the software and must be considered when planning for personnel resources.

Facilities

Special or dedicated facilities normally provide support for maintenance of software. The support for software for a specific system should be limited to a single facility in order to control the configuration of the operational software. Such facilities must be adequate to support both the technical and administrative efforts required for software maintenance. The identification and justification of software maintenance facilities should be accomplished in the same manner as for hardware maintenance facilities.

Training

Software maintenance requires properly trained personnel. This training must be planned and conducted using the same methods as those for training personnel to accomplish hardware maintenance. Many times this training requirement is overlooked because it is a new and unfamiliar area. Lack of adequate training will negatively affect software support when the system becomes operational.

Supply Support

Spare parts requirements to support software are limited in scope, but they can cause problems if not properly addressed. The spares normally consist of the media to transport and store software codes, e.g., magnetic tapes, disks, or card strips, or piece parts that have imbedded firmware. Most commonly used media are available through normal supply channels and should not pose a problem. However, many different types could be used; and as technological changes occur, the future availability of these items must be considered for the long-term support of the system.

Computer chips that contain imbedded, or burned-in, firmware are controlled in two steps. First, there is the acquisition of the chip, and then the software must be transferred to the chip to produce the firmware. This process is normally controlled through a specification control drawing that identifies the basic chip, the software, and the process to be used to produce the firmware. When the system is provisioned, both the basic chip and the firmware chip must be identified. However, the SMR code of the basic chip must indicate that it can be requisitioned only by the activity responsible for software maintenance.

DOCUMENTATION

Several documents are prepared by contractors that provide or plan for operational support of deliverable software. This documentation allows identification of support requirements and provides instructions for operation and maintenance of the software. These documents should reflect support resource planning conducted by ILS disciplines.

Computer Resources Integrated Support Document

The single document prepared and submitted by contractors that identifies the resources necessary to maintain operational software is the Computer Resources Integrated Support Document (CRISD). Included in the CRISD is a detailed description of the recommended resources that will be needed to establish and operate a software support facility. The purpose of the document is to provide the government with a complete set of requirements that must be addressed during the planning stages for operational support. This report can be compared with an LSAR summary for software support.

Figure 22-6 is an outline for a typical CRISD. Note that the CRISD is an all-encompassing document that describes every phase of support for the operational software. Each ILS discipline should have input to this document based on the areas of expertise required.

Technical Manuals

The contractor is normally required to prepare and submit one or more technical manuals specifically designed for the system software. These manuals include (1) computer system operator manual, (2) software user manual, (3) programmer manual, (4) computer system diagnostic manual, and (5) firmware manual. The number and types of manuals required depends on the application and complexity of the software. It is not uncommon for the ILS technical manual discipline to be ignored when these manuals are produced because that discipline has traditionally been associated with the preparation of hardware manuals. However, several ILS disciplines, including technical manuals, reliability, maintainability, safety, human engineering, and training, should be contributing participants in the preparation of any software technical manuals.

As previously stated, the ILS organization has evolved over the years with the goal of planning and providing support for hardware. The tremendous growth in use of software and its impact on the overall supportability of many weapon systems dictates that the ILS disciplines develop the processes and procedures necessary to address both hardware and software as a total integrated system.

1.0 Scope

1.1 Identification

"This computer resources integrated support document provides information required to support all the CSCIs in the XXXXXX system," (or identify the specific portions of the system).

1.2 Purpose

State the purpose of the system and identify the functions of the software to be supported.

Fig. 22-6. Computer resource integrated support document (outline per DI-MCCR 80024).

1.3   Introduction

Summarize the contents of the document.

2.0   Referenced documents

List all applicable documents, specifications, etc., used in preparing the document.

3.0   Support information

3.1   Support environment

Describe the environment where the support will occur to include interrelationships among each component of the environment.

3.1.1   Software required

Identify all the software that will be required to support the operational software. Show the use of each software item.

3.1.2   Hardware required

Identify all the hardware that will be required to support the software. Show the use of each hardware item with relationship to the software identified in 3.1.1, above.

3.1.3   Facilities required

Identify the facilities that will be required to support the operational software, including purpose, recommended location, predicted utilization rates, and special requirements.

3.1.4   Personnel requirements
Identify the personnel required, including number of personnel, skills, skill levels, training, experience, and security clearance requirements.

3.1.5   Other required resources

Identify unique items not addressed above.

3.2   Operations

3.2.1   General usage instructions

Describe general procedures for initiation, operation, and monitoring in the support environment.

3.2.1.1   Initiation of the support environment

Provide instructions for preparation and setup of the support environment.

Fig. 22-6. Computer resource integrated support document (outline per DI-MCCR 80024). (Continued from page 373.)

3.2.1.2 General operation of the support environment

Provide standard instructions for routine operation of the support environment.

3.2.1.3 Monitoring operation of the support environment

Describe the requirements for monitoring the support environment, including trouble and malfunction indicators and possible corrective actions.

3.2.2 Administration

Describe the management and control functions that are applicable to the support environment, including access procedures, security provisions, and access to storage of information.

3.2.3 Software modification

Identify the procedures necessary to modify operational software.

3.2.4 Software integration and testing

Identify procedures for integration and testing of modified software.

3.2.5 System and software generation

Identify procedures for generating new operational software.

3.2.6 Software quality evaluation

Identify methods and equipment necessary to implement a software quality assurance program.

3.2.7 Corrective action system

Describe the recommended method for closed loop identification and resolution of operational software problems.

3.2.8 Configuration management

Describe the procedures to be used to maintain strict configuration management of operational software, both modified and unmodified.

3.2.9 Simulation

Identify any simulation software or hardware that is required to support software maintenance.

3.2.10 Emulation

Identify any emulation software or hardware that is required to support software maintenance.

3.2.11 Reproduction

Identify any software or hardware that is required to reproduce copies of operation software for distribution.

3.2.12 Operational distribution

Identify all available modes for distributing operational software.

3.3 Training

Identify the specific training requirements for personnel to manage and support software maintenance.

3.4 Predicted level of change

Identify the predicted level of changes to operational software, including system modifications, enhancement of operational capabilities, deficiency correction, and frequency and magnitude of changes.

Fig. 22-6. Computer resource integrated support document (outline per DI-MCCR 80024). (Continued from page 375.)

Chapter 23

Using Computers

Computer technology offers a great potential for ILS in terms of both efficiency and timely analysis. The use of computers by engineering and manufacturing in the areas of computer-aided design (CAD), computer-aided engineering (CAE), and computer-aided manufacturing (CAM) has proven invaluable in the design and production of equipment for the military. ILS application of this technology has lagged behind other disciplines, which has caused a gap in the ability of logisticians to effectively participate in the design process.

While other engineering activities have moved toward automated processes wherever possible, the logistics community has continued to rely on manual methods for accomplishing many analyses that should affect the design process. The result is a growing inability to provide any input to the design process except an after-the-fact analysis of what should have been done rather than timely inputs to the design process that enhances the supportability of the equipment being designed.

The emergence of computer-aided logistics support (CALS) is an effort to bring the logistics community up to par with other engineering disciplines. The purpose of CALS is to integrate the efforts of ILS with other disciplines to achieve the goals of timely analysis of designs to identify supportability shortfalls and cost drivers early in the design process so that they can be addressed and corrected before the equipment design is finalized.

This application of computer technology should have two basic benefits. First, there is the inherent capabilities of computer technology to save time and money through efficient data handling. For years, logisticians have relied on manual methods for recording data. This practice is inefficient and costly. Second, the timely availability of data provided by computers allows logisticians to spend their time conducting analyses and providing inputs to the design process based on real-time information.

APPLICATIONS

Computer technology has been applied to several areas of ILS in recent years. These applications have been directed toward automating a specific function to reduce the time required to perform analyses and to record data. In each case, the results have proven extremely satisfactory.

Reliability

Reliability is one of the first areas in which computer technology applications have enjoyed widespread acceptance. This has given reliability engineering the advantage of completing required analyses early and, more importantly, the ability to conduct trade-off analyses rapidly and accurately to affect the design process with the resulting information. Currently available software packages for reliability engineering applications include part stress analysis and part count analysis models, in accordance with MIL-HDBK 217, and data collection and manipulation for conducting a FMECA. There are several different approaches to these analyses, and the specific application should dictate which is most useful.

Figure 23-1 illustrates how a typical part stress analysis model uses existing data on discrete parts plus data on new items to develop reliability predictions for an item of equipment. As the model is used on subsequent programs, the part library is expanded, which provides even greater efficiency. Some models also store predictions for lower-level assemblies that can be used to build analyses on new items having similar parts and functions early in the concept phase when the detailed makeup of an item is still unknown. The flexibility provided by this type of modeling is invaluable for "what if" trade-off analyses.

Maintainability

Computer models for maintainability analyses function similarly to those for reliability predictions. Application of these models is more dependent on the equipment design being modeled since there are several different modeling methods that can be used. Models for mechanical items are normally less complex than those for electronic items. The provisions of MIL-STD 470 must be considered when choosing which model is appropriate.

Calculation of the maintainability predictions requires more manual inputs than the reliability predictions discussed above because the times that result from the prediction models reflect maintenance times rather than the inherent reliability of the equipment. This, in turn, requires more interface between the analyst and the software; however, standard times for common operations aid the modeling process. Many of the maintainability predictions require reliability predictions as an input factor for computation, so modeling packages that allow an interaction between reliability and

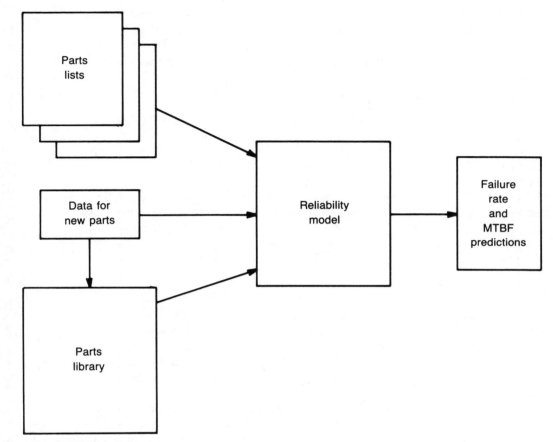

Fig. 23-1. Reliability model.

maintainability predictions prove to be efficient in handling the exchange of applicable data.

Reliability-Centered Maintenance

Software packages for reliability-centered maintenance are limited in availability and flexibility. The types of software commonly available are limited to data collection and data retrieval. While this application might seem minor, the efficiency of simply automating the handling of data reduces the overall cost of performing this type of analysis.

Repair-Level Analysis

The complexity of repair-level analysis computations necessitates the use of computers. Manual computations are laborious, time consuming, and do not provide the capability of performing sensitivity analyses in a timely manner. In most cases, each military service has a standard computer model that is available to contractors for performing this type of analysis when required by contract. Use of the government-furnished model enables the generation of RLA data compatible with the service's unique requirements.

Life Cycle Cost

Models for life cycle cost are probably the most complex of any used by logisticians. The range of variables and combination of factors possible for use in these models require an extensive program. Sensitivities built into the programs are dependent on the purposes for the programs and the amount of valid data available when the programs are used.

LCC predictions require tremendous amounts of data points and computations within the framework of the programs. Use of a life cycle cost model requires a comprehensive knowledge of the types of data involved and the interrelationships among categories of information. An LCC model also requires a large data storage and processing capability that must be considered when using this type of model.

Logistic Support Analysis

Software packages for storing and processing LSA data have proven the applicability of computer technology to this type of analysis. Most programs provide the capability to transfer common data between data records and print hardcopy data sheets as required. The biggest advantage of using LSAR software is having the ability to determine total resource requirements based on the maintenance task analysis for all support functions.

Using the LSAR as a commonly shared data base for all ILS disciplines allows each organization to enter and extract applicable data for analysis and to see the changes as they occur based on the analyses of other groups. This type of computer application to the analysis process aids in identifying total support resources as early as possible in equipment design and assesses the impact of changes to the hardware or the support concept with regard to resource requirements.

Technical Manuals

Computers have been used to generate technical manuals for several years. The combination of text and illustrations using a single system streamlines manual production and decreases the time required for initial preparation and subsequent updates. Further refinement of this process includes transmittal of the final manuals by means of magnetic tape or other modes rather than printed copy. Several software packages are available for this application, each having its own merits and limitations that must be considered.

Common Logistics Data Base

There are many existing applications for computer technology in the area of ILS. Many of these applications are based on the abilities of computers to eliminate redundant paperwork and enhance responsiveness. Future applications include consolidation of one or more of these uses into an ultimate single logistics data base that will serve as the single data base for all source data and analyses performed by ILS disciplines. Such a data base, although not now in existence, would combine all pertinent data into one data base that could service all requirements for logistics applications. Inputs to this data base could be obtained from compatible CAD, CAM, or CAE data systems, or be combined with these systems to form one data system for complete system design, manufacturing, and support planning. Figure 23-2 illustrates this concept.

CHOOSING A COMPUTER SYSTEM

When choosing the correct computer system to use for logistics applications, many variables must be considered, including hardware, software, data, methods of transmission, volume, and cost effectiveness. A contractor might be buying an initial computer system for logistics applications or adding to an existing system to include logistics capabilities.

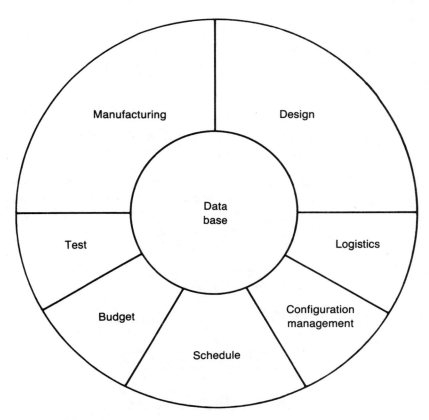

Fig. 23-2. Single data base concept.

Another point to consider is whether ILS will be using the system as a stand-alone operation that is separate from other uses or if the system will be used on a time-share basis with other users. Planning for what type of system to install should be based on a logical evaluation of the requirements that the contractor intends to meet. Figure 23-3 illustrates the planning process that should be used.

The first points to consider are what outputs are required, how the output is to be provided to customers or other organizations, and the anticipated volume of the output and the data to be stored. The next step is to identify the software requirements necessary to provide the outputs required to process the volume of data in a timely manner. The final step is selecting the hardware configuration necessary to support the required software packages.

Output Requirements

A common mistake when choosing a computer system is to base decisions on the merits of a portion of the system rather than on the total system. There are many different types of hardware and software available that have good and bad points based on the intended application. Sometimes an inexpensive system might prove more useful than an expensive one that does not meet the basic requirements of producing the required outputs. In other words, if you don't know what you want out of a computer, it is hard to make the right decision as to what system is appropriate for your needs. Know, in detail, what is needed before choosing a system. The output requirements are the key to this decision.

Software

After the output requirements have been identified, the next step in choosing a computer system is to determine the right software to produce the outputs. Several areas should be considered. First is the source of the software. Off-the-shelf software packages are available for the applications indicated above. The problem is determining which package fits the detailed output requirements of the contractor. Many of

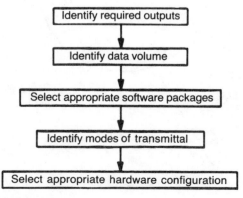

Fig. 23-3. Choosing a computer system.

the commercially available packages are designed to perform specific functions, which they do rather well; however, much of the time they do not have built-in flexibility that allows tailoring of the package to fit a unique program requirement. Should this occur, the software package might be more of a liability than an asset.

Some contractors elect to create their own software packages for ILS applications. This might or might not be a wise decision. Creation of an ILS software package that has broad applications can be a very expensive and time-consuming undertaking. If a contractor has the time and money to invest in building a software package, and can use it for multiple applications, the decision is probably warranted. However, if the contractor does not have the time to design and create the software or needs to start using it before one can be completed, then it may not be advisable.

A compromise solution might be to use existing off-the-shelf packages as a foundation and enhance the programs with extra software that tailors the package to a contractor's requirements. This personalized software package approach has proven very acceptable in acquiring the necessary capabilities to produce the required outputs of a computer system. Areas that must be considered when choosing this approach are the flexibility afforded by the software and the compatibility among packages used in the total system.

Often overlooked is the need for operating system software for the computer hardware. The operating system software and the program software must be compatible; otherwise, the application software package will not function. Another consideration is whether the programs operate in a batch mode or on-line mode.

Finally, and surely not the least important, is how easily the software can be operated by the intended user. The term "user friendly" is supposed to indicate that a computer-literate person should be able to competently operate the software with minimal training. This is not always the case. Many software packages are very rigid in requiring specific commands for operation, but they do not always provide sufficient information for easy operation. The ease of operation might be the key to whether the package actually increases productivity or creates more problems than it provides solutions.

Data Requirements

An area that is often overlooked when choosing a computer system is the volume of data that must be processed and stored. This can cause a twofold problem. First, software packages are written with an inherent limit on the volume of data that can be processed efficiently. When this limit is exceeded, the software programs experience a decrease in processing speed at an alarming rate.

Demonstrations of commercial software packages usually consist of a limited amount of data, which might not accurately show the real speed that the user will experience. Second, the data storage requirements of a contractor may exceed the optimum capacity of the software data handler. Both of these occurrences will degrade the operation of the computer system. Additionally, when two or more software packages are required to work in tandem, they must be compatible in data storage language, data record layout, and data record length. If an incompatibility exists, an additional data-handling software package will be required.

The last data consideration is the total anticipated volume of application software programs, operating system software, data files, and working file space that must be stored on the hardware. These questions should be addressed before choosing the correct computer hardware.

Data Transmittal

The method of transmittal of required outputs must also be considered when choosing a computer system. The common modes of transmittal (Fig. 23-4) are printed hardcopy, magnetic tape, magnetic disks, or electronic via modem. Hardcopy transmittal is the traditional method of providing information to the government. It is the most expensive and time-consuming mode. Computer technology is not being used efficiently if a contractor continues to use this method of transmitting data after choosing a computer system. The government is normally willing to accept other modes of transmittal if the contractor meets the data compatibility requirements of the receiving government agency.

Transmittal using magnetic tape or magnetic disks allows large amounts of data to be provided in a small package that can be prepared in minimal time. It also gives the recipient the flexibility of manipulating the data in whatever method necessary. The problem with this type of transmittal is compatibility between the computer system that generates the data and the receiving system. If the systems are different, then

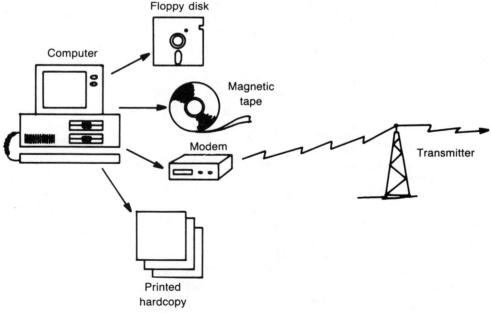

Fig. 23-4. Data transmittal modes.

the recipient may not be able to use the data without undue translation of the data into a usable format or language.

The use of modems for direct access to data is the most desirable method of transmitting data. Electronic transmittal provides recipients with timely information or even direct access to a contractor's data system which may negate the requirement for a contractor to generate any reports. Computer systems should be capable of providing the government direct modem access to data for review or use, as appropriate.

Hardware

The final step in choosing a computer system is to determine the hardware configuration that best supports the software and data transmittal requirements. There are many different brands and sizes of computers, each having unique advantages and disadvantages. The typical computer hardware configuration includes the computer processer unit (CPU), disk drives for data storage, tape deck, computer terminals, video displays, a printer, and communication interface capability. Sizes of computer hardware systems include large mainframes capable of supporting tremendous amounts of software and data files, minicomputers that have similar processing capabilities as mainframes but on a much smaller scale, microcomputers that have limited capabilities, and desktop computers that are oriented toward single-user, single-application use. Each has its place in meeting ILS requirements. There is no simple answer to which one should be selected for a specific application.

During the selection process, the contractor must consider which hardware configuration provides sufficient short-term and long-term support to the unique needs of the company. Flexibility and potential for growth must be considered. The acquisition of a computer system represents a significant investment of capital that must be justified through continued application to support ILS efforts. A cost-versus-utility trade-off analysis should be conducted to determine the optimum return on investment that the contractor will realize if the computer system is acquired.

MANAGEMENT

Acquisition of the right computer system does not assure proper use to fulfill requirements. The system will require continual management to be effective. There are several areas that should be considered, including training, security, maintenance, allocation, and modes of use.

Training

Proper user training is critical. The training program should include initial formal training for all anticipated users when the computer system becomes operational. The training program should address the use of hardware and specific software operating instructions. Follow-up training for new employees and periodic refresher training should also be provided to maintain proficiency.

Security

Security of the computer system is necessary to prevent unauthorized access to the data. Lack of adequate security might result in erroneous data being entered into the system or accidental loss of data. Most computers have a security feature that requires preassigned passwords for use. Additional software is available that increases overall security. The most common security programs limit access to users by identifying those permitted to actually enter or change data as opposed to those allowed to only read or review existing data.

Maintenance

Computer system maintenance covers three areas: hardware maintenance, software maintenance, and data file maintenance. Hardware maintenance might be provided as a service contract with an equipment manufacturer or service company. The cost for an annual maintenance contract might prove to be insignificant for the services required if complex hardware is acquired. Initial maintenance services are sometimes included as a part of the purchase of hardware.

Software maintenance must be performed by a programmer conversant in the language of the programs. Larger companies maintain a staff of programmers for this purpose. However, software purchased from commercial sources must contain the source code for the programs, or the software cannot be changed except by the supplier. Software purchases also include provisions for maintenance support, which should be considered during the decision-making process.

Data maintenance can be performed by a person who is familiar with the software and associated data files. Such file maintenance is normally required on larger computer systems to reduce file space requirements and to keep the usable disk working space as free as possible. Use of the computer system must be closely and continuously monitored to ensure that usable access time is maximized.

Allocation

The purpose of allocation is to ensure that the users who really need computer time receive it. Users tend to monopolize computer access, and certain procedures might be required to allocate computer use time. This can be done through control of security passwords and limiting or increasing the number of computer terminals as necessary. The allocation can be based on program time or data access time.

Computers offer a tremendous potential for ILS to increase the efficiency of the logistics process. Proper selection of the appropriate computer system coupled with necessary management is the key to realizing the benefits of computer technology.

Appendix A

Abbreviations and Acronyms

AAL—additional authorized list
ACO—administrative contracting officer
ADM—advanced development model
AFLC—Air Force Logistics Command
AFSC—Air Force specialty code
ALC—Air Logistics Center
AMC—Army Materiel Command
APO—acquisition program office
ATE—automatic test equipment

BAFO—best and final offer
BCS—baseline comparison system
BII—basic issue item
BIT—built-in test
BITE—built-in test equipment
BOA—basic ordering agreement
BOM—bill of material

CA—criticality analysis
CAD—computer-aided design
CAE—computer-aided engineering
CAGE—commercial and government entity
CALS—computer-aided logistic support

CAM—computer-aided manufacturing
CBD—*Commerce Business Daily*
CBIL—common bulk items list
CCB—configuration control board
CDR—critical design review
CDRL—contract data requirements list
CECOM—communications electronics command
CFE—contractor-furnished equipment
CM—configuration management
COEI—components of end item
CPFF—cost plus fixed fee
CPIF—cost plus incentive fee
CPM—critical path method
CPU—computer processor unit
CRISD—computer resources integrated support document
CRT—cathode-ray tube
CSC—computer software component
CSCI—computer software configuration item
CSD—computer software documentation
CSI—contractor source inspection

D—depot
DCAS—Defense Contract Administration Service
DCN—design change notice
DEMVAL—demonstration validation
DID—data item description
DLA—Defense Logistics Agency
DLSC—Defense Logistics Service Center
DM—data management
DMWR—depot maintenance work requirements
DSARC—Defense Systems Acquisition Review Council

EAC—estimate at completion
ECP—engineering change proposal
ECP—engineering change proposal
EDM—engineering development model
EOQ—economic order quantity
ESML—expendable supplies and materials list
ESS—environmental stress screening
ETC—estimate to complete

FAR—federal acquisition regulations
FCA—functional configuration audit
FFIP—firm fixed incentive price

FFP—firm fixed price
FMEA—failure modes and effects analysis
FMECA—failure modes, effects, and criticality analysis
FMS—foreign military sales
FQR—formal qualification review
FRACAS—failure reporting, analysis, and corrective action system
FRB—failure review board
FSC—federal stock classification
FSCM—federal supply code for manufacturers
FSD—full scale development
FSED—full scale engineering development
FTE—factory test equipment

G&A—general and administrative expenses
GFE—government-furnished equipment
GFP—government-furnished property
GIDEP—government/industry data exchange program
GPETE—general purpose electronic test equipment
GPTE—general purpose test equipment
GSA—General Services Administration
GSI—government source inspection

HWCI—hardware configuration item

I—intermediate
ICD—installation control drawing
ICD—interface control drawing
ID—interface device
ILS—integrated logistics support
ILSMRT—ILS management review team
ILSMT—integrated logistics support management team
ILSP—integrated logistics support plan
IPB—illustrated parts breakdown
ISIL—interim support items list
ISP—integrated support plan
ISSPP—integrated system safety program plan

LCC—life cycle cost
LLCSC—lower-level computer software component
LLTIL—long lead-time items list
LO—lubrication order
LORA—level-of-repair analysis
LSA—logistic support analysis
LSACN—logistic support analysis control number

LSAP—logistic support analysis plan
LSAR—logistic support analysis record

MAC—maintenance allocation chart
MCRL—master cross reference list
MHE—material handling equipment
MICOM—Missile Command
MIL-HDBK—military handbook
MIL-SPEC—military specification
MIL-STD—military standard
MILSTAMP—military standard transportation and movement procedure
MILSTRIP—military standard requisition and issue procedure
MMH/MA—mean manhours per maintenance action
MMH/OH—mean manhours per operating hour
MOS—military occupational specialty
MPTA—manpower, personnel, and training analysis
MTBF—mean time between failures
MTTR—mean time to repair

NATO—North Atlantic Treaty Organization
NAVAIR—Naval Air Systems Command
NAVSEA—Naval Sea Systems Command
NETT—new equipment training team
NHA—next higher assembly
NICP—national inventory control point
NIIN—national item inventory number
NRLA—network repair level analysis
NSN—national stock number
NTE—not to exceed

O—organization
O&S—operation and support
ODC—other direct costs
OJT—on-the-job training
OMB—office of management and budget
ORLA—optimum repair-level analysis
OST—order ship time

PCA—physical configuration audit
PCO—procurement contracting officer
PDM—preliminary development model
PDR—preliminary design review
PERT—program evaluation and review technique
PHS&T—packaging, handling, storage, and transportability

PLISN—provisioning list item sequence number
PLT—production lead time
PM—program manager
PMCS—preventative maintenance checks and services
PMO—program management office
PPDS—preservation and packaging data sheet
PPL—provisioning parts list
PPP—personnel performance profile
PPS—provisioning performance schedule
PPSL—program parts selection list
PRAT—production reliability acceptance test
PRS—provisioning requirements statement
PTD—provisioning technical documentation

QA—quality assurance
QC—quality control
QPEI—quantity per end item
QQPRI—qualitative and quantitative personnel requirements information
QVL—qualified vendor list

R&D—research and development
R&M—reliability and maintainability
RAM—reliability, availability, and maintainability
RCM—reliability-centered maintenance
RD/GT—reliability development/growth testing
RFI—request for information
RFP—request for proposal
RFQ—request for quotation
RIL—repairable items list
RLA—repair-level analysis
ROP—reorder point
RPSTL—repair parts and special tools list
RQT—reliability qualification test

SCD—source control drawing
SCD—specification control drawing
SDL—software development library
SDR—system design review
SDRL—subcontract data requirements list
SE—support equipment
SERD—support equipment recommendations data
SMR—source, maintenance, and recoverability
SOW—statement of work
SPETE—special-purpose electronic test equipment

SPS— software procurement specification
SPTD— supplementary provisioning technical documentation
SPTE— special-purpose test equipment
SRR— system requirements review
SSC— skill specialty code
SSP— system safety program
SSPP— system safety program plan
SSR— software specification review
SSWG— system safety working group
STE— special test equipment

T&M— time and material
TACOM— Tank Automotive Command
TDA— table of distribution and allowances
TDP— technical data package
TLCSC— top-level computer software component
TM— technical manual
TMDE— test, measurement, and diagnostic equipment
TO— technical order
TOE— table of organization and equipment
TP— test program
TPI— test program instruction
TPS— test program set
TRD— test requirements document
TRR— test readiness review
TTEL— tools and test equipment list

UUT— unit under test

WBS— work breakdown structure

Appendix B

References

The following are military and DoD publications.

MILITARY STANDARDS

MIL-STD 12	Abbreviations for Use on Drawings, Specifications,Standards, and in Technical Documents
MIL-STD 129	Marking for Shipment and Storage
MIL-STD 137	Material Handling Equipment
MIL-STD 280	Definitions of Item Levels, Item Exchangeability,Models, and Related Terms
MIL-STD 335	Technical Manuals: Repair Parts and Special Tools List
MIL-STD 470	Maintainability Program for Systems and Equipment
MIL-STD 471	Maintainability Verification/Demonstration/Evaluation
MIL-STD 482A	Configuration Status Accounting Data Elements and Related Features
MIL-STD 680A	Contractor Standardization Program Requirements
MIL-STD 756B	Reliability Modeling and Prediction
MIL-STD 781	Reliability Design Qualification and Production Acceptance Tests: Exponential Distribution
MIL-STD 785B	Reliability Program for Systems and Equipment Development and Production

MIL-STD 794	Procedures for Packaging of Parts and Equipment
MIL-STD 810D	Environmental Test Methods and Engineering Guidelines
MIL-STD 839	Selection and Use of Parts with Established Reliability Levels
MIL-STD 881A	Work Breakdown Structures for Defense Materiel Items
MIL-STD 882B	System Safety Program Requirements
MIL-STD 965A	Parts Control Program
MIL-STD 1309C	Definition of Terms for Test, Measurement, and Diagnostic Equipment
MIL-STD 1319A	Item Characteristics Affecting Transportability and Packaging and Handling Equipment Design
MIL-STD 1345B	Preparation of Test Requirements Document
MIL-STD 1364	Standard General Purpose Electronic Test Equipment
MIL-STD 1365	General Design Criteria for Handling Equipment Associated with Weapons and Weapon Systems
MIL-STD 1366	Definition of Transportation and Delivery Mode Dimensional Constraints
MIL-STD 1367	Packaging, Handling, Storage, and Transportability Program Requirements for Systems and Equipments
MIL-STD 1379C	Military Training Programs
MIL-STD 1387	Preparation and Submission of Data for Approval of Non-Standard General Purpose Electronic Test Equipment
MIL-STD 1388-1A	Logistic Support Analysis
MIL-STD 1388-2A	DoD Requirements for a Logistic Support Analysis Record
MIL-STD 1390	Level of Repair
MIL-STD 1472	Human Engineering Design Criteria for Military Systems, Equipment, and Facilities
MIL-STD 1510	Procedures for Use of Container Design Retrieval System
MIL-STD 1519	Preparation of Test Requirements Document
MIL-STD 1521B	Technical Reviews and Audits for Systems, Equipments, and Computer Software
MIL-STD 1552	Uniform Department of Defense Requirements for Provisioning Technical Documentation (superceded by MIL-STD 1388-2A)
MIL-STD 1556A	Government/Industry Data Exchange Program
MIL-STD 1561	Uniform Department of Defense Provisioning Procedures
MIL-STD 1567A	Work Measurement
MIL-STD 1629	Failure Modes, Effects, and Criticality Analysis
MIL-STD 1635	Reliability Growth Testing
MIL-STD 2073-1A	Procedures for Development and Application of Packaging Requirements
MIL-STD 2073-2	Packaging Requirement Codes
MIL-STD 2076	General Requirements for Unit Under Test Compatibility with Automatic Test Equipment
MIL-STD 2077	General Requirements for Test Program Sets
MIL-STD 2068	Reliability Development Tests

MIL-STD 2173   Reliability-Centered Maintenance Requirements for Naval Aircraft, Weapon Systems and Support Equipment.

MILITARY HANDBOOKS

MIL-HDBK 217	Reliability Prediction of Electronic Equipment
MIL-HDBK 220B	Glossary of Training Device Terms
MIL-HDBK 259	Life Cycle Cost in Navy Acquisitions
MIL-HDBK 338	Electronic Reliability Design Handbook
MIL-HDBK 472	Maintainability Prediction
MIL-HDBK 759A	Human Factors Engineering Design for Army Materiel
MIL-HDBK 63038-1	Technical Manual Writing Handbook
MIL-HDBK 63038-2	Technical Writing Style Guide

MILITARY SPECIFICATIONS

MIL-A-8421	General Specification for Air Transportability
MIL-D-26239A	Qualitative and Quantitative Personnel Requirements Information (QQPRI) Data
MIL-G-29011	Preparation of Guides for Operation and Maintenance of Training Aids
MIL-H-46855B	Human Engineering Requirements for Military Systems, Equipment and Facilities
MIL-M-38784B	Technical Manuals: General Style and Format Requirements
MIL-M-38807A	Technical Manuals: Preparation of Illustrated Parts Breakdown
MIL-M-63036A	Preparation of Operator's Technical Manual
MIL-M-63038B	Technical Manual: Organizational or Aviation Unit, Direct Support or Aviation Intermediate, and General Support Maintenance
MIL-M-63041C	Technical Manuals: Preparation of Depot Maintenance Work Requirements
MIL-M-81919A	Preparation of Support Equipment Technical Manuals
MIL-P-116	Methods of Preservation
MIL-P-9024G	Packaging, Handling, and Transportability in System/Equipment Acquisition
MIL-T-23991E	General Specification for Military Training Devices
MIL-T-29053A	Training Requirements for Aviation Weapon Systems

DOD STANDARDS

DOD-STD 100C	Engineering Drawing Practices

DOD-STD 480A   Configuration Control - Engineering Changes, Deviations, and Waivers

DOD-STD 1685   Comprehensibility Standards for Technical Manuals

DOD-STD 2167   Defense System Software Development

DOD HANDBOOKS

DOD-HDBK 743   Anthropometry of U.S. Military Personnel

DOD DIRECTIVES

DOD-D-1000B   Drawings, Engineering, and Associated Lists

Glossary

EXTRACTED FROM MIL-STD 1388-1A

availability—A measure of the degree to which an item is in an operable and commitable state at the start of a mission when the mission is called for at an unknown (random) time.

baseline comparison system (BCS)—A current operational system, or a composite of current operational subsystems, which most closely represents the design, operational, and support characteristics of the new system under development.

comparative analysis—An examination of two or more systems and their relationships to discover resemblances or differences.

computer resources support—The facilities, hardware, software, and manpower needed to operate and support embedded computer systems. One of the principal elements of ILS.

constraints—Restrictions or key boundary conditions that impact overall capability, priority, and resources in system acquisition.

contract data requirements list (CDRL), DD form 1423—A form used as the sole list of data and information which the contractor will be obligated to deliver under the contract, with the exception of that data specifically required by standard Defense Acquisition Regulation (DAR) clauses.

corrective maintenance—All actions performed as a result of failure to restore an item to a specified condition. Corrective maintenance can include any or all of the following steps: Localization, Isolation, Disassembly, Interchange, Reassembly, Alignment, and Checkout.

data item description (DID), DD form 1664—A form used to define and describe the data required to be furnished by the contractor. Completed forms are provided to contractors in support of and, for identification of, each data item listed on the CDRL.

design parameters—Qualitative, quantitative, physical, and functional value characteristics that are inputs to the design process, for use in design trade-offs, risk analyses, and development of a system that is responsive to system requirements.

end item—A final combination of end products, component parts, and/or materials which is ready for its intended use; e.g., ship, tank, mobile machine shop, aircraft.

facilities—The permanent or semipermanent real property assets required to support the materiel system, including conducting studies to define types of facilities or facility improvements, locations, space needs, environmental requirements, and equipment. One of the principal elements of ILS.

failure modes, effects, and criticality analysis (FMECA)—An analysis to identify potential design weaknesses through systematic, documented consideration of the following: all likely ways in which a component or equipment can fail; causes for each mode; and the effects of each failure (which may be different for each mission phase).

goals—Values, or a range of values, apportioned to the various design, operational, and support elements of a system which are established to optimize the system requirements.

government-furnished material (GFM)—Material provided by the Government to a contractor or comparable Government production facility to be incorporated in, attached to, used with or in support of an end item to be delivered to the Government or ordering activity, or which may be consumed or expended in the performance of a contract. It includes, but is not limited to, raw and processed materials, parts, components, assemblies, tools and supplies. Material categorized as Government Furnished Equipment (GFE) and Government Furnished Aeronautical Equipment (GFAE) are included.

integrated logistics support (ILS)—A disciplined approach to the activities necessary to (1) cause support considerations to be integrated into system and equipment design, (2) develop support requirements that are consistently related to design and to each other, (3) acquire the required support; and (4) provide the required support during the operational phase at minimum cost.

logistic support analysis (LSA)—The selective application of scientific and engineering efforts undertaken during the acquisition process, as part of the system engineering and design process, to assist in complying with supportability and other ILS objectives.

logistic support analysis documentation—All data resulting from performance of LSA tasks conducted under this standard pertaining to an acquisition program.

logistic support analysis record (LSAR)—That portion of LSA documentation consisting of detailed data pertaining to the identification of logistic support resource requirements of a system/equipment. See MIL-STD-1388-2A for LSAR data element definitions.

maintainability—The measure of the ability of an item to be retained in or restored to specified condition when maintenance is performed by personnel having specified skill levels, using prescribed procedures and resources, at each prescribed level of maintenance and repair.

maintenance levels—The basic levels of maintenance into which all maintenance activity is divided. The scope of maintenance performed within each level must be commensurate with the personnel, equipment, technical data, and facilities provided.

maintenance planning—The process conducted to evolve and establish maintenance concepts and requirements for a materiel system. One of the principal elements of ILS.

manpower—The total demand, expressed in terms of the number of individuals, associated with a system. Manpower is indexed by manpower requirements, which consist of quantified lists of jobs, slots, or billets that are characterized by the descriptions of the required number of individuals who fill the job, slots, or billets.

manpower and personnel—The identification and acquisition of military and civilian personnel with the skills and the grade required to operate and support a materiel system at peacetime and wartime rates. One of the principal elements of ILS.

objectives—Qualitative or quantitative values, or range of values, apportioned to the various design, operational, and support elements of a system which represent the desirable levels of performance. Objectives are subject to trade-offs to optimize system requirements.

operating and support (O&S) costs—The cost of operation, maintenance, and follow-on logistics support of the end item and its associated support systems. This term and "ownership cost" are synonymous.

operational concept—A statement about intended employment of forces that provides guidance for posturing and supporting combat forces. Standards are specified for deployment, organization, basing, and support from which detailed resource requirements and implementing programs can be derived.

operational scenario—An outline projecting a course of action under representative operational conditions for an operational system.

operational suitability—The degree to which a system can be satisfactorily placed in field use, with consideration being given availability, compatibility, transportability, interoperability, reliability, wartime usage rates, maintainability, safety, human factors, manpower supportability, logistics supportability, and training requirements.

optimization models—Models which accurately describe a given system and which can be used, through sensitivity analysis, to determine the best operation of the system being modeled.

packaging, handling, storage, and transportation—The resources, processes, procedures, design considerations and methods to ensure that all system, equipment, and support items are preserved, packaged, handled, and transported properly including: environmental considerations and equipment preservation requirements for short- and long-term storage, and transportability. One of the principal elements of ILS.

personnel—The supply of individuals, identified by specialty or classification, skill, skill level, and rate or rank, required to satisfy the manpower demand associated with a system. This supply includes both those individuals who support the system directly (i.e., operate and maintain the system) and those individuals who support the system indirectly by performing those functions necessary to produce the maintain the personnel required to support the system directly. Indirect support functions include recruitment, training, retention, and development.

preventive maintenance—All actions performed in an attempt to retain an item in specified condition by providing systematic inspection, detection, and prevention of incipient failures.

provisioning—The process of determining and acquiring the range and quantity (depth) of spares and repair parts and support and test equipment required to operate and maintain an end item of materiel for an initial period of service.

readiness drivers—Those system characteristics that have the largest effect on a system's readiness values. These may be design (hardware or software), support, or operational characteristics.

reliability—(1) The duration or probability of failure-free performance under stated conditions. (2) The probability that an item can perform its intended function for a specified interval under stated conditions.

reliability centered maintenance—A systematic approach for identifying preventive maintenance tasks for an equipment end item in accordance with a specified set of procedures and for establishing intervals between maintenance tasks.

repair parts—Those support items that are an integral part of the end item or system which are coded as nonreparable.

requiring authority—That activity (government, contractor, or subcontractor) which levies LSA task or subtask performance requirements on another activity (performing activity) through a contract or other document of agreement.

scheduled maintenance—Preventative maintenance performed at prescribed points in the item's life.

sensitivity analysis—An analysis concerned with determining the amount by which model parameter estimates can be in error before the generated decision alternative will no longer be superior to others.

site survey—An examination of potential locations and supporting technical facilities for capability to base a system.

source, maintenance, and recoverability (SMR) codes—Uniform codes assigned to all support items early in the acquisition cycle to convey maintenance and supply instructions to the various logistic support levels and using commands. They are assigned based on the logistics support planned for the end item and its components. The uniform code format is composed of three, 2-character parts: Source Codes, Maintenance Codes, and Recoverability Codes, in that order.

spares—Those support items that are an integral part of the end item or system which are coded as reparable.

standardization—The process by which member nations achieve the closest practicable cooperation among forces; the most efficient use of research, development, and production resources; and agree to adopt on the broadest possible basis the use of (1) common or compatible operational, administrative, and logistics procedures; (2) common or compatible technical procedures and criteria; (3) common, compatible, or interchangeable supplies, components, weapons, or equipment; and (4) common or compatible tactical doctrine with corresponding organizational compatibility.

supply support—All management actions, procedures, and techniques required to determine requirements for, acquire, catalog, receive, store, transfer, issue, and dispose of secondary items. This includes provisioning for initial support as well as replenishment supply support. One of the principal elements of ILS.

supportability—The degree to which system design characteristics and planned logistics resources including manpower meet system peacetime operational and wartime utilization requirements.

supportability assessment—An evaluation of how well the composite of support considerations necessary to achieve the effective and economical support of a system for its life cycle meets stated quantitative and qualitative requirements. This includes integrated logistic support and logistic support resource-related O&S cost considerations.

supportability factors—Qualitative and quantitative indicators of supportability.

supportability-related design factors—Those supportability factors that include only the effects of an item's design. Examples include inherent reliability and maintainability values, testability values, transportability characteristics, etc.

support concept—A complete system-level description of a support system, consisting of an integrated set of ILS element concepts, which meets the functional support requirements and is in harmony with the design and operational concepts.

support equipment—All equipment (mobile or fixed) required to support the operation and maintenance of a materiel system. This includes associated multiuse end items, ground handling and maintenance equipment, tools, metrology and calibration equipment, communications resources, test equipment and automatic test equipment, with diagnostic software for both on- and off-equipment maintenance. It includes the acquisition of logistics support for the support and test equipment itself. One of the principal elements of ILS.

support plan—A detailed description of a support system covering each element of ILS and having consistency among the elements of ILS. Support plans cover lower hardware indenture levels and provide a more detailed coverage of maintenance level functions than support concepts.

support resources—The materiel and personnel elements required to operate and maintain a system to meet readiness and sustainability requirements. New support resources are those that require development. Critical support resources are those that are not new but require special management attention due to schedule requirements, cost implications, known scarcities, or foreign markets.

support system—A composite of all the resources that must be acquired for operating and maintaining a system or equipment throughout its life cycle.

system engineering process—A logical sequence of activities and decisions transforming an operational need into a description of system performance parameters and a preferred system configuration.

system/equipment—The item under analysis, be it a complete system or any portion thereof being procured.

system readiness—A measure or measures of the ability of a system to undertake and sustain a specified set of missions at planned peacetime and wartime utilization rates. System readiness measures take explicit account of the effects of system design (reliability and maintainability), the characteristics and performance of the support system, and the quantity and location of support resources. Examples of typical readiness measures are sortie rate, mission capable rate, operational availability, and asset ready rate.

tailoring—The process by which the individual requirements (sections, paragraphs, or sentences) of the selected specifications and standards are evaluated to determine the extent to which each requirement is most suitable for a specific materiel acquisition and the modification of these requirements, where necessary, to assure that each tailored document invoked states only the minimum needs of the Government.

technical data—Recorded information regardless of form or character (e.g., manuals, drawings) of a scientific or technical nature. Computer programs and related software are not technical data; documentation of computer programs and related software are. Also excluded are financial data or other information related to contract administration. One of the principal elements of ILS.

testability—A design characteristic that allows the status (operable, inoperable, or degraded) of an item and the location of any faults within the item to be confidently determined in a timely fashion.

thresholds—Values, or a range of values, apportioned to the various design, operational, and support elements of a system that impose a quantitative or qualitative minimum essential level of performance. Thresholds are usually associated with a goal.

trade-off—The determination of the optimum balance among system characteristics (cost, schedule, performance, and supportability).

training—The structured process by which individuals are provided with the skills necessary for successful performance in their job, slot, billet, or specialty.

training and training devices—The processes, procedures, techniques, and equipment used to train active and reserve personnel to operate and support a materiel system. This includes individual and crew training, new equipment training, and logistic support for the training devices themselves. One of the principal elements of ILS.

transportability—The inherent capability of material to be moved with available and projected transportation assets to meet schedules established in mobility plans, and the impact of system equipment and support items on the strategic mobility of operating military forces.

unscheduled maintenance—Corrective maintenance required by item conditions.

Index